Dairy Farming and Livestock Production

Dairy Farming and Livestock Production

Edited by **Christian Snider**

SYRAWOOD
PUBLISHING HOUSE

New York

Published by Syrawood Publishing House,
750 Third Avenue, 9th Floor,
New York, NY 10017, USA
www.syrawoodpublishinghouse.com

Dairy Farming and Livestock Production
Edited by Christian Snider

© 2016 Syrawood Publishing House

International Standard Book Number: 978-1-68286-146-2 (Hardback)

Printed in the United States of America.

Contents

Preface

The world is advancing at a fast pace like never before. Therefore, the need is to keep up with the latest developments. This book was an idea that came to fruition when the specialists in the area realized the need to coordinate together and document essential themes in the subject. That's when I was requested to be the editor. Editing this book has been an honour as it brings together diverse authors researching on different streams of the field. The book collates essential materials contributed by veterans in the area which can be utilized by students and researchers alike.

Dairy farming is one of the core areas of agriculture. It is an essential economic activity and a source of livelihood for many. This book discusses in detail various aspects of dairy farming as well as livestock production through lucid elaborations on topics like methods to enhance milk production, herd management, animal welfare, etc. The extensive content of this text provides the readers with a thorough understanding of the subject. It is ideal for students who are looking for an elaborate reference text on dairy farming and livestock production. This book will also be beneficial for the researchers and academicians pursuing dairy sciences and related fields of study.

Each chapter is a sole-standing publication that reflects each author's interpretation. Thus, the book displays a multi-facetted picture of our current understanding of applications and diverse aspects of the field. I would like to thank the contributors of this book and my family for their endless support.

Editor

Seroprevalence and risk factors influenced transmission of *Toxoplasma gondii* in dogs and cats in dairy farms in Western Thailand

Pipat Arunvipas[1]*, Sathaporn Jittapalapong[2], Tawin Inpankaew[2], Nongnuch Pinyopanuwat[2], Wissanuwat Chimnoi[2] and Soichi Maruyama[3]

[1]Department of Large Animal and Wildlife Clinical Sciences, Faculty of Veterinary Medicine, Kasetsart University, Kamphaengsaen Campus, Thailand.
[2]Department of Parasitology, Faculty of Veterinary Medicine, Kasetsart University, Thailand.
[3]Laboratory of Veterinary Public Health, College of Bioresource Sciences, Nihon University, Japan.

The objectives of this study were to determine seroprevalence of *Toxoplasma gondii* infection in dogs and cats in dairy farms of the western provinces of Thailand, and to evaluate the risk factors for the infection. Dogs and cats from positive dairy farms were collected as case and pets from neighboring farms were randomly selected as control; 40 herds in total. 114 dogs and 36 cats sera sample was tested in the Western provinces, and examined for antibodies against *T. gondii* infections by latex agglutination test. Seven sera samples from dogs (6.1%, 6/114) and 3 sera samples from cats (8.3%, 3/36) were found to have titer of *T. gondii* ranging from 1:64 to 1:256 in dogs and from 1:256 to 1:2048 in cats. A toxoplasmosis-positive herd with the presence of a pet was tended to have a 4.6 times (OR = 4.6, 95%CI: 0.95, 22.5; P = 0.06) higher chance of fecal contamination of *T. gondii* infections compared to a herd without the presence of pets. In herds with positive pets on the farm, the prevalence was 6.9 times (OR = 6.9, 95% CI; 1.06-50.8; P = 0.01) more likely to have positive cows in the herd.

Key words: *Toxoplasma gondii*, seroprevalence, risk factors, dogs, cats.

INRODUCTION

Toxoplasma gondii is a ubiquitous protozoan parasite of warm-blooded animals (Dubey, 2010). It has an extremely broad host range including birds, wild animals, pets, livestock and humans. *T. gondii* is prevalent in most areas of the world. Environmental contamination with *T. gondii* oocysts extends even into the oceans. High *T. gondii* seroprevalence rates were recently demonstrated in a variety of marine mammals (Dubey, 2004). *T. gondii* is of veterinary and medical importance, because it may cause abortion or congenital disease in its intermediate hosts. It is estimated to infect 30% of the population worldwide (Flegr and Striz, 2011). Infections in most immunologically normal humans are asymptomatic or result in an influenza-like illness, which often goes undiagnosed as Toxoplasmosis (Tenter et al., 2000). Women who become infected for the first time during pregnancy and immunosuppressed patients are at serious risk from toxoplasmosis. *Toxoplasma* infection can be life threatening in congenitally infected and immunosuppressed patients (Chintana et al., 1998). *T. gondii* is recognised worldwide as a major cause of morbidity and mortality in AIDS patients, primarily as a result of encephalitis (Luft et al., 1984). While *T. gondii* can be transmitted directly by animal-human contact or through contact with contaminated feces, soil and herbage, it can also be transmitted through contaminated food and water. Humans usually become infected with *T. gondii* by ingesting tissue cysts from undercooked meat

*Corresponding author. E-mail: fvetpia@ku.ac.th.

or by ingesting food or water contaminated with oocystes from infected cat feces (Kijlstra and Jongert, 2008). In animals, *T. gondii* infection not only results in significant reproductive and hence economic losses, but also has implications for public health since consumption of infected meat or milk can facilitate zoonotic transmission (Hill et al., 2005).

Cats play an important role in the spread of toxoplasmosis as they are the only animals that excrete resistant oocysts into the environment (Dubey, 2006). On farms this can result in significant reproductive problems in dairy cows (Arunvipas et al., 2012) and economic losses in livestock from abortion and neonatal mortalities (Dubey, 2010).

The standard diagnosis of toxoplasmosis recommended by the OIE for cats is based on coprological diagnosis. The results are often difficult to interpret since the numbers of oocysts of *T. gondii* in cat feces are usually low (Dubey and Beattie, 1988). The latex agglutination test is now widely available as a useful tool for the serological diagnosis of toxoplasmosis and was used by this study as detecting antibodies ensured more accurate results. Serological surveys are good indicators of the occurrence of *T. gondii* infection in cats which is important as serologically positive cats undoubtedly shed oocysts (Dubey and Thulliez, 1989).

Serological studies in Thailand have found evidence of widespread in humans with 1.2 to 4.6% (Maruyama et al., 2000), cats with 11% and dogs with 9.4% (Jittapalapong et al., 2007), goats with 27.9% (Jittapalapong et al., 2005), dairy cows with 7% (Arunvipas et al., 2012). The overall seroprevalence of *T. gondii* was found in 21 of 136 (15.4%) captive felids (Thiangtum et al., 2006), and elephants with 25.6% (Tuntasuvan et al., 2001). There is no evidence on the association between dogs and cats within dairy farms in Thailand. Therefore, the purpose of this study was to determine seroprevalence and the risk factors transmitted of Thailand *T. gondii* in dogs and cats in dairy farms in Western part Thailand.

MATERIALS AND METHODS

Animals and blood collections

Blood samples of dogs and cats were collected from 13 positive dairy farms. Dogs and cats from 27 nearby farms were also randomly collected as control in negative herds between June 2007 and May 2008. A total of 114 dogs and 36 cats from 40 dairy herds were examined in this study. Questionnaire for pet owners included information on age, sex, feeding habits (drinking raw milk), pet born in farm, illness history, and owners data (eating raw meat, pregnancy abnormality) and were obtained during farm visits. The blood was centrifuged at 1000 x g for 10 min and the serum separated and stored at -20˚C until serological analysis.

Serological assay

All sera were examined with commercial latex agglutination test (LAT) kits (TOXOCHECK-MT; Eiken Chemical, Tanabe, Japan). This test was evaluated as a screening serologic test for toxoplasmosis in animals (Tsubota et al., 1977a, b). The procedure was described in a previous report (Jittapalapong et al., 2005). Twenty-five microliters of latex agglutination buffer was added to each well of a U-shaped 96 well cluster plate. Then 25 μl of 1:8 diluted sera was mixed with the buffer in the first well. Serial two-fold dilutions were performed in all wells and the final 25 μl was discarded. Then, 25 μl of *T. gondii*-antigen-coated latex beads were added to each well. The plate was shaken gently and incubated at room temperature overnight. The cut-off titer for this test was 1:64 in accordance to the manufacturer's instructions.

Statistical analysis

Descriptive statistics were performed and unconditional associations between disease status and each variable were determined. Eight variables used in pet analysis included sex (male, female), age (≤ 1 year, >1 year), dogs and cats born on the farm, pets illness during year one of the study, pet dying during year one of the study, feces contamination in feed, habit of owners drinking raw milk, and pregnancy abnormality in owners. The variables which P-values < 0.10 from univariate analysis were analyzed with multiple logistic regression (Backward stepwise) to identify risk factors associated with *T. gondii* infection at P <0.05 (two-sided), based on the likelihood ratio chi-square test. Goodness-of-fit of the final model was examined to assess how well the model fit the observed data. All analyses were conducted using the statistical software package STATA (version 8.2, Stata Corp., College Station, TX (2003))

RESULTS

A total of 13 toxoplasmosis-positive herds, with 27 toxoplasmosis-negative farms randomly selected as the control were used. The numbers of pets from positive and negative herds are shown in Table 1. Overall seroprevalence of *T. gondii* infections in dogs and cats were 6.1% (7/114) and 8.3% (3/36), respectively. The positives ranged from 1:64 to 1:256 in dogs and from 1:256 to 1:2,048 in cats (Table 2). The seropositive rates of dogs and cats in Kanchanaburi, Nakhon Pathom and Ratchaburi were 14.8% (4/27), 5.1% (2/39), and 4.8% (4/84), respectively. There were two variables significant in unconditional association but not significant in final model. Toxoplasmosis-positive herds with the presence of a pet have a 4.6 times (OR = 4.6, 95%CI: 0.95, 22.5; P = 0.06) higher chance of fecal contamination compared to herds without the presence of pets (Table 3). In toxoplasmosis-positive pet households, one case of pregnancy abnormality was found. However, no documented proof was recorded to show a direct link to the disease. Herds with positive pets, were 6.9 times (OR = 6.9, 95% CI; 1.06-50.8; P = 0.01) more likely to have positive cows (Table 4).

DISCUSSION

The prevalence of dogs and cats in dairy farm was lower

Table 1. Number of pets in positive and negative herds in the case-control study.

Species	Positive herds	Negative herds	Total
Dogs	42	72	114
Cats	11	25	36
Total	53	97	150

Table 2. Antibody titer to *T. gondii* infections of cats and dogs.

Species	No. tested	Frequency of the antibody titer					
		1:64	1:128	1:256	1:512	1:1024	1:1028
Cats	36	0	0	1	1	0	1
Dogs	114	1	1	5	0	0	0

Table 3. Factors related to *T. gondii* infection in pets, OR and P value.

Variable[a]	No. of seronegative	No. of seropositive	OR	95% CI	P value
Sex			0.75	[0.20,2.76]	0.66
Male	74	6			
Female	66	4			
Age			1.88	[0.51,6.98]	0.34
≤ 1 year	78	4			
>1 year	62	6			
Born in farm			0.83	[0.22,3.09]	0.78
Yes	90	6			
No	50	4			
Illness			0.90	[0.22,3.66]	0.89
Yes	45	3			
No	95	7			
Death			0.86	[0.23,3.19]	0.83
Yes	61	4			
No	79	6			
Fecal contamination			4.61	[0.95,22.5]	0.06
Yes	65	8			
No	75	2			
Drinking raw milk			3.73	[0.45,30.4]	0.22
Yes	99	9			
No	41	1			
Pregnancy abnormality			7.67	[0.64,92.7]	0.10
Yes	9	1			
No	138	**2**			

[a]Variable comprises of sex (male, female), age (≤ 1 year, >1 year), pet born on farm or not, illness status, pet died during one year period, feces contamination in feed, owner drinking raw milk, and pregnancy abnormality in owner.

than street dogs. Jittapalapong et al. (2007) reported the prevalence of *T. gondii* in stray dogs in Bangkok was 9.4% (40/427) and *T. gondii* antibodies were detected in 65 (11.0%) of the 592 cats.

In this study, the sex of dogs was not significantly associated with seroprevalence, as has been reported by other researchers (Jittapalapong et al., 2007). No significant differences were observed in the seroprevalence of *T. gondii* in both the sexes of cats as similar to study in Japan (Maruyama et al., 2003).

Table 4. Numbers of positive and negative pets in positive and negative dairy herds.

Species	Positive herds	Negative herds	Total
Positive pet	6	3	9
Negative pet	7	24	31
Total	13	27	40

For the seropositive pets, fecal transmission was the major cause of transmission by 4.6 times more likely to be positive in fecal contamination herds. Cats are the natural reservoir of *T. gondii* and excrete the resistant oocysts to environments. Therefore, cats should be not allow in farm areas and should be kept in the house.

In positive households, one case of pregnancy abnormality may be interpreted as demonstrating the risk as being 7.7 times more likely than of negative households. However, no documented proof was recorded to show a direct link to the disease since the details of pregnancy abnormality were not investigated in this study. Women infected with *T. gondii* for the first time during early pregnancy and immunosuppressed patients are at serious risk for clinical toxoplasmosis (Luft et al., 1984; Chintana et al., 1998). In humans, *T. gondii* is transmitted by several routes such as direct animal contact, eating raw or undercooked meat, and unpasteurised milk containing the parasites in the infective stage (Dubey, 2010). *T. gondii* is not simply another protozoan that ought to be considered in the differential diagnosis of protozoal abortion in cattle. It has great implications concerning public health since the consumption of infected meat or milk can facilitate zoonotic transmission.

This study showed the association between seropositive dogs and cats and positive dairy cows. In herds with positive pets on the farm, the prevalence was 6.9 times more likely to have positive cows in the herd and was highly associated with reproductive problems in dairy herds. Most dairy farms in the Western part Thailand had cats and dogs as pets and no boundary between dairy farm and their houses. Cats are capable of roaming all areas including the food storage and even in stalls. Therefore, the spreading of toxoplasmosis in dairy farms might rely on the number of dogs and cats roaming in the farm. The increasing number of pet will increase the risk of toxoplasmosis among animals in dairy farms.

Conclusions

In conclusions, our results showed that *T. gondii* infection was present in dogs and cats in dairy farms in western province. In herds with positive pets on the farm, the prevalence of having positive cows in the herd was 6.9 times higher. Fecal contamination of feed was found to be a major source of transmission in the study. It is essential to control the number of pets on the farms in order to reduce the transmission of toxoplasmosis in both animals and humans.

ACKNOWLEDGEMENTS

Financial support of this study was provided by Thailand Research Fund (TRF). The authors are also very grateful for the cooperation of all dairy farmers who participated in the study.

REFERENCES

Arunvipas P, Inpankaew T, Jittapalapong S (2012). Seroprevalence and Risk Factors of *Toxoplasma gondii* Infection among Dairy Cows in the Western Provinces of Nakhon Pathom, Ratchaburi and Kanchanaburi, Thailand. J. Kasetsart Vet. 22:76-84.

Chintana T, Sukthana Y, Bunyakai B, Lekkla A (1998). *Toxoplasma gondii* antibody in pregnant woman with and without HIV infection. Southeast Asian J. Trop. Med. Public Health 29:383-386.

Dubey JP (2004). Toxoplasma – a waterborne zoonosis. Vet. Parasitol. 126:57-72.

Dubey JP (2006). Comparative infectivity of oocysts and bradyzoites of *Toxoplasma gondii* for intermediate (mice) and definitive (cats) hosts. Vet. Parasitol. 140:69-75.

Dubey JP (2010). Toxoplasmosis of Animals and Humans. CRC Press, Beltsville, Maryland, USA. pp. 1-313.

Dubey JP, Beattie CP (1988). Toxoplasmosis of Animals and Man. CRC Press, Boca Raton, FL; pp.1-200.

Dubey JP, Thulliez P (1989). Serologic diagnosis of toxoplasmosis in cats fed *Toxoplasma gondii* tissue cysts. J. Am. Vet. Med. Assoc. 194:1297-1299.

Flegr J, Striz I (2011). Potential immunomodulatory effects of latent toxoplasmosis in humans. BMC Infect. Dis. 11:274-281.

Hill DE, Chirukandoth S, Dubey JP (2005). Biology and epidemiology of *Toxoplasma gondii* in man and animals. Anim. Health Res. Rev. 6:41-61.

Jittapalapong S, Nimsupan B, Pinyopanuwat N, Chimnoi W, Kabeya S, Maruyama S (2007). Seroprevalence of *Toxoplasma gondii* antibodies in stray cats and dogs in the Bangkok metropolitan area,Thailand. Vet. Parasitol. 145:138-141.

Jittapalapong S, Sangvaranond A, Pinyopanuwat N, Chimnoi W, Khachaeram W, Koizumi S, Maruyama S (2005). Seroprevalence of *Toxoplasma gondii* infection in domestic goats in Satun Province, Thailand. Vet. Parasitol. 127:17-22.

Kijlstra A, Jongert E (2008). Control of the risk of human toxoplasmosis transmitted by meat. Int. J. Parasitol. 38: 1359-1370.

Luft BJ, Brooks RJ, Couley FK, McCabe RE, Remington JS (1984). Toxoplasmic encephalitis in patients with acquired immune deficiency syndrome. J. Am. Vet. Med. Assoc. 252:913-917.

Maruyama S, Boonmar S, Morita Y, Sakai T, Tanaka S, Yamaguchi F, Kabeya H, Katsube Y (2000). Seroprevalence of Bartonella henselae and Toxoplasma gondii among healthy individuals in Thailand. J. Vet. Med. Sci. 62: 635-637.

Maruyama S, Kabeya H, Nakao R, Tanaka S, Sakai T, Katsube Y, Mikami T (2003). Seroprevalence of *Bartonella henselae, Toxoplasma gondii*, FIV and FeLV infections in domestic cats in Japan. Microbiol. Immunol. 47:147-153.

StataCorp. (2003). Stata Statistical Software: Release 8.2College Station, TX:Stata Corporation.

Tenter AM, Heckeroth AR, Weiss LM (2000). *Toxoplasma gondii*: from animals to humans. Int. J. Parasitol. 30:1217-1258.

Thiangtum K, Nimsuphun B, Pinyopanuwat N, Chimnoi W, Tunwattana W, Tongthainan D, Jittapalapong S, Rukkwamsuk T, Maruyama S (2006). Seroprevalence of *Toxoplasma gondii* in captive felids in Thailand. Vet. Parasitol. 136:351-355.

Tsubota N, Hiraoka K, Sawada Y, Watanabe T, Ohshima S (1977a). Studies on latex agglutination test for Toxoplasmosis. Jpn. J.

Seroprevalence and risk factors influenced transmission of Toxoplasma gondii in dogs and cats in dairy farms...

5

Parasitol. 26:286-290.

Tsubota N, Hiraoka K, Sawada Y, Oshino M (1977b). Studies on latex agglutination test for toxoplasmosis. III. Evaluation of the microtiter test as a serologic test for toxoplasmosis in some animals. Jpn. J. Vet. Sci. 26:291-298.

Tuntasuvan D, Mohkaew K, Dubey JP (2001). Seroprevalence of *Toxoplasma gondii* in elephants (*Elephus maximus indicus*) in Thailand J. Parasitol. 87:229-230.

Cross sectional study of bovine trypanosomosis and major clinical signs observed in Diga District, Western Ethiopia

Bekele Dinsa[1], Moti Yohannes[1], Hailu Degefu[1]* and Mezene Woyesa[2]

[1]College of Agriculture and Veterinary Medicine, Jimma University, P. O. Box 307, Jimma, Ethiopia.
[2]School of Veterinary Medicine, Wollega University, Nekemte, P. O. Box 395, Ethiopia.

A cross sectional study was conducted on a total of 410 zebu cattle in selected Kebeles of Diga *Woreda*, Western Ethiopia. The purposes of the study were to determine the prevalence and species of trypanosomes infecting cattle using buffy coat darkground-phase contrast technique and thin blood smear and to assess the associations of common complaints observed by cattle owners with detected trypanosomosis. An overall prevalence of 5.85% was recorded and no statistical significant difference in the prevalence between the Kebeles involved in the present study. The species of trypanosomes identified were *Trypanosoma Congolese* (54.17%) followed by, *Trypanosoma Vivax* (37.5%) *and Trypanosoma Brucei* (8.33%). Trypanosome infection rate based on different age groups was not found statistically significant (P > 0.05). Sex-wise prevalence of 7.47 and 3.55% were recorded in female and male cattle, respectively. Upon case history assessment, the likelihood of cattle with poor body condition to be trypanosome positive was higher when compared to animals with good body condition (OR = 15.82, 95% CI = 5.9- 44.6). Besides cattle with anaemic status were 52.4 times (OR = 52.4, 95% CI = 8.2-216) more likely to have trypanosome infection than non anaemic animals. The mean PCV value of parasitemic animals (20.04%) was lower (P < 0.05) than that of aparasitemic cattle (26.85%). In conclusion, the result of this survey indicated that bovine trypanosomosis is potentially a major constraint to the livestock production and common clinical signs loss of weight and anemia could be considered as one option to keep the disease in check for cattle owners in the area.

Key words: Cattle, clinical signs, Diga *Woreda*, Ethiopia, prevalence, trypanosomosis.

INTRODUCTION

Ten million square kilometers of Africa are infested by the tsetse fly which transmits animal trypanosomosis. Animals lose condition and become progressively emaciated. Milk yield has been reported to decrease in dairy animals (CFSPH, 2009). Available data indicated that bovine trypanosomosis is one of the major impedements to livestock development and agricultural production in Ethiopia, contributing negatively to the overall development in general and to food self-reliance efforts of the nation in particular (Ministry of Agriculture and Rural Development (MoARD, 2004; Abebe, 2005). Anemia, generalized enlargement of the superficial lymph nodes, lethargy and progressive loss of condition are the major signs of bovine trypanosomosis (Radostitis et al., 2007, Blood et al., 1989).

The most prevalent trypanosome species in tsetse infested areas of Ethiopia are *Trypanosoma congolense* and *Trypanosoma vivax* (Abebe and Jobre, 1996; Abebe, 2005). The reported prevalence of bovine trypanosomosis

*Corresponding author. E-mail: hailu.degefu@ju.edu.et.

varies from locality to locality depending on agro climatic conditions, seasons, and activities which were intended to control the impact of the disease. Abebe and Jobre (1996) indicated the prevalence of 58.5, 31.2 and 3.5% for *T. congolense*, *T. vivax*, and *Trypanosoma brucei*, respectively in southwest Ethiopia. A similar report was also recorded for Awi zone in northwestern part of the country (Kebede and Animuit, 2009), Bekele et al. (2010), recently reported prevalence of 70, 20 and 8% in cattle for *T. congolense*, *T. vivax*, and *T. brucei*, respectively, in southern rift valley of the country. Furthermore, an overall prevalence of 4.4% bovine trypanosome infection, with abundance of 36.36, 18.18 and 9.09% for *T. congolense, T. vivax* and *T. brucei*, respectively was also reported by Tadesse and Tsegaye (2010) from Bench Maji zone, South Western Ethiopia. Specific status regarding bovine trypanosomosis is not known for Diga Woreda of East Wollega zone, which is located very near to high risk area for tsetse transmitted trypanosomosis in the country. Therefore the objectives of the present study were to determine the prevalence of infection, to identify the trypanosomes species involved for the infection and to assess the association of cattle trypanosomosis with observed major clinical signs and complaints by cattle owners.

MATERIALS AND METHODS

Study area

The present study was carried out in selected villages of Diga Woreda which is found in Oromia National Regional state, Western Ethiopia. The Woreda is located at 343 km away from Addis Ababa and has a total population of 58,826 and land area of about 40,788 ha. The study area has an altitude range of 1380 to 2300 m above sea level and receives an average annual rain fall of 1416 mm. The temperature range is 14.6 to 30.4°C and the annual average is 22.5°C. According to the Diga District agriculture and rural development office, agro-climatic classification of the *Woreda* is low land (58%), mid-altitude (42%) and no high land coverage (0%). The farming practice in the area is mixed where crop production and all classes of live stock are kept by farmers with the estimated population of cattle 57,586, sheep 11, 220, goats 6,091, horses 136, mules 46, donkeys 2,948 and poultry 31, 241(Central Statistic Authority (CSA), 2009)

Study animals

The study was conducted on local zebu cattle. These animals were raised in different villages of Diga district. The animals examined in this particular study were representing different Kebeles (the smallest administrative unit), sex, body conditions and age groups (calf, young and adult) and reared in extensive management system.

Study design

A cross sectional study was conducted from November 2010 to March 2011, by selecting 11 Kebeles out of 23 Kebeles found in the Woreda based on their accessibility and environmental variations.

A total of 410 animals were sampled, the number of animals sampled from each village was based on proportional weighting. At the Kebele level, animals were selected by simple random sampling using lottery method. The number of cattle sampled from a particular herd in a given Kebele depended on proportional weighting; but a minimum of 20 cattle per herd was fixed. The study animals were classified in different age groups. Animals less than one year (calf), 1 to 6 years and greater than 6 years. Prior to taking the blood sample, information about the history and the clinical signs for each sampled animals were discussed with the cattle owners and recorded and summarized. In addition, the body condition of each animal was also assessed by the amount of fat covering the rump, loin and degree of depression in tailhead area (Nicholson and Butterwarth, 1986).

Sample size

The sample size was calculated using expected prevalence of 50% and desired absolute precision of 0.05 as per the standard procedure described by Thursfield (1995). An estimated minimum sample size of 384 cattle was obtained, but we were able to involve 410 cattle for our study.

Sample collection

Cattle were properly restrained and following aseptic procedure, the marginal ear vein was pricked with the tip of sterile lancet to let blood in to capillary tube. Blood samples were collected in to two capillary tubes from each animal (Murray et al., 1977).

Parasitological examination

Buffy coat darkground-phase contrast technique (BCT)

Blood was collected in two heparinized microhematocrit capillary tubes filled up to 3/4th of their volume from each animal. One end of the tube was sealed with crystal sealant. The tube was placed in the microhematocrit centrifuge with sealant at the outer end. The blood in capillary tube was centrifuged at 10,000 rpm for 5 min. The tubes were taken from the hematocrit centrifuge and placed on the microhematocrit reader to determine the PCV. Animals with PCV value below 24% were considered to be anaemic (Murray et al, 1983). After reading PCV, the capillary tubes were cut 1 mm below the buffy coat including the top layer of the RBC and the contents were expressed onto slide and examined by dark ground-phase contrast microscope for detection of trypanosome (Murray et al., 1983)

Thin blood smear and Giemsa staining

For preparation of the thin smear, first the slide was polished with dry and clean cloth. The blood in microhematocrit capillary tube was expressed approximately 20 mm away from one end on the slide. The spreader (another slide) was placed on a head of the drop of the blood approximately at an angle of 45°. The spreader slide was drawn back to make contact with blood. Then, the blood was allowed to run to both ends of the spreader slide and spread the blood along the slide with steady motion. The slide was dried by waving it in the air and fixed for 5 min with methyl alcohol. The smear was flooded with Giemsa staining solution for 45 min. Excess stain was drained and washed off by using distilled water and allowed dry for examination. Then microscopic examination was made under oil emersion objective (Losos, 1986; Losos and kede, 1972).

Table 1. Prevalence of trypanosome infection in different villages of Diga Wereda, Western Ethiopia,

Kebeles	Numbers of animal examined	Numbers of animal positive (%)	Species of Trypanosame		
			Trypanosoma congolense (%)	*Trypanosoma vivax* (%)	*Trypanosoma brucei* (%)
Adugna	46	1(2.17)	-	1 (2.17)	-
Bikila	49	4(8.16)	2(4.1)	2(4.1)	-
Damaksa	26	1(2.17)	-	1(3.85)	-
Fromsa	51	3(5.88)	3(5.88)	-	-
Garuma	35	1(2.86)	1(2.86)	-	-
Gemechis	28	2(7.14)	1(3.57)	-	1(3.57)
Gudisa	66	4(6.06)	3(4.55)	1(1.5)	-
Ifa	60	4(6.66)	1(1.66)	3(5.0)	-
Jirata	49	4(8.16)	2(4.08)	1(2.04)	1(2.04)
Total	410	24(5.85)	13(3.17)	9(2.19)	2(0.49)

$\chi^2 = 3.34$; P = 0.99.

Table 2. Sex-wise prevalence of trypanosome infection.

Sex	Numbers of animal examined	Numbers infected and prevalence (%)	95% CI*	OR (95%CI)
Female	241	18(7.47)	4.15 - 10.79	
Male	169	6(3.55)	0.76 - 6.34	2.8(0.81-6.91)
Total	410	24(5.85)	3.58 - 8.12	

CI* = Confidence interval; χ^2 = 2.768, P = 0.096.

Data management and analysis

Raw data generated for this study was stored in Microsoft Excel and the prevalence of bovine trypanosomosis in different age groups and sexes were analyzed by using STATA version 8.0; chi-square was used to compare the prevalence of trypanosome infection in different variables and to determine association between variables and the disease. Odds ratios (OR) were used to assess the association of trypanosomosis and observed clinical signs in sampled animals. The mean PCV values of parasitemic animals against that of parasitemic animals were compared using Paired t-test. In all cases differences between parameter were tested for significance at probability levels of 0.05.

RESULTS

Parasitological findings

The overall prevalence of bovine trypanosomosis was found to be 5.85% for Diga Woreda (95% CI: 3.58 to 8.12). The prevalences recorded for each Kebeles are summarized in Table 1. The highest (8.16%) prevalences were observed at Bikila and Jirata Kebeles and the lowest prevalence (2.17%) in Adugna Kebele. However, statistically there was no significant difference in the infection rate among the different Kebeles (P>0.05). In the present study, among 24 cattle detected positive13 (54.17%) were found to be infected by *T. congolese*, 9

(37.50%) by *T. vivax* and 2 (8.33%) cattle were infected by *T. brucei* Trypanosome infection prevalence between male and female animals showed no statistical difference (P > 0.05), even though there were slightly more infections in female animals (Table 2).

The age distribution of bovine trypanosomosis is presented in Table 3. Infections were observed in all the four age groups studied, with no statistically significant variations (P > 0.05)

The results showed that trypanosome infection prevalences among good and poor body condition animals showed statistically significant difference (P> 0.05). Higher prevalence was observed in poor body condition animals (Table 4).

The results of the monthly trypanosome infection showed higher and low prevalence rates in March and November respectively as it is illustrated in Table 5. There was no statistically significant difference in infection rate among the different months studied (P>0.05).

Hematological findings

The PCV value in the animals examined for trypanosome infection showed that the mean PCV value for the parasitemic cattle was 20.04 ± 0.46 while the mean PCV

Table 3. Trypanosome infection in different age groups.

Age group	Number of animals examined	Number infected and prevalence (%)	95% CI*
Calf (< 1 years)	27	1(3.70)	3.42 - 10.82
1-6 years	118	6(5.08)	1.12 - 9.04
>6 years	265	17(6.42)	3.47 - 9.37
Total	410	24(5.85)	3.58 - 8.12

CI* = Confidence interval; χ^2 = 0.504, P = 0.777.

Table 4. Trypanosome infection based on body condition.

Body condition	Number of animals examined	Number infected and prevalence (%)	95% CI*	OR(95%CI)
Good	364	8(2.20)	0.69 - 3.71	
Poor	46	16(34.78)	21.02 - 48.54	15.82(5.9-44.6)
Total	410	24(5.85)	3.58 - 8.12	

CI* = Confidence interval; χ^2= 78.682, P= 0.000.

value for the aparasitemic cattle was observed to be 26.86 ± 0.46 and the over all mean PCV value was 26.46 ± 0.23. There was statistically significant difference in the mean PCV value between the infected (20±0.22) and non infected animals (t = 7, P = 0.000).

In the present study, cattle owners were asked to describe the major clinical signs he or she observed on each sampled animals. Accordingly, the majority of them responded the occurrence of weight loss, milk reduction among adult cows and other health problems listed in Table 6. However the majority of complaints made by owners were not positively associated with trypanosome infection.

Cattle anaemic status (with low PCV) was positively associated with trypanosome infection (OR = 52.4, 95% CI = 8.2 - 216). However cattle with low PCV (anaemic) were 52.4 times more likely to be trypanosome positive than non anaemic cattle (Table 6).

DISCUSSION

The overall prevalence of 5.85% bovine trypanosomosis recorded in the present study is an indicator of the disease as a limiting factor to cattle production in the study area; in addition, this find is in agreement with a similar research conducted in Ethiopia NTTICC (2005) and Garoma (2009) found prevalence of 4.2 and 4.15% in Kenaf and Gari settlement areas of East Wellega , 4.2% was also recorded in South Achefer district in Amhara regional state by Denbarga et al. (2012). However our result is very low when compared to the reports of Afework et al., (2001) and Molalegen et al. (2011), from northwest of the country and 19% prevalence found by Abiy (2002) for Goro district of

Southwest Ethiopia. The relatively low prevalence found in this study may be attributed to the frequent use of chemotherapeutic drugs, an increase in agricultural investment and decreased tsetse challenge in the area. In addition, the parasitological test has been reported to be less sensitive when compared with molecular techniques (Paris et al., 1982; Geysen et al., 2003).

In the present study, trypanosome infection rate observed in female was higher than male animals but not statistically significant. This observation coincides with the findings of Tefera (1994) in Arbaminch districts and Adane (1995) in and around Bahir Dar who reported no significant difference in prevalence between the two sexes. Similarly the infection rates of trypanosome in different age groups showed no statistically significant difference in the present study. This finding supports the result of the previous works by Sinshaw (2004), Tadess et al. (2011), Bitew et al. (2011) and Denbarega et al. (2012), who concluded that there is no significant difference in infection rate between different age groups. The possible explanation for the low prevalence of trypanosome infection observed in calves in the current study may be due to the fact that claves mostly remain confined around home and there is less exposure to tsetse fly whereas the older age groups might have faced the vector challenge during grazing in the field and at watering points.

In the present study, *Trypanosoma congolense* was the predominant species in the study area. The predominance of *T. congoloense* (54.17%) in the present study is in agreement with the previous works that documented 66.1% (Mereb et al., 1999), 60.9% (Afework et al., 2001) and 63.4% in southern part of Ethiopia (Terzu, 2006). In contrast, higher *T. congolense* proportions of 75% at Kone settlement (Tewelde et al., 2004), 85.2% at

Table 5. Prevalence of trypanosome infection in different months.

Months	Number of animals examined	Number infected and prevalence (%)	95% CI*
November	81	1(1.23)	0.03 - 6.60
December	123	9(7.32)	3.40 -13.10
January	112	10(8.93)	1.1 - 26.00
February	74	2(2.70)	0.3 - 9.20
March	20	2(10)	1.20 - 31.10
Total	410	24(5.85)	3.58 - 8.12

CI* = Confidence interval; χ^2 = 7.493, P = 0.112.

Table 6. The relationship between trypanosomosis and some selected clinical signs observed or witness by the owners in cattle of Diga District , Western Ethiopia.

Case history		No. Examined	No. positive (%)	OR	95% CI
Weight loss	Yes	163	3(1.84)	4.71	1.45 - 15.11
	No	247	21(8.5)*		
Decrease in milk yield	Yes	160	8 (4.96)	3.8	1.42 - 10.12
	No	54	9 (18.6)*		
Loss of weight and appetite	Yes	128	3(2.34)	3.3	1.03 - 10.56
	No	282	21(7.4)*		
Loss of weight and coughing	Yes	115	8(6.9)*	1.3	0.55 - 3.06
	No	295	16(5.4)		
Diarrhoea and weight loss	Yes	16	2(12.5)*	2.4	0.0 - 10.5
	No	394	22(5.6)		
Dry feces and coughing	Yes	85	6(7.1)*	1.28	0.51 - 3.25
	No	325	18 (5.5)		
Anemia status (based on PCV)	Anaemic*	134	23 (17.91)*	52.4	8.2 - 216
	Non anaemic	276	1 (0.36)		

*The reference group for comparison.

Arbaminch Zuria districts (Woldeyes and Aboset, 1997) and 84% in Ghibe Valley (Rowlands et al., 1993) were documented. These high ratios of *T. congolense* suggest that the major cyclical vectors or Glossina species are more efficient transmitters of *T. congolense* than *T. vivax* in east Africa (Langridge, 1976).

According to Abebe (2005), *T. congolese* and *T. vivax* are the most prevalent trypanosomes that infect cattle in tsetse infested and tsetse free areas of the Ethiopia, respectively. In the tsetse infested areas of the country, though the prevalence of *T. congolense* was found to be high; a considerable number of examined animals were also harboring *T. vivax* infection which supports the result of the present study.

An assessment of the difference between mean PCV value of parasitemic and aparasitemic cattle was found to be 20.04 and 26.86, respectively and there was statistically significant difference between both groups (P < 0.05). This result agrees with the report of Haile (1996) who reported that the mean PCV value of parasitemic animals were significantly (P < 0.05) lower than that of

aparasitemic animals.

Analysis of trypanosome infection prevalence on the basis of body condition revealed that cattle in poor body condition has 15.8 times more association with positivity to trypanosome when compared with cattle having good body condition. This result is in agreement with previous work by Mussa (2002) who also documented significant difference in infection rate between poor and good body condition cattle.

In the current study, anaemia was more frequently (OR: 52.4, 95% CI: 8.2 - 216) observed among trypanosome positive animals than trypanosome negative cattle. The positive relationship between anemia and trypanosomosis is of course a well established fact (Monga et al., 2004; Radostitis et al., 2007). So clinical manifestations particularly anaemia, could be taken into consideration with ecological conditions, might provide sufficient grounds for a putative diagnosis for trypanosomosis and its impact. In relation with the anemia one should also consider other factors such as the presence of nutritional deficiency and other diseases which are common in

Ethiopia (Mukasa et al., 1989; Berhanu, 2002) could also contribute for anaemia and poor body conditions we observed in this study.

The reported weight loss condition and reduction in milk by the animal owners in the examined animals were found to have no statistical positive associations with animals' which were infected with trypanosome. However a Delphi technique study done on expert opinion on key signs for clinical diagnosis of bovine trypanosomosis in some other East African countries considered weight loss as one of the important clinical sign for bovine trypanosomosis (Magona et al., 2004). Further, more the absence of positive association between trypanosome infection and loss of weight, a reduction in milk yield and other reported clinical signs (Table 6) could also depend on many factors such as stages of the disease, seasons and the type of study design (crossectional) we followed. In addition, it is difficult to attribute some common clinical signs to a single disease, which can be observed as a result of infection of many other diseases, which is very common in extensive cattle keeping system in country like Ethiopia. However previous reports (Kristjanson et al., 1999; FAO, 2000; Radostits et al., 2007), indicated that clinical signs especially weight loss and milk reduction are common observations in trypanosomosis.

In conclusion, bovine trypanosomosis is potentially a constraint to the live stock production and the disease was commonly associated with loss of body condition and anemia in the study area. These signs could be used as an important input in creating awareness in the area of treatment, control and prevention of bovine trypanosomosis.

ACKNOWLEDGEMENT

The authors would like to acknowledge the financial support of Jimma University, College of Agriculture and veterinary medicine to undertake the study.

REFERENCES

Abebe G (2005). Trypanosomosis in Ethiopia. Ethiopia J. Biol. Sci. 4:75-121.

Abebe G, Jobre M (1996). Trypanosomosis: A Threat to cattle production in Ethiopia. Revue. Med. Vet. 147:897-902.

Abiy M (2002). Prevalence Of bovine trypanosomosis in Goro woreda, southwest Ethiopia. DVM Thesis. Addis Ababa University, Debre Zeit, Ethiopia.

Adane M (1995). Survey on the prevalence of bovine trypanosomosis I and around Bahir Dar. DVM Thesis. Addis Ababa University, Debre Zeit, Ethiopia.

Afework Y, Caulosen PH, Abebe G, Tilahun G, Dieter M (2001). Appearance of Multiple drug resistant trypanosome population in village cattle of Metekel District, North west Ethiopia. Livestock community and environment. Veterinary Medicine. Copenhagh, Denmark. pp 1-11.

Bekele J, Asmare K, Abebe G, Ayelet G, Esayas G (2010). Evaluation of Deltamethrin applications in the control of tsetse and trypanosomosis in the southern rift valley areas of Ethiopia. Vet.

Parasitol. 168:177-184.

Berhanu A (2002). Animal health and poverty reduction strategies. In: proceedings of the 16th Annual Conference of the Ethiopian Veterinary Association (EVA), held at Addis Ababa, Ethiopia. pp: 117-137.

Blood DC, Radostitis OM, Henderson JA (1989). Veterinary Medicine: A textbook of Diseases of Cattle, sheep,pigs, goats and horses, 7th Edition, Bailliere Tindall, pp. 1012-1015.

Bitew M, Amedie Y, Abebe A, Tolosa T (2011). Prevalence of bovine trypanosomosis in selected areas of Jabi Tehenan district, West Gojam of Amhara regional state, Northwestern Ethiopia. Afr. J. Agric. Res. 6-1:140-144

CSA (2009). Central statistical agency, Federal democratic republic of Ethiopia, agricultural sample survey. Statistical Bulletin 446, Addis Ababa.

CFSPH (2009). African animal trypanosomiasis. The Center for Food Security and Public Health (CFSPH), Iowa State University. www.cfsph.iastate.edu.

Denbarga Y, Ando O, Abebe R (2012). Trypanosoma Species Causing Bovine Trypanosomosis in South Achefer District, Northern Ethiopia. J. Vet. Adv. 2(2):108-113.

FAO (2000) Impacts of trypanosomosis on African agriculture, by B.M. Swallow. PAAT Technical andScientific Series p. 2. Rome.

Garoma D (2009). prevalence of bovine trypanosomosis in Gari Settlementarea of east wollega zone. DVM Thesis. Jimma University, Jimma, Ethiopia.

Geysen D, Delespaux V, Geerts S (2003). PCR-RFLP using Ssu-rDNA amplification as an easy method for species-specific diagnosis of Trypanosoma species in cattle. Vet. Parasitol. 110:171–180.

Haile C (1996). Bovine trypanosommosis in North Omo: prevalence and assessment of drug Efficacy . DVM Thesis. Addis Ababa University, Debre Zeit, Ethiopia.

Kebede N, Animut A (2009). Trypanosomosis of cattle in selected districts of Awi zone, northwestern Ethiopia. Trop. Anim. Health Prod. 41(7):1353-1356.

Kristjanson PM, Swallow BM, Rowlands GJ, Kruska RL, De Lew PP (1999). Measuring the costs of African animal trypanosomosis, the potential benefits of control and returns to research. Agri. Syst. 59:79-98.

Langridge WP (1976). A tsetse and trypanosomosis survey of Ethiopia.Addis Ababa, Ethiopia, Minstry of Over seas Development of British and Ministry of Agriculture of Ethiopia. p. 97.

Losos GJ (1986). Infectious tropical diseases of domestic animals. Longman, UK. p. 186

Losos GJ, Ikede BO (1972). Review of pathology of disease in domestic and laboratory animals caused by T. vivax, T. brucei, T. rhodesiense and T. gambiense. Vet. Pathol. Suppl. 9:1-17.

Magona JW, Walubengo J, Olaho-Mukani W, Revie CW, Jonsson NN, Eisler MC (2004). A Delphi survey on expert opinion on key signs for clinical diagnosis of bovine trypanosomosis, tick-borne diseases and helminthoses. Bull. Anim. Health. Prod. Afr. 52:130-140.

Ministry Of Agriculture and Rural Development of the Government of Ethiopia (MoARD) (2004). Tsetse and trypanosomiasis prevention and control strategies. Paper presented on Farming In Tsetse Controlled Areas (FITCA), Ethiopia final workshop held at Adama, Ethiopia

Mukasa-Mugerwa E. Bekele E, Tessema T (1989). Type and productivity of Indigenous cattle in central Ethiopia. Trop. Anim. Health Prod. 21:120-120.

Murray M., Murray PK, Mmcintyre WI (1977). An improved parasitological technique for the diagnosis of African Trypanosomosis Trans. R.S.C Trop . Med . Hyg. 71:325-326.

Murray M., Trial TCM, Stephen LE (1983). Livestock productivity and Trypanosomosis, ILCA, Addis Ababa Ethiopia. pp. 4-10

Mussa A (2002). Prevalence of bovine Trypanosomosis in Goro Worda South west Ethiopia . DVM Thesis. Addis Ababa University, Debre Zeit, Ethiopia

Nicholson MJ, Butterworth MH (1986). A guide to condition scoring of zebu cattle. International Livestock Centre for Africa (ILCA).

NTTICC (2005). National Tsetse and Trypanosomosis Investigation and control centre. report for the period 7th June 2003 to 6th July 2004 .Bedelle, Ethiopia. p. 1

Paris J, Murray M, McOdimba F (1982). A comparative evaluation of the parasitological techniques currently available for the diagnosis of African trypanosomiasis in cattle. Acta Trop. 39:307–316.

Radostitis OM, Gay CC, Hinchcliff KW (2007). Veterinary Medicine, A text book of the disease of Cattle, Horses, Sheep, Pigs and Goats, tenth Ed. Saunders Elsevier, Baillier Tindall, London, Philadelphia, New York.

Rowlands GJ, Mulatu W, Authié D'Ieteren GDM, Leak SGA, Nagda SM, Peregrine AS (1993). Epidemiology of bovine trypanosomiasis in the Ghibe valley, Southwest Ethiopia.. Factors associated with variations in trypanosome prevalence, incidence of new infections and prevalence of recurrent infections. Act. Trop. 53:135-150.

Sinshaw A (2004). Prevalence of trypanosomosis of cattle in three woreda of Amhara Region. Msc thesis , Addis Ababa University, Debrezeit Ethiopia.

Tadesse A, Hadgu E, Mekbib B, Abebe R, Mekuria S (2011). Mechanically transmitted bovine trypanosomosis in Tselemty Woreda, Western Tigray, Northern Ethiopia. Agric. J. 6 (1):10-13.

Tadesse A, Tsegaye B. (2010). Bovine trypanosomosis and its vectors in two districts of Bench Maji zone, South Western Ethiopia. Trop. Anim. Health Prod. 42 (8):1757-1762.

Tefera S (1994) prevalence of bovine trypanosomes in Arba Minch districts. DVM Thesis .Addis Ababa University, Debrezeit, Ethiopia

Terzu D (2006). The distribution of tsetse fly and trypanosomosis in Southern Region of Ethiopia. In: Tsetse and trypanosomosis control approaches, and Southern tsetse and trypanosomosis control actors' forum establishment, proceeding of Regional Workshop, Hawassa; Ethiopia . pp. 1-8.

Tewelde N, Abebe G, Eisler M, MacDermott J, Greiner M, Afework Y, Kyule M Munsterman S, Zessin KH, Clausen PH (2004). Application o field methods to assess isomethamidium resistance of trypanosomosis in cattle in Western Ethiopia. Acta Trop. 90:63-170.

Thrusfield M (2005). Veterinary Epidemiology , 3rd ed. Black Science Ltd., Oxford , Great Britain, PP. 182-184.

Woldeyes G. Aboset G (1997). Tsetse and trypanosomosis distribution, identification and assessment of socio- economic viabilities of the new vector control approaches in Arbaminch Zuria Woreda. In proceeding of the11th conference of Ethiopia veterinary association held at Addis Ababa, Ethiopia. pp. 143-154.

Factors affecting milk market outlet choices in Wolaita zone, Ethiopia

Berhanu Kuma[1], Derek Baker[2], Kindie Getnet[3] and Belay Kassa[4]

[1]EIAR, Holetta Agricultural Research Center, P. O. Box 2003, Addis Ababa, Ethiopia.
[2]Research Economist, International Water Management Institute, P. O. Box 5689, Addis Ababa, Ethiopia.
[3]Senior Agricultural Economist and Agricultural Marketing Program Leader, ILRI, Nairobi, Kenya.
[4]Haramaya University, P. O. Box 138, Dire Dawa, Ethiopia.

The study was undertaken with the objective of assessing factors affecting milk market outlet choices in *Wolaita* zone, Ethiopia. Using farm household survey data from 394 households and Multinomial Logit Model, milk market outlet choices were analyzed. Multinomial Logit model results indicate that compared to accessing individual consumer milk market outlet, the likelihood of accessing cooperative milk market outlet was lower among households who owned large number of cows, those who considered price offered by cooperative lower than other market outlets and those who wanted payment other than cash mode. The likelihood of accessing cooperative milk market outlet was higher for households who were cooperative member, who owned large landholding size, who had been in dairy farming for many years and who received better dairy extension services. Compared to accessing individual consumer milk market outlet, the likelihood of accessing hotel/restaurant milk market outlet was lower among households who were at far away from urban center and higher among households who accessed better dairy extension services and who owned large number of dairy cows. As one of the key factors to boost milk market outlet choices, dairy extension services should be strengthened through redesigning or reforming implementation strategies or improving/strengthening existing policy. It should be strengthened to enable farmers produce surplus milk for markets and should devise means to reduce local milking cow numbers by replacing them with crossbred cows. Moreover, governments should strengthen milk processing cooperatives and improve infrastructure facilities.

Key words: Factors, milk market, participation, volume of supply, *Wolaita* zone.

INTRODUCTION

Development policy of Ethiopia has placed an emphasis on increasing agricultural production to serve as a base for rural development. Even though there have been an increase in agricultural production, its attempt experienced drawbacks in the absence of household's market participation. The lack of market participation that many agricultural households face is considered to be a major constraint to combating poverty (Best et al., 2005). This shows that an efficient, integrated and responsive market that is marked with good performance is of crucial importance for optimal allocation of resources and stimulating households to increase output (FAO, 2003). Thus facilitating market participation of households as well as developing chain competitiveness and efficiency are valuable preconditions to improve livelihoods (Lundy et al., 2004; Padulosi et al., 2004). Unless farm households adjust to rapidly changing markets which are characterized by quality and food safety, vertical integration, standards and product traceability, reliability of supply, there will be a risk of competitiveness and inefficiency for the entire value chain (Vermeulen et al., 2008). Household market participation is an important strategy for poverty alleviation and food security in developing countries (Heltberg and Tarp, 2002). Moreover, increasing household participation in markets is a key factor to lifting rural households out of poverty in Africa countries (Delgado, 1995).

The literature on market outlet choices has been thin, especially in developing countries where significant frictions make this question most salient. Goetz (1992) studied participation of Senegalese agricultural households in grain markets. He used probit model to analyze household's discrete decision either to participate in a market or not which was followed by a second-stage regression model to analyze the extent of market participation. Key et al. (2000) developed a structural model to estimate structural supply functions and production thresholds for Mexican households' participation in maize market, based on a censoring model with an unobserved censoring threshold. Holloway et al. (2005) used a Bayesian double-hurdle model to study participation of Ethiopian dairy farmers in milk market when non negligible fixed costs lead to non zero censoring, as in Key et al. (2000), but distinguishing between discrete participation and continuous volume marketed, as in Goetz (1992). Some others studied livestock and livestock products marketing in parts of Ethiopia (Holloway et al., 2000; Yigezu, 2000; Muriuki and Thorpe, 2001; Tsehay, 2001; Mohammed et al., 2004; Woldemichael, 2008). However, none of past studies identified factors affecting milk market outlet choices in *Wolaita* zone, Ethiopia.

Wolaita zone is one of the potential milk production and marketing areas in Ethiopia. In the zone, it is common to see household choices among milk market outlets. Then, what motivate households to choose among milk market outlets available in the study area? Systematic identification of factors faced by households in market outlet choice is increasingly seen by agricultural research as important component of any strategy for reaching the millennium development goals (Giuliani and Padulosi, 2005). Given *Wolaita* zone's potential for milk production, processing, marketing and consumption, results of the study become essential to provide vital and valid information for effective research, planning and policy formulation. Therefore the study provides an empirical basis for identifying options to increase milk market outlet choices of households. In doing so, the study attempts to contribute to filling the knowledge gap by assessing factors affecting milk market outlet choices in *Wolaita* zone, Ethiopia.

METHODOLOGY

A multistage random sampling procedure was used to select representative households from the study area. In the first stage, *Wolaita* zone was selected purposively as it is one of the potential milk production, processing, marketing and consumption areas of the country. Within the zone, four rural districts/*weredas* (*Sodo zuria, Bolosso Sore, Ofa* and *Damote Gale*) and one town (*Wolaita Sodo*) were selected purposively on the basis of milk production, marketing and consumption potential. Then 33 peasant associations/*kebeles* from the *weredas* and the town were selected purposively on the basis of milk production and market participation potential. Sample frame of the *kebeles* was updated and sample size was determined using a simplified formula provided by Yamane (1967). Out of the total 32,972 households, 398 households were selected using simple random sampling methods. However, four households with inappropriately filled questionnaire and missing data were dropped and the data set to 394 households were analyzed.

Both quantitative and qualitative data types were used in the study under investigation. In order to generate these data types, both secondary and primary data sources were used. Secondary sources include reports of line ministries, journals, books, Central Statistical Authority (CSA) and internet browsing, national policies, zonal and *wereda* reports, among others. Primary data sources include zonal and *weredas* Agricultural and Rural Development Offices, zonal and *weredas* Agricultural Marketing Offices, *Wolaita Sodo* Cattle Breeding and Multiplication Center and dairy households. The major data collection methods used includes discussions, rapid market appraisal, observation, formal survey and visual aids. Survey questionnaires were prepared and pre-tested for households operating within the study area. Using the questionnaire, interviews were conducted to gather data on household characteristics, socioeconomic and demographic characteristics, farm information, input utilization, and access to services such as extension, credit and information, technology use, milk production, milk market outlets, among others. Trained and experienced enumerators collected data from households during July and August, 2010.

Two types of data analysis, namely descriptive statistics and econometric models were used to analyze the data collected from households. Descriptive method of data analysis included the use of ratios, percentages, means and standard deviations in the process of comparing socioeconomic, demographic and institutional characteristics of households. To identify factors affecting milk market outlet choices, multinomial logit model was used. If there are a finite number of choices (greater than two), multinomial logit estimation is appropriate to analyze the effect of exogenous variables. The multinomial logit model has been widely used by researchers such as Schup et al. (1999) and Ferto and Szabo (2002). It is a simple extension of the binary choice model and is the most frequently used model for nominal outcomes that are often used when a dependent variable has more than two choices. The results revealed that households accessed milk market outlets such as individual consumer, cooperative, hotel/restaurant and combinations thereof. However, due to mutually inclusiveness of choices, fewer representation and similar collection and operation practices, only households who had access to individual consumer, cooperative and hotel/restaurant milk market outlets were considered in multinomial logit regression. For estimation purpose, the base category used was access to individual consumer; thus the model assessed the effects of various independent variables on the odds of two market outlets versus access to individual consumer market outlet. The general form of the Multinomial Logit model is (McFadden, 1973; Long, 1997):

$$P_{ki} = \frac{\exp(x_i' \beta_k)}{\sum_{K=1}^{J} \exp(x_i' \beta_j)}$$

(1)

$$for\ i = 1,2 - - -, N; K = 1,2, - - -, J$$

where P is the likelihood that a household i chooses to access J milk market outlet from K milk market outlet choices; x is explanatory variable vector that contains the set of factors about household attributes and socioeconomic and demographic characteristics; and β_j is a vector of parameters relating explanatory variables to the valuation of K outlets (K = 1, 2, 3).

The marginal effects are obtained from the logit regression results by the following equation:

$$\frac{\partial P_{ji}}{\partial X_{ji}} = P_{ji}\left(\beta_j - \sum P_{ki}\beta_k\right)$$

(2)

Where β and P represent the parameter and likelihood, respectively, of one of the choices. Marginal likelihood gives better indications and represents changes in dependent variable for a given change in a particular explanatory variable whereas holding the other explanatory variables at their sample means. The models are estimated under maximum likelihood procedures, which yield consistent, asymptotically normal and efficient estimates.

The data covered information necessary to make household level indices of social, economic, demographic and institutional indicators comparable across different categories of households and milk market outlets. In order to identify factors affecting household milk market outlet choices, continuous and discrete variables were identified based on economic theories and empirical studies as follows.

Market outlets (ACCESS)

This is a categorical dependent variable that represents milk market outlets of the study area. The results revealed that households had three milk market outlets and combinations thereof. However, due to mutually inclusiveness of outlets, fewer representation and similar collection and operation practices, only households who had access to individual consumer, cooperative and hotel/restaurant milk market outlets were considered in the regression. Accordingly, dependent variables were created from the data, which indicated sales to (1) individual consumer, (2) cooperative and (3) hotel/restaurant. For estimation purpose, the base category used was access to individual consumer; thus the model assessed the effects of various independent variables on the odds of two milk market outlets versus access to individual consumer milk market outlet.

Mode of milk sale (PAY)

This is a dummy independent variable that takes the value 1 if mode of milk sale is in cash and 0 otherwise. Most households need cash from milk sale to purchase household needs such as soap, salt, food, etc and want payment to be made in cash. Staal et al. (2006) found out that cash mode of payment negatively and significantly affected accessing cooperative and private trader milk market channel selection as compared with accessing individual consumer milk market channel. Therefore, cash based mode of payment is hypothesized to affect accessing individual consumer milk market outlet positively as compared with accessing cooperative and hotel/restaurant milk market outlets.

Milk price by market outlet (PRICE)

This is a continuous independent variable that is measured in Ethiopian birr. It is the actual price received by a household per liter of milk sold to milk market outlets. Staal et al. (2006) found out that the better the price offered by milk market channel, the more a household prefers that outlet for accessing and selling milk. They found out that price offered per liter of milk by individual consumer was lower than price offered by private trader and cooperative and thus households accessed these market outlets than accessing individual consumer milk market outlet. Therefore, the variable is hypothesized to affect accessing individual consumer milk market outlet positively as compared with accessing cooperative milk market outlet and negatively as compared with accessing hotel/restaurant milk market outlet.

Size of milk output (YIELD)

This is a continuous independent variable measured in liter. Past studies revealed that milk yield per day significantly and positively affected marketed surplus of milk (Singh and Rai, 1998; Woldemichael, 2008). Therefore, the variable is hypothesized to affect accessing hotel/restaurant milk market outlet positively than others because of hotel/restaurant capacity to purchase large volume of milk.

Distance to the nearest urban center (DIST)

This is a continuous independent variable measured in kilometer. The closer a household to the nearest urban center, the lesser would be transportation costs, loss due to spoilage and better access to market information and facilities. Berhanu and Moti (2010) found out negative relationship between market participation and distance to the nearest urban market center. Therefore, households who are at far away from urban center are hypothesized to affect the likelihood of accessing cooperative milk market outlet positively as compared with accessing other milk market outlets.

Education of household head (EDU)

This is a dummy independent variable that takes the value 1 if a household head had attended formal schooling and 0 otherwise. Literate households are expected to have better skills and better access to information and ability to process information. Education plays an important role in adoption of new technologies and believed to improve readiness of a head to accept new ideas and innovations. It also enables a head to get updated demand and supply information. Therefore, formal education of household head is hypothesized to affect accessing hotel/restaurant milk market outlet choice positively as compared with accessing other milk market outlets.

Age of household head (AGE)

This is a continuous independent variable that is measured in years. Tshiunza et al. (2001) identified age of a household head as a major household characteristic that significantly affected the proportion of cooking banana plant for markets. They found out that young aged household heads tended to produce and sell more cooking banana than older aged household heads. Therefore, being young aged household head is hypothesized to affect accessing hotel/restaurant milk market outlet choice positively as compared with accessing other milk market outlets.

Sex of household head (SEX)

This is a dummy independent variable that takes the value 1 if the head of a household is male and 0 otherwise. Female contribute more labor in the area of feeding, cleaning of bans, milking, butter and cottage cheese making and sale of dairy products. However, such constraints as lack of capital and poor access to institutional credit and extension service, may affect female participation in dairy production and markets (Tanga et al., 2000). Due to their potential dairy production advantages over female headed households, male headed households are expected to be more market oriented. Therefore, being male headed household is hypothesized to affect accessing hotel/restaurant milk market outlet choice positively as compared with accessing other milk market outlets.

Household size (HSIZE)

This is a continuous independent variable that is measured in the number of members in a household. Household size increases domestic consumption requirements and may render households more risk averse. Families with more household members tend to consume more milk which in turn decreases milk market participation and marketed milk surplus. Hence, controlling for labor supply, larger households are expected to have lower market participation. Heltberg and Trap (2002), Lapar et al. (2003), Edmeades (2006) and Berhanu and Moti (2010) found out negative relationship between household size and market participation of households. It is therefore hypothesized to affect accessing cooperative milk market outlet choice positively as compared with accessing other milk market outlets.

Access to dairy extension services (EXT)

This is a dummy independent variable taking the value 1 if a household had access to dairy extension services and 0 otherwise. It is expected that dairy extension service widens household knowledge with regard to use of improved dairy technologies. Agricultural extension services are expected to enhance households' skills and knowledge, link households with technology and markets (Lerman, 2004). The number of extension agent visits improves household's intellectual capitals and helps in improving dairy production and impacts milk market outlet choices. Past studies revealed that extension agent visits had direct relationship with market outlet choices (Holloway and Ehui, 2002; Rehima, 2006). Thus access to dairy extension service is hypothesized to affect accessing hotel/restaurant milk market outlet choice positively as compared with accessing other milk market outlets.

Access to market information (INFOM)

This is a dummy independent variable taking the value 1 if a household had access to market information services and 0 otherwise. Households marketing decision is based on market price information. Poorly integrated markets may convey inaccurate price information leading to inefficient product movement. Study conducted by Goetz (1992) on food marketing behavior showed that better market information significantly raised likelihood of market participation of households. Therefore, the variable is hypothesized to affect accessing hotel/restaurant milk market outlet

choice positively as compared with accessing other milk market outlets.

Milking cow ownership (COW)

This is a continuous independent variable measured in the number of milking cows owned by a household in TLU. As the number of dairy cows owned increases, milk production increases and percentage share of consumption declines and milk sales increase (Holloway and Ehui, 2002). Past studies indicated that the variable showed positive and significant relationship with market participation and marketable milk volume (Holloway and Ehui, 2002; Gizachew, 2005). Therefore, the variable is hypothesized to affect accessing hotel/restaurant milk market outlet choice positively as compared with accessing other milk market outlets.

Presence of children under six years of age (CHILD)

This is a dummy independent variable taking the value 1 if a household had at least a child less than six years of age and 0 otherwise. There is a competition between milk for child requirement and the amount needed for market. Staal et al. (2006) included the variable in probit model and found out that the variable revealed negative relation to milk market outlet choices. Therefore, households with at least a child under age six are hypothesized to affect accessing cooperative milk market outlet choice positively as compared with accessing other milk market outlets.

Dairy farming experience (EXP)

This is a continuous independent variable measured in the number of years a household has been engaged in dairy farming. Households who have been in dairy farming for many years are expected to have rich experiences regarding opportunities and challenges of dairy production, processing and marketing. Staal et al. (2006) included the variable in probit model and found out that the variable revealed positive relation to milk market participation and market outlet choice. Therefore, the variable is hypothesized to affect accessing cooperative milk market outlet choice positively as compared with accessing other milk market outlets.

Landholding size (LAND)

This is a continuous independent variable measured in hectare. As input for dairy production, land is very important for forage and pasture development to feed dairy cows. It is expected that as the size of land increases, the proportion of land allocated for feed development and improvement increases. According to Staal et al. (2006) the variable has shown negative relationship with milk market participation and market outlet choice. However, in this study the variable is hypothesized to affect accessing cooperative milk market outlet choice positively as compared with accessing other milk market outlets.

Membership to cooperative (MEMB)

This is a dummy independent variable that takes the value 1 if a household has a membership to cooperative and 0 otherwise. Households who are member to cooperative are supposed to sell milk to milk processing cooperative rather than selling to individual consumer and hotel/restaurant. Therefore, membership to cooperative is hypothesized to affect accessing cooperative market

Table 1. Mean household characteristics by milk market outlets

Variable	Mean (Standard deviation) of market outlets		
	Individual consumer (N=118)	Cooperative (N=46)	Hotel/restaurant (N=118)
Age of household head (year)	44.4(10.83)	45.3(13.04)	43.51(8.96)
Household size (number)	5.86(2.11)	6.39(2.40)	5.58(1.87)
Distance to the nearest urban market (km)	2.27(1.61)	3.36(2.16)	1.78(1.39)
Dairy cow in TLU	2.47(1.36)	1.91(1.31)	2.97(1.81)
Milk yield per day (liter)	10.02(3.03)	7.54(1.74)	10.44(3.31)
Dairy farming experiences (year)	8.7(3.81)	19.46(3.25)	7.02(3.77)
Milk price by outlet per liter (Birr)	5.40(1.21)	4.50(0.51)	5.27(0.97)
Land holding size (ha)	0.96(0.07)	1.41(1.45)	0.48(0.31)

Source: Authors collection, July and August 2010.

outlet positively as compared with accessing other milk market outlets.

RESULTS AND DISCUSSION

Mean household characteristics by milk market outlets

The mean household characteristics by milk market outlets are provided in Table 1. The mean household size by milk market outlets was 5.9, 6.4 and 5.6 with individual consumer, cooperative and hotel/restaurant, respectively. The mean household size for households who accessed cooperative milk market outlet was higher than the mean household size (6.0 people) in the rural areas of southern Ethiopia (CSA, 2007). The mean age of household heads that had access to individual consumer, cooperative and hotel/restaurant milk market outlets was 44, 45 and 43.5 years, respectively. The mean dairy cow ownership of households who had access to cooperative, individual consumer and hotel/restaurant milk market outlets was 1.9, 2.5 and 3.0 TLU, respectively. This indicates that households that owned large dairy cows accessed hotel/restaurant milk market outlet because of hotel/restaurants' capacity to purchase large amount of milk.

On average 10, 7.5 and 10.4 L of milk per day was accessed by individual consumer, cooperative and hotel/restaurant market outlets, respectively. The mean dairy farming experience was highest for households who had access to cooperative (19.5 years) milk market outlet and lowest to households that had access to hotel/restaurant (7 years) market outlet. This indicates that households who had access to cooperative milk market outlet were engaged in crop-livestock production whereas others may be peri-urban households. The mean landholding size was highest for households that had access to cooperative (1.41 ha) milk market outlet and lowest for households who had access to hotel/restaurant (0.48 ha) milk market outlet. The average

distance travelled to the nearest urban milk market was highest to households who had access to cooperative (3.36 km) milk market outlet and lowest to households that had access to hotel/restaurant (1.8 km) milk market outlets. However, the average price offered by cooperative market outlet was 4.5[1] birr which is lower than price offered by other market outlets.

Proportion of household characteristics by milk market outlets is given in Table 2. About 29, 46 and 31% of households that had access to individual consumer, cooperative and hotel/restaurant milk market outlets, respectively had at least a child under the age of six. About 60, 54 and 69% of household heads who had access to individual consumer, cooperative and hotel/restaurant milk market outlets, respectively attended formal schooling. 75, 78 and 77% of households that had access to individual consumer, cooperative and hotel/restaurant milk market outlets, respectively were headed by male. About 31, 50 and 40% of households who had access to individual consumer, cooperative and hotel/restaurant milk market outlets, respectively accessed dairy extension services.

About 76, 85 and 81% of households that had access to individual consumer, cooperative and hotel/restaurant milk market outlets, respectively accessed milk market information services. Households that had access to cooperative milk market outlet received relatively better of the service than others because cooperative were established by government. This was because they were given due attention by government extension services to ensure quality supply, support processing and to access better markets as compared to other outlets. Households who had access to cooperative milk market outlet replied that they did not have any other options as they are far from accessing urban market. About 43, 42 and 17% of households that had access to individual consumer, hotel/restaurant and cooperative milk market outlet, respectively received payment to their sales in cash.

---

[1] US$ 1 = Birr 13.632 during the survey period. Birr is the currency unit of Ethiopia.

Table 2. Proportion of household characteristics by milk market outlets

Variable	Category	Proportion (%)		
		Individual consumer (N=118)	Cooperative (N=46)	Hotels (N=118)
Sex of household head	Male	75	78	77
	Female	25	22	23
Education level of head	Formal	60	54	69
	Otherwise	40	46	31
Presence of at least a child under 6 years	Yes	29	46	31
	No	71	54	69
Mode of payment	Cash	43	17	42
	Others	57	83	58
Membership to cooperative	Yes	15	85	25
	No	85	15	75
Access to market information	Yes	76	85	81
	No	24	15	19
Access to dairy extension services	Yes	31	50	40
	No	69	50	60

Source: Authors collection, July and August 2010.

About 85% of households who had access to cooperative market milk outlet were cooperative members. All the households that had access to cooperative milk market outlet replied that they had not received payment for sales made for two months before data collection.

Factors affecting milk market outlet choices

The multinomial logit model has been estimated by the maximum likelihood method. The overall model was significant at 0.01 significance level indicating 99% confidence level that the explanatory variables included in the model assessed the effects on the odds of two market outlets versus sales to individual consumer as indicated by the log pseudo likelihood value of -198.34. Moreover, based on the pseudo R^2 of 0.314, the model appears to have a good fit to the data (Table 3).

The results indicated that households were less likely to access cooperative and hotel/restaurant milk market outlets as compared to individual consumer milk market outlet. Although search, bargaining and delivery costs for access to individual consumer milk market outlet may be high, the preference for accessing it may be an indication of social values attached with. Out of 15 explanatory variables included in multinomial logit model, seven variables to cooperative milk market outlet and three variables to hotel/restaurant milk market outlet were

found to affect milk market outlet choices as compared with accessing individual consumer milk market outlet.

Compared to accessing individual consumer milk market outlet, the likelihood of accessing cooperative milk market outlet was lower among households who owned large number of cows, who considered price offered by cooperative lower than other market outlets and who wanted payment other than cash mode. Households that had access to cooperative milk market outlet received lower price per liter of milk and their mode of sales was not cash. On the other hand, the likelihood of accessing cooperative milk market outlet was higher for households who were cooperative members, who owned large landholding size, who had been in dairy farming for many years and who received better dairy extension services. These households responded that they bypassed access to relatively profitable market outlet (hotel/restaurant) because they considered the opportunity costs in terms of their labor time and transportation, compared to additional profit they could have obtained.

Compared to accessing individual consumer milk market outlet, the likelihood of accessing hotel/restaurant milk market outlet was lower among households who were at farthest distance to the nearest urban center and higher among households who accessed better dairy extension services and who owned large number of dairy cows. Households who owned large number of dairy cows produced more milk and supplied milk to hotel/

Table 3. Results of Multinomial logit regression on milk market outlet choices.

Symbol	Cooperative	Marginal effect (Coop)	Hotel	Marginal effect (Hotel)
Constant	2.653(2.394)	-	0.875(1.191)	-
AGE	-0.029(0.031)	-0.002(0.002)	0.003(0.016)	0.001(0.004)
SEX	-0.255(0.709)	-0.003(0.054)	-0. 079(0.334)	-0.025(0.081)
EDU	1.071(0.653)	0.017(0.051)	0.257(0.317)	0.070(0.078)
HSIZE	0.033(0.148)	-0.007(0.012)	-0.072(0.080)	-0.016(0.019)
CHILD	-0.212(0.614)	-0.010(0.041)	0.310(0.347)	0.083(0.083)
DIST	-0.062(0.122)	-0.010(0.011)	-0.234(0.101)**	-0.057(0.025)
COW	-0.797(0.350)**	-0.057(0.027)	0.208(0.108)*	0.050(0.025)
EXT	2.107(0.668)***	0.202(0.116)	0.854(0.325)***	0.210(0.075)
YIELD	0.063(0.042)	0.008(0.003)	-0.025(0.015)	-0.005(0.003)
EXP	0.096(0.034)***	0.008(0.004)	-0.006(0.017)	-0.001(0.004)
INFO	0.569(0.863)	0.049(0.036)	0.265(0.338)	0.065(0.082)
LAND	0.658(0.231)***	0.053(0.024)	0.052(0.132)	0.011(0.030)
PRICE	-1.400(0.377)***	-0.084(0.037)	-0.237(0.158)	-0.056(0.036)
MEMB	4.000(0.727)***	0.517(0.091)	0.422(0.375)	0.102(0.088)
PAY	-2.039(0.821)**	-0.101(0.039)	0.075(0.292)	0.015(0.072)

Number of observation = 282; Wald Chi-Square (30) = 80.09; Log pseudo likelihood = -198.357***; Pseudo R square: = 0.314.
Source: Authors collection, July and August 2010. ***, **, and * indicate the significance level of 1, 5 and 10%, respectively.
Numbers in brackets indicate robust standard error.

restaurant as they have capacity to absorb supplied milk. Households who are at farthest to access hotel/restaurant milk market considered transaction costs of travelling as a hindering factor and thus accessed neighborhood individual consumer milk market outlet.

Milking cow ownership

Number of milking cows owned by households negatively and significantly affected accessing cooperative milk market outlet as compared with accessing individual consumer milk market outlet. The marginal effect indicates that the likelihood of accessing cooperative milk market outlet decreases by 5.7% for an increase in ownership of milking cow by a TLU as compared with accessing individual consumer milk market outlet.

Milk price by market outlets

Price offered by milk market outlet per liter of milk significantly and negatively affected accessing cooperative milk market outlet as compared with accessing individual consumer milk market outlet. The marginal effect shows that the likelihood of accessing cooperative milk market outlet decreases by 8.4% for a birr increase per liter of milk as compared with accessing individual consumer milk market outlet.

Access to dairy extension services

Access to dairy extension services such as dairy techno-

logy, information, training, field days, field visits and field tours received by households positively and significantly affected accessing cooperative milk market outlet as compared with accessing individual consumer milk market outlet. The marginal effect shows that the likelihood of accessing cooperative milk market outlet increases by 20.2% as compared with accessing individual consumer milk market outlet for one more member access to dairy extension services.

Dairy farming experiences

Number of years a household has been in dairy farming positively and significantly affected accessing cooperative milk market outlet as compared with accessing individual consumer milk market outlet. The marginal effect indicates that the likelihood of accessing cooperative milk market outlet increases by 0.8% as compared with accessing individual consumer milk market outlet for an increase in dairy farming experiences by a year.

Landholding size

Landholding size of households positively and significantly affected accessing cooperative milk market outlet as compared with accessing individual consumer milk market outlet. The marginal effect of landholding size shows that the likelihood of accessing cooperative milk market outlet increases by 5.3% as compared with accessing individual consumer milk market for a hectare increase in landholding size.

Cooperative membership

Membership to cooperative positively and significantly affected accessing cooperative milk market outlet as compared with accessing individual consumer milk market outlet. The marginal effect indicates that the likelihood of accessing cooperative milk market outlet increases by 51.7% as compared with accessing individual consumer milk market outlet for an addition of a household who has membership to cooperative.

Distance to the nearest urban center

Distance to the nearest urban center negatively and significantly affected accessing hotel/restaurant milk market outlet as compared to accessing individual consumer milk market outlet. The marginal effect indicates that the likelihood of accessing hotel/restaurant milk market outlet decreases by 5.7% as compared with accessing individual consumer milk market outlet for a km distance away from the nearest urban center.

Milking cow ownership

Number of milking cows owned by a household positively and significantly affected accessing hotel/restaurant milk market outlet as compared with accessing individual consumer milk market outlet. The marginal effect indicates that the likelihood of accessing hotel/restaurant milk market outlet increases by 5% as compared with accessing individual consumer milk market outlet for an increase in milking cow ownership by one TLU.

Access to dairy extension services

Access to dairy extension services positively and significantly affected accessing hotel/restaurant milk market outlet as compared with accessing individual consumer milk market outlet. The marginal effect shows that the likelihood of accessing hotel/restaurant milk market outlet increases by 21% as compared with accessing individual consumer milk market outlet for an addition of a household who accessed dairy extension service.

Conclusion

The study was undertaken with the objective of assessing factors affecting milk market outlet choices in *Wolaita* zone, Ethiopia. Using farm household survey data from 394 households and Multinomial Logit Model, milk market

outlet choices were analyzed. Multinomial Logit model results indicate that compared to accessing individual consumer milk market outlet, the likelihood of accessing cooperative milk market outlet was lower among households who owned large number of cows, who considered price offered by cooperative lower than other market outlets and who wanted payment other than cash mode. The likelihood of accessing cooperative milk market outlet was higher for households who were cooperative members, who owned large landholding size, who had been in dairy farming for many years and who received better dairy extension services. Compared to accessing individual consumer milk market outlet, the likelihood of accessing hotel/restaurant milk market outlet was lower among households who were at far away from the nearest distance to the nearest urban center and higher among households who accessed better dairy extension services and who owned large number of dairy cows. As one of the key factor to boost milk market outlet choices, dairy extension services should be strengthened through redesigning or reforming implementation strategies or improving/strengthening existing policy. It should be strengthened to enable farmers produce surplus milk for markets and should devise means to reduce local milking cow numbers by replacing them with crossbred cows. Moreover, governments should strengthen milk processing cooperatives and improve their infrastructure facilities.

REFERENCES

Berhanu G, Moti J (2010). Commercialization of smallholders: does market orientation translate into market participation? Improving Productivity and Market Success (IPMS) of Ethiopia farmer project working paper 22. Nairobi Kenya, ILRI.

Best R, Ferris S, Schiavine A (2005). Building linkages and enhancing trust between small-scale rural producers, buyers in growing markets and suppliers of critical inputs. In: F.R. Almond and S.D. Hainsworth (eds.). Beyond Agriculture-making markets work for the poor: Proceedings of an international seminar, 28 February-1 March 2005. Westminster, London, UK. Crop Post Harvest Program (COHP), Natural Resources International Limited, Aylesford, Kent and Practical Action, Bourton on Dunsmore, Warwickshire, UK. P. 176.

Central Statistical Authority (CSA) (2007). Summary and statistical report of 2007 population and housing census. Federal Democratic Republic of Ethiopia population and census commission.

Delgado C (1995). Africa's changing agricultural development strategies: Past and Present Paradigms. As a Guide to the Future, IFPRI.

Edmeades S (2006). Varieties, attributes and marketed surplus of a subsistence crop: Banana in Uganda. Paper presented at international association of agricultural economists association, Gold Coast, Australia, August 12-18.

FAO (2003). FAO action program for the prevention of food loses. Milk and dairy products, post harvest loses and food safety in sub-Saharan Africa and the near east. Regional approaches to national challenges. Synthesis report. ILRI, Nairobi, Kenya.

Ferto I, Szabo G (2002). The choice of supply channels in Hungarian fruit and vegetable sector. In: Economics of Contracts in Agriculture, Second Annual Workshop, Annapolis, MD, 21-23 July.

Giuliani A, Padulosi S (2005). Enhancing the value chain for markets for smallholder producers of (neglected and underutilized) aromatic, vegetables and fruit species in the Near East: A pilot study in Syria. In: Proceedings of ICARDA International Conference on: Promoting community-driven conservation and sustainable use of dry land agro biodiversity, 18-21 April 2005. Aleppo, Syria.

Gizachew G (2005). Dairy marketing patterns and efficiency: The Case of Ada' Liben District, Eastern Oromia. M.Sc. Thesis, Alemaya University, Ethiopia.

Goetz S (1992). A selectivity model of farmer food marketing behavior in sub-Saharan Africa. Am. J. Agric. Econ. 74:444-452.

Heltberg G, Tarp F (2002). Agricultural supply response and poverty in Mozambique. Food Pol. 27:103-124.

Holloway G, Barrett CB, Ehui S (2005). The Double-Hurdle Model in the Presence of Fixed Costs. J. Int. Agric. Trade Dev. 1:17-28.

Holloway G, Ehui S (2002). Expanding market participation among smallholder livestock producers: A collection of studies employing Gibbs sampling and data from the Ethiopian highlands. Socio-economic and Policy Research Working Paper 48. ILRI, Nairobi, Kenya. P. 85.

Holloway G, Nicholson C, Delgado C, Staal S, Ehui S (2000). Agro industrialization through organizational innovation: Transaction costs, cooperatives and milk market development in the east African highlands. Agric. Econ. 23:279-288.

Key N, Sadoule E, de Janvry A (2000). Transaction costs and agricultural farmer supply response. Am. J. Agric. Econ. 82:245-245.

Lapar ML, Holloway G, Ehui S (2003). Policy options promoting market participation among smallholder livestock producers: A case study from Philippines. Food Pol. 28(2003):187-211.

Lerman Z (2004). Policies and institutions for commercialization of subsistence farms in transition countries. J. Asian Econ. 15:461-479.

Long J (1997). Regression Models for Categorical and Limited Dependent Variables. Advanced Quantitative Techniques in Social Sciences. Series 7, SA gE, London.

Lundy M, Gottret M, Cifuentes W, Ostertag C, Best R, Peters D, Ferris S (2004). Increasing the competitiveness of market chains with smallholder producers. Field Manual 3. The Territorial Approach to Rural Agro-enterprise Development. Centro Internacionale de Agricultura Tropical, Cali, Colombia.

McFadden D (1973). Conditional logit analysis of qualitative choice behavior. In: Zarembka P. (ed.): Frontiers in Econometrics. Acad. Press, New York. pp. 105-142.

Mohammed A, Ahmed M, Ehui S, Yemesrach A (2004). Milk development in Ethiopia. EPTD Discussion Paper No. 123. International Food Policy Research Institute, NW Washington, D.C, U.S.A.

Muriuki HG, Thorpe W (2001). Smallholder dairy production and marketing. Constraints and opportunities. P. Smith. Princeton, New Jersey: Princeton University Press. pp. 206-247.

Padulosi S, Noun J, Giuliani A, Shuman F, Rojas W, Ravi B (2004). Realizing the benefits in neglected and underutilized plant species through technology transfers and human resources development. In: Proceedings of the Norway/UN Conference on Technology Transfer and Capacity Building, 23-27 May, 2004. Trondheim, Norway. pp. 117-127.

Rehima M (2006). Pepper marketing chains analysis: the case of Alaba and Siraro Districts, Southern Ethiopia. M.Sc. Thesis, Haramaya University, Ethiopia.

Schup A, Gillepsie J, Reed D (1999). Consumer choice among alternative red meats. J. Food Distrib. Res. 29(3):35-43.

Singh V, Rai KN (1998). Economics of production and marketing of buffalo milk in Harayana. Indian J. Agric. Econ. 53(1):43-52.

Staal SJ, Baltenweck I, Njoroge L, Patil BR, Ibrahim MNM, Kariuki E (2006). Smallholder dairy farmer access to alternative milk market channels in Gujarat. IAAE Conference, Brisbane, Australia.

Tanga FK, Jabbar MA, Shapario BI (2000). Gender roles and child nutrition in livestock production systems in developing countries: A critical review. Socioeconomics and policy research paper 27. ILRI, Nairobi Kenya. P. 64.

Tsehay R (2001). Small scale milk marketing and processing in Ethiopia. In: D. Rangnekar and W. Thorpe (eds), smallholder dairy production and marketing-opportunities and constraints. Proceedings of a South–South workshop held at Anand, India, 13-16 March 2001. National Dairy Development Board, Anand, India, and ILRI, Nairobi, Kenya. pp. 352-367.

Tshiunza M, Lemchi J, Tenkouano A. (2001). Determinants of market production of cooking banana in Nigeria. Afr. Crop Sci. J. 9(3):537-547.

Vermeulen S, Woodhill J, Proctor FJ, Delnoye R (2008). Chain-wide learning for inclusive agro food market development: A guide to multi-stakeholder processes for linking small scale producers with modern markets. International Institute for Environment and Development, London, UK, and Wageningen University and Research Centre, Wageningen, The Netherlands.

Woldemichael S (2008). Dairy marketing chains analysis: The case of Shashaname, Hawassa and Dale District's milk shed, Southern Ethiopia. M.Sc. Thesis, Haramaya University, Ethiopia.

Yamane T (1967). Statistics, an Introductory Analysis, 2nd ed., New York: Harper and Row.

Yigezu Z (2000). Dairy development experience in milk collection, processing and marketing. The role of village dairy cooperatives in dairy development. Smallholder dairy development project proceeding. Addis Ababa, Ethiopia.

4

Teat perimeters for South African Boer and Nguni goats: Use of these measurements to predict milk production potential and kids' growth

V. M. Mmbengwa[1]*, J. A. Groenewald[1], H. D. van Schalkwyk[1] and P. J. C. Greyling[2]

[1]North-West University (NWU), Potchefstroom Campus, Potchefstroom, Republic of South Africa.
[2]University of Free State (UFS), Bloemfontein Campus, Bloemfontein, Republic of South Africa.

The aim of this trial was to assess the possibility of using teat perimeter in estimating milk production potential and kids' growth rate in Nguni and Boer goats of South Africa. Traditionally, the productive capacity of goats doe, are judged to a large extent by the physical phenotypic characteristics. In this trial a 2 × 2 factorial design (two goat breeds and two nutritional management systems) were used. Thirty six (36) recently kidded does (18 Boer goats and 18 Nguni goats) were also used in this experiment. The investigation was conducted between seven (7) and hundred (100) days post-partum. Half (n = 9) the Boer goat does and half (n = 9) the Nguni does were allocated randomly within breed to an extensive (veld) feeding treatment, and the other half per breed to an intensive (high energy) feeding treatment (this is traditionally practiced by profit driven intensive farmers). The results revealed that teat perimeter measurement could be used in both breeds under an intensive feeding regime. The results further revealed that in the extensive feeding regime, the prediction of milk potential through teat perimeter can only be carried out in the Nguni goat breed. These results indicate that the teat perimeter measurement can be used to measure milk production of the Nguni goats based on the two feeding systems. These results appear to suggest that teat perimeters in the Nguni goat could be the best tool for the resource poor farmers.

Key words: Milk production, teat, perimeter, Boer goat, Nguni goat.

INTRODUCTION

This article attempts to uncover the possible of using the teat perimeters in measuring the milk production potential of indigenous South African goat does. In South Africa, indigenous goats such as Boer and Nguni are kept and reared solely for meat production in the communal areas (Mammabolo et al., 1998; Greyling et al., 2004; Simela and Merkel, 2008; Pambu, 2011). In these communal poor remote rural areas, milk and milk products are sold far from these communities (and for expensive prices), goat owners usually resort to the indigenous goats as their source for milk and its products (King, 2009). Most of the goat farmers use these indigenous goats for multipurpose production in the locality (Linzell, 1966; Devendra and Burns, 1970, Pambu, 2011) despite the fact that they receive little or no supplementary feeding during lactation period to boost their milk production potential. According to Pambu (2011), these livestock are not bred for milk production nor selected for such purpose. Intuitively, rural goat owners randomly select these nursing goats for the supply of milk and milk products for their households (Peris et al., 1997; Fourie et al., 2002). In view of these roles and stress imposed on these livestock, it will be necessary for farmers to select which of these livestock could be involved in this multi-purpose use without adversely affecting their growth and survival of their kids. Studies have revealed that the

*Corresponding author. E-mail: vmmmbengwa@gmail.com.

Table 1. Nutritional composition of the prepared pelleted feed used in the experiment.

Nutrients	Minimum (g/kg)	Maximum (g/kg)
Total protein	130	340.92
Urea	-	10
Moisture	-	120
Fibre	-	150
Fat	25	-
NH$_4$Cl	10	10
Calcium	10	15
Phosphorus	3	-

Minimum and maximum represent the range at which the ration was formulated.

physical features of the udder have been used in other goat and sheep breeds to measure the milk potential (Gall, 1981; Mavrogenis et al., 1989; Godfrey et al., 1997; Maiwashe, 2000; Ahuya et al., 2009; Pambu, 2011). However, the use of udder characteristics in the indigenous goat of South Africa as measure of their milk production potential has not been fully explored (Mmbengwa, 2009). Therefore, the aim of this study is the use of teat perimeter as an alternative technique to measure milk production potential of indigenous goats for resource poor communities in South Africa.

MATERIALS AND METHODS

The data were obtained from two different locations. An extensive and intensive group of animals were kept at Paradys experimental farm and a small stock on the campus of University of Free State, respectively. Each breed type was equally represented in the experimental locations. The management practices on each location differ on the feeding management. The feeding manage-ment of the animals in each feeding regime were designed to simulate the winter and summer season. The intensive feeding regime was designed to resemble a balance diet where nutritional factors are assumed to play a maximum role in enhancing the productivity of the livestock.

The extensive feeding regime was designed to resemble natural pasture during winter period where nutrient deficiencies prevail. Hence, this treatment is often called unfavourable nutritional environment for livestock. In this environment, livestock often lose their condition and some often die due to inadequate nutrition. In this trial, these feeding regimes mimic the communal farming environment in South Africa, where veld management is not put into practise by poor resource farmers resulting to poor natural pasture and grasslands. This practice is as result of communal land owners where livestock of each and every livestock owners graze at liberty and that the grazing lands are not demarcated for rotational grazing practises. Both Boer and Nguni goats were subjected to this treatment. Each individual animal performance was recorded and an average performance for each lactation period was determined. Therefore, the breed in the extensive feeding regime was kept on the farm and was allowed to feed on the natural pasture eight hours per day. The weather condition in the farm was dry cold winter with

a temperature range of 0 to 14°C. On the other hand, those animals kept in the intensive feeding regime were kept in each cage, where they were provided with feeds every day. The pelleted feeds were provided twice a day. The animals were not exposed to temperature outside the building. The temperature inside the building ranges from 15 to 30°C making it a perfect simulation of summer temperature for the area. The experimental farm is situated approximately 20 km south from Bloemfontein. It is situated at latitude 28.34° south, longitude 25.89° east and an altitude of 1412 m above sea level. The small stock facility is situated west of the Faculty of Agriculture building of the main campus (Bloemfontein). This location is 1412 meters above sea level. This study was carried out during winter season with low quality or poor nutrition of the natural pasture. A 2 × 2 factorial design (two goat breeds and two nutritional management systems) were used in this trial. Thirty six (36) recently kidded does (18 Boer goats and 18 Nguni rural goats were used. The 50% of Boer goats gave birth to twins, whilst only 17% of Nguni goats had gave birth to twins. The amounts of milk production per feeding regime were measures on individual doe. The investigation was conducted between seven (7) and hundred (100) days post-partum. Half (n = 9) of the Boer goat does and half of (n = 9) the Nguni does were allocated randomly within breed to an extensive (veld) feeding treatment, and the other half per breed to an intensive (high energy) feeding treatment (Table 1).

The does were adapted to the pelleted feed for two weeks prior to the collection period. During the experimental period, clear fresh water was always available ad libitum. These goats were fed pelleted lamb-ram-ewe diet twice per day in a paddock, while the extensive treatment group was allowed to graze freely on natural pastures which was located at Paradys University experimental farm. In this farm, no supplementary feeding was provided to these groups. The pastures consisted of 80% red grass (Themeda triandra), 15% of finger grass (Digitaria eriantha) and weeping love grass (Eragrostis species) and 5% of other minor species. In this group, feed intake was ad libitum. Water was freely available to all the animals. Milk production, teat perimeters (teat width, teat volume and and teat length), body weights and daily feed intake were measured. Two milking periods were undertaken during the trial. Prior (5 min before) to the commencement of the first milking, each doe was intra-muscularly injected with 1ml of oxytocin (Fentocin).

The first milking was not recorded. The second milking which was recorded was done after an interval of two hours. , The does were again injected with 1ml of oxytocin 5 min prior the second milking. During the period of the second milking, each teat milk output was measured individually and this milk yield was used to measure the milk production by each doe. The total milk yield for the milking was calculated by adding the output of both teats, that is, the output of the left and right teat (MPL and MPR respectively), while the total milk yield was recorded as total milk production (TOTLR). Milk production after a 2-h period was extrapolated to 24 h, to give the daily milk production.

The measurements of the teat perimeter were done prior the commencement of each milking. These measurements comprised of teat volume, length and width for both left and right teats and were measured with the aid of a calliper. The teat volume was measured (ml) by displacement of water. This was done by inserting the teat into a glass beaker filled with water (37°C). The water displaced was recovered by 5 L container, measured and recorded as the volume of the teat. Regardless of the feeding regime, all kids were separated from their dams and were only given access to them 5 min before and after each milking times. This practice was done to stimulate milk release and feeding of the kids. All the kids and dams were weighed (kg) weekly throughout the experimental period. These body weight measurements were

Table 2. Mean (±S.D.) variables measured from Boer goats in an intensive feeding regime.

Variable	Early lactation (0-28days)			Mid-lactation (28-56days)			Late lactation (56-84days)		
	Mean (n=30)	(±S.D)	Mean (%)	Mean (n=30)	(±S.D)	Mean (%)	Mean (n=30)	(±S.D)	Mean (%)
Milk production (L/day)	2.97	1.31	32.07343	3.39	1.73	36.60907	2.9	1.23	31.31749
Feed intake (g/kg)	1.49	0.5	42.32955	1.36	0.61	38.63636	0.67	0.49	19.03409
Doe weight(kg/day)	46.8	9.49	34.92016	44.13	8.89	32.92792	43.09	7.35	32.15192
Kids weights (Kg/day)	4.7	0.96	17.49814	8.20	1.66	30.52867	13.96	3.91	51.97319
Teat volume (ml)	4.00	1.18	30.76923	4.14	1.30	31.84615	5.86	2.92	45.07692
Teat width (mm)	4.8	0.50	34.09091	4.82	0.70	34.23295	4.46	0.13	31.67614
Teat length (mm)	18.03	3.90	32.13904	18.8	5.60	33.51159	19.27	6.00	34.34938

Values within each row for each parameter differ significantly ($P<0.01$).

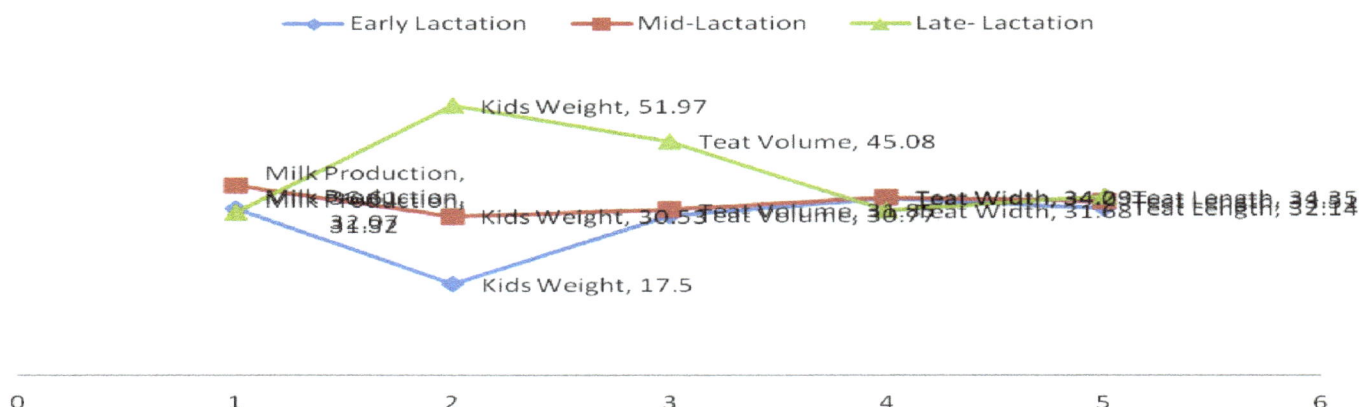

Figure 1. Mean of the milk production, teat perimeters and kids weights in the intensively managed Boer goat for three lactation periods.

recorded prior each milking. The samples were collected weekly from five ($n = 5$) different animals per breed per feeding system. The mean and standard deviations of variables were analysed using the one-way ANOVA with treatments in a 2 × 2 factorial design. Data analysis was carried out using the General Linear Models Procedures of SAS (1991).

RESULTS AND DISCUSSION

Intensive goat production

Performance of Boer goats group in an intensive regime

The results of the performance of the Boer goats in the intensive feeding regime are presented in Table 2. Figure 1, illustrates the results of the performance of milk production, teat perimeters and kids weights in various lactation periods. According to the results, milk production shows an increase of 4.54% (32.07 to 36.61%) trend in the early to mid-lactation period and a decline of 5.29% in the mid to late lactation period. The results of

the feeding intake showed that there was a declining trend during the entire lactation period. On the other hand, the kid weights increased throughout the entire lactation period. In relation to teat perimeter, the results revealed that the teat volume increased throughout the entire lactation period which had similar trend with that of the kid weights. In view of these results of teat volume and kid weights, it appears that teat volume could be the best teat perimeter in predicting kid growth potential as compared to other measurements.

The results further revealed that the teat width and teat length have slightly similar trends with that of milk production (where they all increased in the early to mid-lactation and declined in the mid to late lactation periods). In view of these similarities, it appeared that both teat width and length of Boer goat in the intensive group may be the best measure of milk production. In actual practice, poor resource goat farmers in South Africa do not measure teat width. On the other hand goat farmers in rural poor areas of South Africa estimate the milk production of goat using teat length (as this is important in the milking processes). In view of the traditional milking

Table 3. Mean (±S.D.) variables measured from Indigenous (Nguni) goats in an intensive feeding regime.

Variable	Early lactation (0-28days)			Mid-lactation (28-56days)			Late lactation (56-84days)		
	Mean (n=30)	(±S.D)	Mean (%)	Mean (n=30)	(±S.D)	Mean (%)	Mean (n=30)	(±S.D)	Mean (%)
Milk production (L/day)	1.32	1.56	29.60	1.64	4.65	36.77	1.50	6.48	33.63
Feed intake (g/Kg)	0.85	0.45	37.12	0.71	0.50	31.00	0.73	0.57	31.88
Doe weight(kg/day)	31.99	5.46	33.12	31.52	6.51	32.63	33.09	6.92	34.26
Kids weights (Kg/day)	4.05	0.79	20.81	6.54	2.16	33.61	8.87	2.97	45.58
Teat volume (ml)	4.40	1.90	38.60	2.60	1.87	22.81	2.40	1.83	21.05
Teat width (mm)	5.00	1.70	55.25	3.00	150	33.15	1.01	0.10	11.16
Teat length (mm)	2.20	3.20	47.64	12.10	2.80	28.54	10.10	2.00	23.82

*Values within each row for each parameter differ significantly (P<0.01).

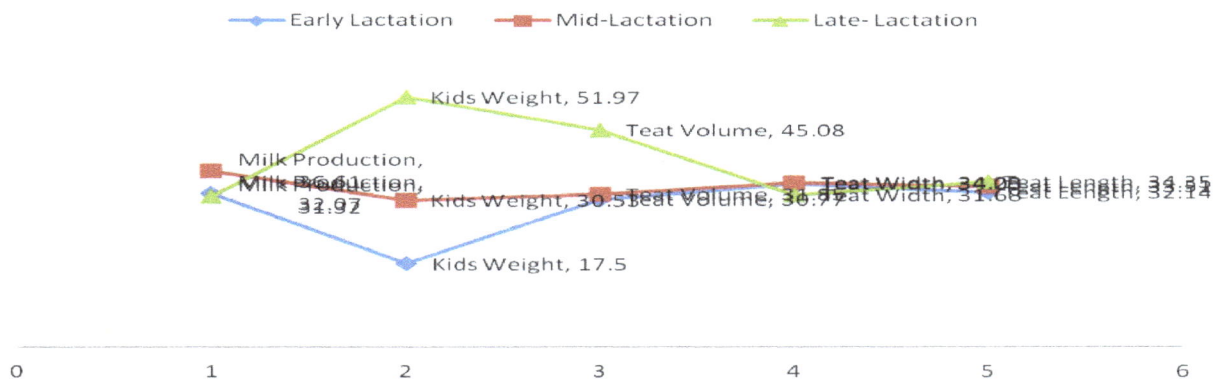

Figure 2. Mean of the milk production, teat perimeters and kids weights for Nguni goats in intensive feeding regime during three lactation period.

practices and the outcome of the trial, it appears that teat length could be recommended as the best measure for milk production potential for the resource poor rural farmers in an intensive system. Another advantage of the teat length over teat width is that it is easier to estimate the teat length visually (without the actual measurements) than teat width. Although, the method used in this experiment to measure teat width was practical and easy, it may be handy for farmers who are well educated.

Performance of Nguni goats group in an intensive feeding regime

Table 3 present the performance of the Nguni goats group in an intensive feeding regime. Figure 2 illustrates the mean milk production, teat perimeters and kid weights for three lactation periods in the intensively managed Nguni goats. The results obtained appeared to be slightly similar in trend compared to that of the Boer goat in the intensive feeding system.

The only observed difference is the results on the teat length, where it was found that there was a slight increase

on the outcome of the measurement during the mid to late lactation period compared to a decrease observed in the Boer goat used in the intensive system (Figure 2). Hence, it is concluded that teat perimeters in this group could be used to predict similar variables as in the Boer goat within the intensive group. Therefore, it is deduced that regardless of breeds' difference, teat length and volume could used to measure milk production and kid weight gains respectively.

Table 4 provides the results of a comparative mean (%) of both Boer and Nguni goats. According to the results, Boer goats are superior in all outcomes of variables measured in this trial. These results appear to indicate that the teat perimeters could be easily measured in the Boer goat group as compared to the Nguni counterpart. This was evident during the measuring processes where it was not as easier to measure the teat perimeters of the Nguni goat as it was when measuring the teat perimeters of the Boer goat due to the sizes difference between the two breeds. However, it is known that Boer goat has a higher genetic potential for milk production traits, compared to the Nguni goat (Greyling et al., 2004). Therefore, those resource poor farmers using Boer goats

Table 4. Comparative mean performance (%) for both Boer and Nguni goats in an intensive feeding regime.

Boer goat	Early lactation Mean (%)	Mid-lactation Mean (%)	Late- lactation Mean (%)
Milk production	2.97 (32.07)	3.39 (36.61)	2.90 (31.32)
Kids weight	4.70 (17.50)	8.20 (30.53)	13.96 (51.97)
Teat volume	4.00 (30.77)	4.14 (31.85)	5.86 (45.08)
Teat width	4.80 (34.09)	4.82 (34.23)	4.46 (31.68)
Teat length	18.03 (32.14)	18.8 (33.51)	19.27 (34.35)
Nguni goat	**Mean (%)**	**Mean (%)**	**Mean (%)**
Milk production	1.32 (29.60)	1.64 (36.77)	1.50 (33.63)
Kids weight	4.05 (20.81)	2.16 (33.61)	8.87 (45.58)
Teat volume	4.40 (38.60)	1.87 (22.81)	2.40 (21.05)
Teat width	5.00 (55.25)	1.50 (33.15)	1.01 (11.16)
Teat length	2.20 (47.64)	2.80 (28.54)	10.10 (23.82)

Table 5. Mean (±S.D.) variables measured from Boer goat s in an extensive feeding regime.

Variable	Early lactation (0-28days)			Mid-lactation (28-56days)			Late lactation (56-84days)		
	Mean (n=30)	(±S.D)	Mean (%)	Mean (n=30)	(±S.D)	Mean (%)	Mean (n=30)	(±S.D)	Mean (%)
Milk production (L/day)	0.67	0.88	31.46	0.87	0.58	40.85	0.89	0.67	41.78
Kids weights (Kg/day)	4.52	1.52	22.77	7.13	1.97	35.92	8.20	2.51	41.31
Teat volume (ml)	5.60	1.31	43.41	3.4	1.13	26.36	2.70	1.90	20.93
Teat width (mm)	4.07	1.40	33.36	4.06	1.35	33.28	4.07	1.40	33.36
Teat length (mm)	13.46	3.20	34.51	12.61	2.53	32.33	13.03	3.01	33.41

*Values within each row for each parameter differ significantly (P<0.01).

compared to Nguni goats are in a slightly advantageous position should they wish to use teat perimeter measurements as the measure of the kid weights and milk production potential due to the above-mentioned advantage.

Extensive goat production

Performance measurements of Boer goats in an extensive feeding regime

The results of the performance of the Boer goat in the extensive feeding regime are shown in Table 5 and illustrated graphically in Figure 3 respectively. According to these results, milk production increased marginally in all phases of lactation periods. On the other hand, the results also show that the kids growth show higher increase throughout the entire lactation period.

In relation to the teat perimeters, none of them show any similar trend with that of the milk production and kid growth. Therefore, it can be deduced that in the extensive feeding system, it is not possible for the poor resource farmer who own Boer goat to use teat perimeters as a predictor for both milk production and kid growth potentials.

Performance measurements of Nguni goats in an extensive feeding regime

Table 6 and Figure 4 presents and illustrate graphically the results of the Nguni goats in the extensive feeding regime respectively. According to these results, milk production shows a declining trend throughout the entire lactation period. On the contrary, kid weight gain shows an increasing trend. In relation to the teat perimeter measurements, the results revealed that all teat perimeter measured followed a trend similar to that of the milk production.

In view of the above results, it appears that all the teat perimeter measurements could be used to predict milk production potential as opposed to kid weight gain. This appears to suggest that poor resource farmers who are

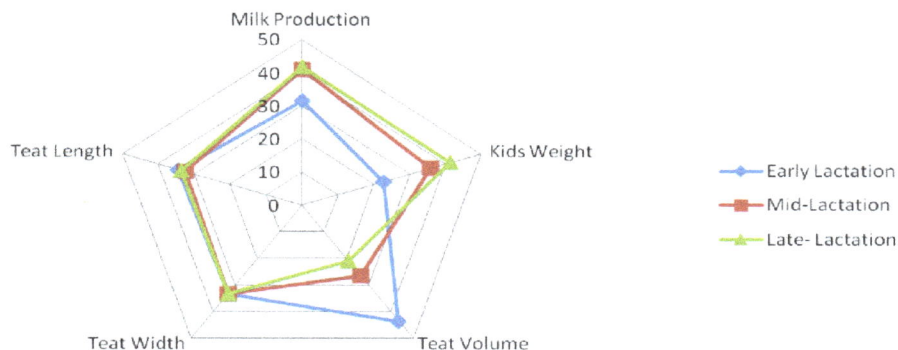

Figure 3. Illustration of the performance of Boer goats in the extensive feeding regime.

Table 6. Mean (±S.D.) variables measured from Indigenous (Nguni) goats in an extensive feeding regime.

Variable	Early lactation (0-28days)			Mid-lactation (28-56days)			Late lactation (56-84days)		
	Mean (n=30)	(±S.D)	Mean (%)	Mean (n=30)	(±S.D)	Mean (%)	Mean (n=30)	(±S.D)	Mean (%)
Milk production (L/day)	0.60	0.54	33.52	0.59	0.24	32.96	0.57	0.79	31.84
Kids weights (Kg/day)	3.85	0.84	51.40	8.20	1.66	10.94	5.76	11.31	16.90
Teat volume (ml)	4.57	1.53	40.62	3.75	1.42	33.33	2.93	1.30	26.04
Teat width (mm)	4.01	1.09	36.45	3.70	1.0	33.33	3.39	1.05	30.54
Teat length (mm)	12.02	3.70	34.44	11.70	3.4	33.52	11.18	2.8	32.03

*Values within each row for each parameter differ significantly (P<0.01).

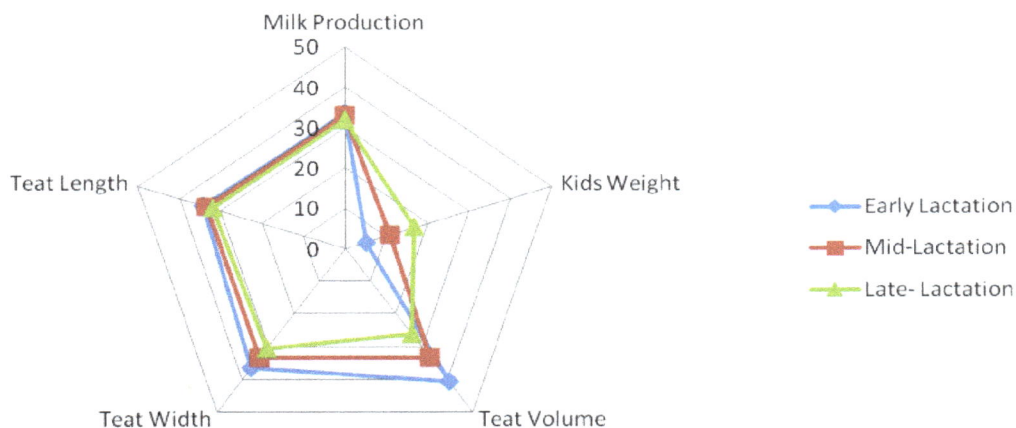

Figure 4. Illustration of the performance of Boer goats in the extensive feeding regime.

farming with Nguni goat in the extensive management system could use any of the teat perimeters to predict milk production. This makes this breed more suitable for milk production by resource poor farmers due to their less or no demand in requiring sophisticated tools for predicting milk production.

Comparative analysis of performance of both Boer and Nguni in the extensive system

Table 7 and Figure 5 present results of the comparative analysis of both the breeds in the extensive system. In view of the results, it appears that milk production of the

Table 7. Comparative mean performance (%) for both Boer and Nguni goats in extensive feeding regime.

Boer goat	Early lactation	Mid-lactation	Late- lactation
	Mean (%)	Mean (%)	Mean (%)
Milk production (L/day)	0.67 (31.46)	0.87 (40.85)	0.89 (41.78)
Kids weight (kg/day)	4.52 (22.77)	7.13 (35.92)	8.20 (41.31)
Teat volume (ml)	5.60 (43.41)	3.40 (26.36)	2.70 (20.93)
Teat width (mm)	4.07 (33.36)	4.06 (33.28)	4.07 (33.36)
Teat length (mm)	13.46 (34.51)	12.61 (32.33)	13.03 (33.41)
Nguni goat	**Mean (%)**	**Mean (%)**	**Mean (%)**
Milk production (L/day)	0.60 (33.52)	0.59 (32.96)	0.57 (31.84)
Kids weight (Kg/day)	3.85 (51.40)	8.20 (10.94)	11.31 (16.90)
Teat volume (ml)	4.57 (40.62)	3.75 (33.33)	1.30 (26.04)
Teat width (mm)	4.01 (36.45)	3.70 (33.33)	1.05 (30.54)
Teat length (mm)	12.02 (34.41)	11.70 (33.52)	2.80 (32.03)

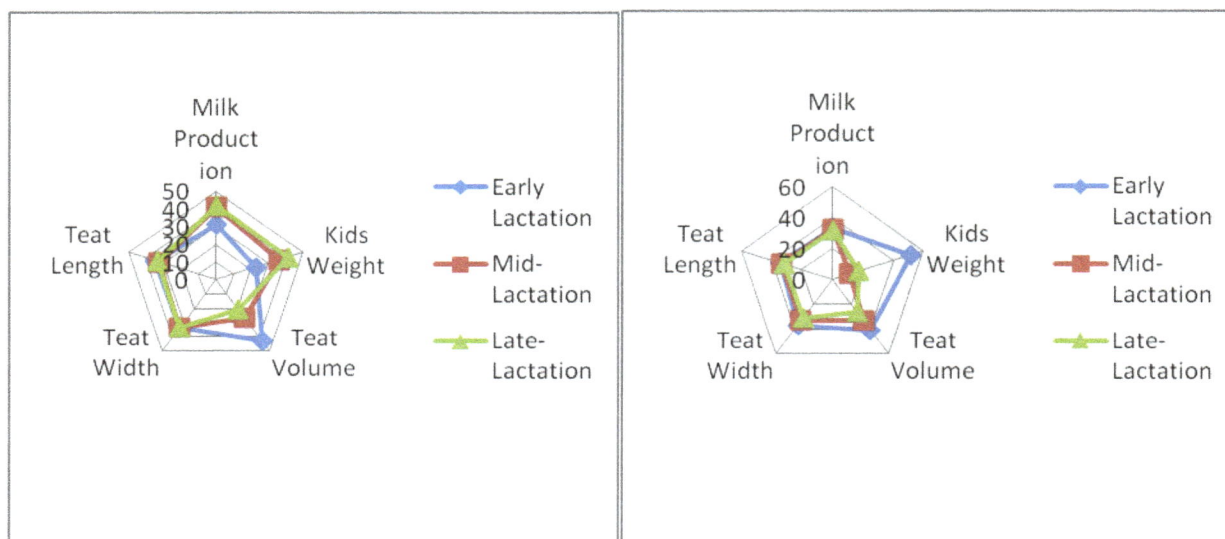

Figure 5. Illustration of the performance of both Boer and Nguni goat groups.

Boer goat increased throughout the entire lactation period in contrast with the milk production of the Nguni goat.

Regarding the results of the kid weight gain, the results revealed that for the Boer goats, the kid weight gain increased throughout the entire lactation period. These results are in contrast with that of the Nguni whose kid weight gain only increased during the mid to late lactation, after its decline in the early to mid lactation periods. In terms of the teat perimeter measurements, Boer goat teat measurements were found to be inconsistent relative to the Nguni goat.

The comparative analysis shows that there is a significant difference (P<0.01) in these breeds under this nutritional regime considered. These results also may be a confirmation that teat perimeters in the Nguni goats can be used in predicting the milk production potential. The results may also suggest that Nguni goats are slightly superior in term of their adaptability in harsh nutritional feeding regime. In this trial, it was also found that the left and right teat measurements were positively correlated (r = 0.21 vs. r = 0.23, r = 0.46 vs. r = 0.36, r = 0.55 vs. r = 0.52) for teat volume, width and length with the milk output. This observation is complementary to the findings by Montaldo and Martinez-Lozano (1993), who found milk production to be correlated significantly (P<0.01) with teat perimeter (r = 0.45). However, data on milk production and kid weight gain to the findings of Montaldo and Martinez-Lozano (1993) mentioned previously. In other words, milk production and teat perimeters do not show a defined correlation. The teat perimeters show clearly small positive correlations. In general, these results may indicate that teat perimeters of goats in a favourable nutritional environment may serve as indicator of milk production through kid weight gain and hence, may be used to measure the kid growth potentials. In unfavorable nutritional environments, the breeds have a significant impact ($P<0.01$) to degree in which teat perimeters may be used to predict kid growth potential.

Conclusion

Teat and udder perimeters are important traits in milk production and growth potential of the kids. This study revealed amongst others that goats possessing higher live weight produce more milk if nutritional conditions are suitable. Similarly, it was also found that does with higher milk production had kids with higher total weaning weights, compared to those with less milk production. This indicates that milk production plays an important role in the daily weight gain of the kids. The teat perimeters (volume, width and length) were found to have small positive correlation with milk production. This implies that teat perimeters can possibly be used as a tool to predict the milk production of does, as compared to evaluating the physical appearance and size of the udder. Teat volume perimeters were found to be the best predictor of kid weights in intensively fed goats. On the other hand, the teat width and length were found to be the best predictor of milk production in the intensive feeding regime for both breeds under consideration.

Feeding management practices

In addition, it was also found that the feeding regime and breeds have a significant impact on the teat perimeter measurements, milk production and kids' growth potential. For instance, it was found that in the extensive feeding regime, Nguni goats are the only goats where teat perimeter measurements could be used to predict milk production in the winter when allowed to graze on natural pasture. Therefore, under harsh nutritional regime, Nguni goats appear to be more adaptable than the Boer goats such that owners could use teat perimeters to predict their milk production.

Seasonal effect on teat perimeter measurement

These results appear to indicate that in an event where a farmer is producing milk in summer and winter respectively from both breeds, he or she can use teat perimeters for both breeds during summer (where nutrition is assumed to be in abundance). However, in winter seasons teat perimeter could be used to predict milk production by the farmers only in the Nguni goats. Result from this experiment is highly suitable for resource poor goat farmers who are farming in any season of the year due to lack of breeding planning and programs.

ACKNOWLEDGEMENTS

The authors would like to thank the Department of Animal, Wildlife and Grassland Sciences at the University of the Free State and the National Research Foundation (NRF) for supporting the project.

REFERENCES

Ahuya CO, Ojango JMK, Mosi RO, Peacock CP, Okeyo AM (2009). Performance of the Toggenburg dairy goats in smallholder production systems of eastern highlands of Kenya. Small Rumin. Res. 83:7-13.

Devendra C, Burns M (1970). Goat Production in the Tropics. Commonwealth Agriculture Bureau, Farnhall, UK. pp. 48–65.

Fourie PJ, Neser FWC, Olivier JJ, van der Westhuizen C (2002). Relationship between production performance, visual appraisal and body measurements of young Dorper rams. S. Afr. J. Anim. Sci. 32(4):256-262.

Gall C (1981). Milk production. In: Gall, C. (Ed.), Goat Production. Academic Press, London, pp. 309–344.

Godfrey RW, Gray ML, Collins JR (1997). Lamb growth and milk production of hair and wool sheep in a semi-arid tropical environment. Small Rumin. Res. 24:77-83.

Greyling JPC, Mmbengwa VM, Schwalbach LMJ, Muller T., (2004). Comparative milk production potential of Indigenous and Boer goats under two feeding systems in South Africa. Small Rumin. Res. 55:97-105.

King FJM (2009). Production parameters for boer goats in South Africa. MSc dissertation, UFS, Bloemfontein, RSA. pp. 1-60.

Linzell JL (1966). Measurement of udder volume in live goats as an index of mammary growth and function. J. Dairy Sci. 49:307.

Maiwashe AN (2000). The value of recording body measurements in beef cattle. MSc (Agric) dissertation, University of the Orange Free Sate, South Africa. pp. 1-50.

Mammabolo MJ, Webb EC, Du Preez ER, Morris SD (1998). Proceeding of the workshop on "Research and strategies for goat production systems in South Africa" EC Webb, PB Cronje and EF Donkin(Ed). pp. 79-85.

Montaldo H, Martinez-lozano FJ (1993). Phenotypic relationship between udder and Milk characteristics, milk production and California mastitis test in goats. Small Rumin. Res. 12:329-337

Mmbengwa VM (2009). Milk production of South African Boer and Indigenous (Nguni) feral goats under intensive and extensive feeding systems. MSc dissertation, UFS, Bloemfontein, RSA. pp. 1-60.

Pambu RG (2011). Effects of goat phenotype score on milk characteristics and blood parameters of indigenous and improved dairy goats in South Africa. PhD Thesis. University of Pretoria, Pretoria, RSA. pp. 1-75.

Peris S, Gaja G, Such X, Casals R, Ferret A, Torre C (1997). Influence of kid rearing systems on milk composition and yield of Murciano-Granadina Dairy Goats. J. Dairy Sci. 80:3249-3255.

SAS (1991). Statistical Analysis System user's guide, Ver. 6. SAS Institute Inc., Cary, N.C. USA.

Simela L, Merkel R (2008). The contribution of chevon from Africa to global meat production. Meat Sci. 80:101-109

5

Performance evaluation of the dairy farmers regarding adoption of precise dairy farming practices in the Punjab, Pakistan

Saleem Ashraf, Muhammad Iftikhar, Ghazanfar Ali Khan, Babar Shahbaz and Ijaz Ashraf

Institute of Agricultural Extension and rural Development, University of Agriculture, 38040- Faisalabad, Pakistan.

Livestock is the dominant cog of economy of Pakistan and also livelihood supporter fulfiller for the subsistent farmers of country. The country has a great potential and prosperous with dairy animals' breeds. Unfortunately, farmers are so far away to cash in the actual potential of animals. Non-adoption of improved practices is the major productivity retarding factor. In this regard, present study was conducted in the different areas of Punjab, Pakistan. Data were collected from the 107 livestock farmers. The sample was a blend of landless, illiterate and young, middle aged and small farmers who were immensely dependent on livestock. Data analysis indicated that cattle were the more likely animal among farmers. Farmers were interested in different management practices regarding feeding, breeding and disease management. However, farmers were adopting the only practice which needs no investment and technicality. Farmers were lacking in technical knowledge to keep animals healthy therefore it got mean value of 2.77 influenced by non-cooperation of livestock extension agents with maximum mean value of 2.39 as a major constraint. Finance shortage ($\bar{x}$=2.41) and high inputs prices ($\bar{x}$=2.81) were other constraints compelling farmers to adopt traditional practices. Correlation analysis indicated highly significant association of age, education, land size and annual income with adoption of scientific dairy practices. On the basis of results, it is suggested that the role of livestock extension should be diversified under strict evaluation. Micro credit schemes with cooperation of government may also boost the adoption. It is also inferred that international organization may also start some projects keeping in mind the ultimate potential of Pakistan dairy sector.

Key words: Livestock, dairy farming, adoption, Punjab.

INTRODUCTION

Livestock is an important part of agriculture along with crop farming having 55.1% stake in agriculture (Government of Pakistan, 2012). Pakistan is bestowed with distinctive geographical location and environment which is supportive for the rearing of multipurpose animals (PBIT, 2011). Among various breeds of animals' castles, buffaloes, goats and sheep are more important. These animals are domestic animals and farmers rely on their products such as milk and meat for their better livelihood.

In world Pakistan ranks 4th milk producer with average production of 37,475 thousand tons per year for human consumption (Government of Pakistan, 2011) and 2nd largest buffalo producer. Buffalo and cattle share 68 and 27% of milk, respectively in the total produces milk in an entire country (Raza and Rabbani, 2012). Livestock sector play a vital role in the economy of Pakistan and dairy animals are the essence of livestock wealth.

Performance evaluation of the dairy farmers regarding adoption of precise dairy farming practices in the Punjab...

31

The livestock sector was also influenced by the massive floods and showed significant declining growth at 3.7% in 2010-2011 as against 4.3% last year (Government of Pakistan, 2011). Globally, Pakistan stands among the 5 largest countries contributing 570 billion annually to the national economy (Kakakhel, 2010). Country is gifted with dairy breeds of various farm animals. Regarding buffalos' nilli-Ravi, Kundhi and regarding cattle, Sahiwal, Red Sindhi, Cholistani are nationally renowned. In case of small ruminants such as goats, beetal, Dera Din Panah, and kamori are famous. These all breeds are well renowned because of their enormous potential. More than 70% dairy animals are owned by small farmers having herd size of 1 to 10 animals.

Livestock farmers are considered as spine of national dairy industry through supply of greater than 80% of market milk for consumption in various sectors. Mostly small or landless farmers are the owners of these significant animals such as cattle, buffaloes and goats (Government of Pakistan, 2011).

Various dairy farming systems are adopted in Pakistan such as rural subsistence smallholdings, rural commercial farms, market-oriented smallholdings and peri-urban commercial dairy farms. The ultimate purpose of these systems is to strengthen their livelihood through earning from different dairy products. Livestock farmers sell milk, meat and other products to earn better livelihoods. Meanwhile, they also use these products at their home for food security. Especially landless and small farmers are connected with this setup. These livestock farmers mostly rely on road side, canal banks and water channel sides grazing for their herds feeding. However, this practice remains insufficient and animals remain underfed. More importantly farmers focus more on lactating animals and they do their best for their furnished diet (Raza et al., 2006).

Despite of the vital importance of livestock and dependency of farmers the productivity is far below than the actual potential. Several factors are responsible for this low production. Ghafoor (2003) and Gillespie et al. (2007) reported poor finance earning, lack of appropriate knowledge, negative attitude of government and poor marketing facilities, as factors responsible for the non-adoption of improved dairy farm practices. Moreover, according to Arif et al. (2013), inadequate feed resources, unawareness of artificial insemination, finance shortage and limited health facilities were the major constraints among farmers. In this context, mostly small farmer or landless farmer suffer more because they fail to get the targeted outcome and in this way their livelihood becomes more diverse. Majority of farmers remain unaware of new practices and low adoption of recommended practices are also cause of poor dissemination of information through information sources (Ahmed et al., 2004). Adoption can be enhances through delivery of accurate extension services regarding all the dairying aspects (SMEDA, 2011). Keep in mind these facts present study was conducted in the different areas of Punjab, Pakistan.

METHODOLOGY

Punjab province is famous for livestock as most of the respondents are dependent on livestock. Three districts as Muzaffargarh, Khanewal and Nankana Sahib were selected purposively for the data collection. All of these districts have unique importance regarding livestock. Total 107 livestock farmers were interviewed from these three selected areas purposively. For the data collection an interview schedule was developed. The interview schedule was based on various aspects of dairying management such as feeding, breeding, disease and management practices and was checked by the livestock experts of Animal Husbandry Department, University of Agriculture Faisalabad, Pakistan. All the suggestions were incorporated in interview schedule.

Later on for further reliability interview schedule was pre tested on 5 livestock farmer. Three point Likert scale was used for data collection. Data were collected from the targeted respondents in selected areas. Farmers were interviewed at their farms and face to face interviews were preferred. Collected data were analyzed through Computer software Statistical Package for Social Sciences (SPSS). Descriptive statistics (Mean, Standard Deviation) and inferential statistics were applied for the interpretation of data analyzed. Ranking of adoption status and constraints faced by livestock farmers was formulated on means values basis. Furthermore, correlation analysis was applied to investigate the association between demographic characteristics of the respondents with adoption status of farmers' regarding scientific practices.

RESULTS AND DISCUSSION

Table 1 is the representation of respondents' personal demographic attributes. According the data mentioned in Table 1, majority of the farmers were of middle aged. Middle aged category appeared prominent with the percentage of 43.9% respondents. Inclination of young and middle aged farmers represents the interest in livestock business. Regarding education 15.9% respondent were found, who never attended the formal education, except to these illiterate people up to primary level education holding respondents were more prominent. This scenario reports the poor literacy level of farmers. More than one third of total respondents were basically landless and were tenants. Generally, overwhelming majority were small farmers with land holding size of up to 12.5 acres. Major income sources observed in the farm level were crop sale and livestock while some were also doing some sort of private business. Total income derived from these different income sources appeared not more than 2 lacs as just 13.1% respondents were earning more than 2 lacs rupees. Mostly, farmers rear dairy animals for better income through milk selling and young animals such as calf and heifers selling. Small rudiments such as goat and sheep are not in good trend nowadays because of low profit. People were having various animals such buffaloes and

Table 1. Demographic characteristics of farmers.

Demographic characteristics of farmers		Frequency
Age (Year)	<35 (Young)	39 (36.4)
	35-50 (Middle age)	47(43.9)
	>50 (Old)	21(19.6)
Education	Illiterate	17(15.9)
	Up to primary	37(34.6)
	Middle-matriculation	24(22.4)
	Above matriculation	29(27.1)
Land holding size	Landless	42(39.3)
	Up to 5 acre	37(34.6)
	5-10 acre	12(11.2)
	Above10 acre	16(15)
Annual income	<1 lac	35(32.7)
	1-2 lac	58(54.2)
	>2 lac	14(13.1)

Source: Field Data 2013. Note: Values in parenthesis are percentages.

cattle for different purpose such as milk purpose. Farmers owned different number of animals but mostly animals on their farm were in range of 2 to 3. Some progressive farmers were also interviewed who were holding a larger herd size. Majority (60.7%) of respondents was holding up to 2 buffaloes in milk. Lesser number of respondents (8.41%) owned buffaloes in milk more than 5. More than one third respondents (39.3%) owned buffaloes (heifers, young and dry). Farmers were more interested in cattle therefore majority of farmers owned cattle including in milk and heifers, dry and young cattle. About half of the respondents owned cattle population up to 2 while one third respondents were found with in milk cattle population of 2 to 5. Population of small rudiments such as goats and sheep was not so good due less profit. During data collection, it was rarely seen that someone is holding goats and sheep except someone who were rearing for domestic purposes. During informal discussion few respondents also stated that goats and sheep did not have impressive benefits, infact they have losses. Mostly these small animals cause damages to crops and home based plantations.

Data depicted in Table 1 revealed that, in general most of the farmers were inclined toward the adoption of scientific dairy practices. The hidden reality behind this adoption was their dependency of livelihoods on livestock. During data collection, it was observed that farmers were so caring and loving to their animals but they were doing in their own way according their available resources and knowledge that they have. Farmers' were enquired regarding breeding, feeding, disease and management practices. Regarding breeding

practices own cross bred cows got the maimum mean value ($=2.13$). This indicates that majority of the farmer were having cattle especially cow. During informal discussion farmers reported the early growth of cows and calf as a reason for more inclination toward cows. Lesser number of respondents was inclined toward artificial insemination therefore it got mean value of 1.85. It was observed that farmers used breeding practices according the feasibility, sometime they breed animals naturally, some time both artificial insemination and natural mating while some one also show interest in their farm produced bull for natural breeding. Same observation was reported by Arif et al. (2013).

Feeding practices were not being adopted according the recommendations as the farmers were not so good in financial condition. oreover, farmers were also lacing in technical nowledge that is wh adoption revealed lower. Farmers were reling more on fodder and roughages ($=2.91$). Diseases cause several losses in animals in form of week health and deaths. In this regard, farmers were mostly caring and were adopting precautionary measures such as regular cleaning, deworming of calves, hygienic milking and vaccination against the diseases. These results in respect of vaccination coincide with Karim and Najeeb (2001).

Colostrums feeding to calves were adopted almost by overwhelming majority because farmers consider calves their net income source assets. These results coincides with Arif et al. (2013) and Fulwider et al. (2007) where they reported many farmers offering 3.8 liters/day milk as first feed to calves. Income availability is the key to avail all the facilities and poor financial condition did not

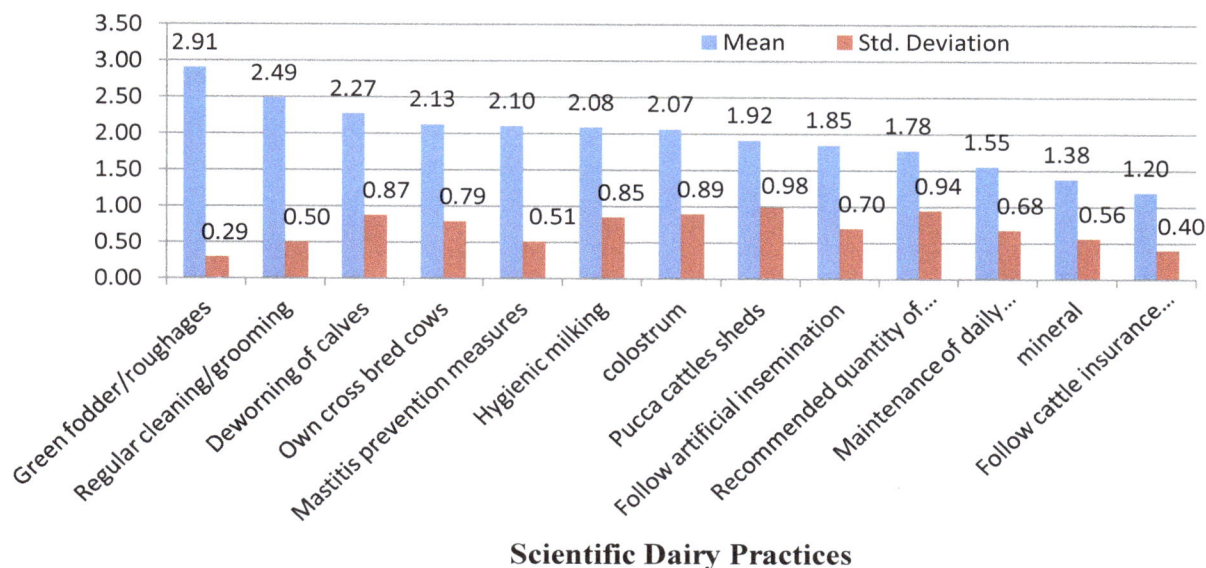

Scientific Dairy Practices

Figure 1. Descending Means in regard of farmers adoption level.

Figure 1. Descending Means in regard of farmers adoption level.

allow many farmers to construct pucca cattle shed for animas. It was observed that farmers were placed inside the house and families were living along with them without any precautions and fear of diseases. Poor literacy was responsible for the low awareness regarding cattle insurance. Only progressive farmers were maintaining the daily management records of their animals. Record keeping system had significant link with increased milk production (Tomaszewski, 1993).

It is clear from Figure 1 that green fodder roughages (2. 1 usage got the ma i mum means value followed b the regular cleaning (2.) and deworming of calves (=2.27). Own cross bred cows attained the 4th highest means value (=2.13). If we look in depth that all the practices that do not need any technicality and expenses got the maximum means. Farmers were doing these practices at their own within the available resources. Other technical aspects such as recommend quantity of concentrates, maintenance of daily records, mineral utilization and cattle insurance go the lower mean values. These aspects are highly technical and need investment, therefore farmers were away of their adoption because of their limited resources and lack of technical knowledge.

From the dissemination to adoption of technology, it is surrounded by various factors which affect its adoption and these factors are simply known as constraints. These constraints enormousl e ist at farmers' level. According to the data mentioned in Table 2, inadequate technical advice by livestock extension department attained the highest mean value and was ranked first. Farmers argued that they have never seen the livestock extension agent here in the area. These comments raise the critics on the livestock extension agents and their deficient role describes their la iness. ac of government support was

ran e d second (2.) followed b the high cost of livestoc inputs (=2.80). Due to these high rates, farmers were not adopting minerals and recommended quantities of concentrates for their animals as they cannot purchase these inputs at high rates.

Naylor et al. (2005) suggested that livestock production can only be increased by reducing the feed prices. Ramsey et al. (2005) also reported that several production practices can increase cow herd returns either by increasing revenue or by reducing costs (Table 3). Farmers were not having technical knowledge and even they were unaware of this technical knowledge. Farmers stated this lack of technical knowledge as the major constraint toward adoption (Table 4). Low quality inputs were another major constraint regarding feeding practices among farmers while provision of balanced rations is vital for better survival of animals (Riley et al., 2004). Lack of training facilities got the mean value ( ) of 2. 1 . ithout training, adoption cannot be enhanced and livestoc e t ension agents are considered trainers among the farmers while the role of these agents is alread negligible. Availabilit of labour (1.) and vaccines (=1.48) were not considered as significant constraints. Vaccines are easily available in the market but their availability was considered as constraints among the farmers living in remote area and where the females were the heads of family.

Association between demographic characteristics and adoption of dairy practices

Table 5 revealed the association checked between demographic characteristics and the adoption level of

Table 2. Status of farmers regarding owned herd size.

Class of animals	No. of animals	Owners (Percentage)	
		f	%
Buffalo (in milk)	< 2	65	60.7
	2-5	4	3.7
	>5	9	8.41
Buffalo (Heifers, dry and young)	< 2	42	39.3
	2-5	8	7.5
	>5	3	2.8
Cattle (in milk)	< 2	48	44.9
	2-5	33	30.8
	>5	4	3.7
Cattle (Heifers, dry and young)	< 2	40	37.4
	2-5	29	27.1
	>5	11	10.2
Adult goat	< 2	13	12.1
	2-5	18	16.8
	>5	-	-
Young goat	< 2	17	15.9
	2-5	19	17.8
	>5	-	-
Sheep	< 2	3	2.8
	2-5	2	1.9
	>5	-	-

Source: Field Data 2013.

of livestock farmers regarding scientific dairy farming practices. Regarding age and adoption mix response was found as with some particulars such as own cross bred cows and regular cleaning positive association was found while with other particulars such as artificial insemination, utilization of recommended concentrated and green forages or roughages positive association was revealed. Positive but negative association of age was found with vaccination and colostrums feeding to the calves. It can be said on the basis of results that as the age increase, interest of livestock farming goes down as potential of farmers does not allow them to do hard work. As compared to these old ages, young aged farmers may have better intensions toward adoption.

Education plays vital role in awareness and adoption of any innovation. In that case, highly significant relation of education was found with most of the particulars except mastitis prevention measures and cattle insurance. Literacy level of the study areas was not so good as 15.9% respondents found were illiterate. On the basis of

results, it can be concluded that because low adoption was influenced by the poor literacy level. Land is another major factor.

Farmers with large land holding have better intentions toward livestock farming and adoption of improved practices. According to correlation analysis, land holding size was highly associated with the adoption as highly significant association as found with most of the particulars except own bred cows, artificial insemination and green fodder or roughages.

Finance is always needed to purchase inputs and financially sound farmers can afford any type of concentrated but financially poor farmers cannot. Income showed highly significant relationship with vaccination against disease and mastitis prevention measures as both of these particulars are money consuming. Age was also significantly associated with adoption of recommended concentration of concentrates. Moreover, income was positively or negatively connected with rest of the particulars.

Table 3. Farmers level of adoption regarding scientific dairy practices.

Scientific dairy practices		Adoption	
		Mean	Std. deviation
Breeding practice	Own cross bred cows	2.13	0.79
	Follow artificial insemination	1.85	0.70
Feeding practice	Recommended quantity of concentrated	1.78	0.94
	Green fodder/roughages	2.91	0.29
	Mineral and vitamin supplements	1.38	0.56
Disease control practice	Regular cleaning/grooming	2.49	0.50
	Vaccination against diseases	2.37	0.61
	Hygienic milking	2.08	0.85
	Deworning of calves	2.27	0.87
	Mastitis prevention measures	2.10	0.51
Management practices	Follow cattle insurance practice	1.20	0.40
	Feed colostrums to newly born calves	2.07	0.89
	Pucca cattles sheds	1.92	0.98
	Maintenance of daily management records	1.55	0.68

Source: Field Data 2013.

Table 4. Ranking of constraints hindering the adoption level.

Constraints	Means	Rank	Std. deviation
Inadequate technical advice by livestock extension department	2.93	1	0.36
Lack of government support	2.87	2	0.46
High cost of livestock inputs	2.80	3	0.56
Lack of technical knowledge	2.77	4	0.51
Lack of awareness about technical knowledge	2.71	5	0.50
Low quality of livestock inputs	2.64	6	0.62
Lack of training facilities	2.61	7	0.66
Literacy	2.51	8	0.81
Lack of financial resources	2.41	9	0.90
Availability of credit/loan	2.38	10	0.88
Marketing of animals	2.33	11	0.80
Less availability of land	2.32	12	0.91
Lack of grazing field	2.32	13	0.94
Shortage of fodder in winter season	2.14	14	0.86
Availability of labour	1.63	15	0.58
Availability of vaccines against infectious diseases	1.48	16	0.77

Source: Field Data 2013.

CONCLUSION AND RECOMMENDATION

It is concluded on the basis of present study that whole livestock husbandry is based on traditional techniques resulting half way productivity than potential due to several factors. Farmers do not have technical knowledge and they lack in resources to create interest toward technicality attainments. More importantly, role of Livestock Extension Field Staff was negligible.

Addressing these factors only on paper will not boost the dairy productivity until the reduction of these factors through proper policies. To improve the economic viability of dairy farming communities, it is essential to modify the role of livestock extension field staff though proper trainings and strict evaluations. It will also be good to create chances for the collaboration of international companies to launch some projects as dairy in Pakistan hold immense potential. Correlation analysis revealed

Table 5. Association between demographic characteristics of livestock farmers and adoption of scientific dairy farming practices.

Particulars	X1	X2	X3	X4	X5	X6	X7	X8	X9	X10	X11	X12	X13	X14
Age	0.006	-0.087	-0.069	-0.074	-0.071	0.122	-0.218(*)	0.068	-0.016	-0.154	0.056	-0.209(*)	-0.012	0.046
	(0.953)	(0.376)	(0.483)	(0.449)	(0.466)	(0.212)	(0.024)	(0.483)	(0.866)	(0.112)	(0.569)	(0.031)	(0.904)	(0.046)
Education	0.403(**)	-0.440(**)	0.650(**)	0.278(**)	-0.222(*)	0.489(**)	0.557(**)	0.259(**)	-0.621(**)	-0.099	0.188	0.455(**)	0.248(**)	0.406(**)
	(0.00)	(0.000)	(0.000)	(0.004)	(0.021)	(0.000)	(0.000)	(0.007)	(0.000)	(0.309)	(0.053)	(0.000)	(0.010)	(0.000)
Land	0.076	-0.112	0.553(**)	0.189	0.403(**)	0.535(**)	0.519(**)	0.620(**)	-0.476(**)	0.293(**)	0.422(**)	0.260(**)	0.419(**)	0.475(**)
	(0.435)	(0.253)	(0.000)	(0.051)	(0.000)	(0.000)	(0.000)	(0.000)	(0.000)	(0.002)	(0.000)	(0.007)	(0.000)	(0.000)
Income	0.142	-0.065	0.204(*)	0.151	-0.102	0.006	0.307(**)	-0.107	0.011	0.316(**)	0.077	0.186	-0.173	0.003
	(0.144)	(0.504)	(0.035)	(0.122)	(0.295)	(0.952)	(0.001)	(0.275)	(0.907)	(0.001)	(0.432)	(0.055)	(0.076)	(0.972)

** Significant at 0.01 level, *Significant at 0.05 level. Note: values given in parenthesis are significance level. X1=Own cross bred cows , X2=Follow artificial insemination, X3=Recommended quantity of concentrated, X4=Green fodder/roughages, X5=Mineral and vitamin supplements, X6=Regular cleaning/grooming, X7=Vaccination against diseases, X8=Hygienic milking, X9=Deworming of calves, X10=Mastitis prevention measures, X11=Follow cattle insurance practice, X12=Feed colostrums to newly born calves, X13=Pucca cattle sheds, X14=Maintenance of daily management records.

highly significant association of education and income with adoption therefore, it is also suggested that government should introduce farmers' trainings strategies and subsidies on inputs along with micro credit schemes to cover up the finance shortage.

REFERENCES

Ahmed AM, Simeon E, Yemesrach A (2004). Dairy development in Ethiopia, International Food Policy Research Institute, 2033 K Street, Washington DC 20006 U.S.A.

Arif M, Iqbal A, Younas M, Khan BB, Sarwarand M, Ahmad S (2013). Prospects and limitations of dairying in Gujranwala district (Punjab-Pakistan). Proc. Int. Workshop Dairy Sci. Park. Nov 21-23, 2011, Agric. Univ. Peshawar, Pakistan. J. Anim. Plant Sci. 23(Sup. 1):2013.

Fulwider WK, Grandin T, Rollin BE, Engle TE, Dalsted NL, Lamm WD (2007). Survey of dairy management practices on one hundred thirteen North Central and Northeastern United States Dairies. J. Dairy Sci. 91:1686-1692

Ghafoor A (2003). WTO regime: livestock scenario. Daily Dawn. Available online on www.dawn.com.pk.

Gillespie J, Kim A, Paudel S (200). h don't producers adopt best management practices? An analysis of the beef cattle industry. J. Agric. Econ. 36(1):89-102.

Government of Pakistan (2011). Ministry of Finance, Economic Advisors Wing, Islamabad.

Kakakhel I (2010). Daily Times. Dated February, 27.

Karim M, Najeeb S (2001). An evaluation of extension activities of LPRI, Bahadurnagar during the year 1999-2000in Okara District. 22nd Annual Report, Livestock Production Research Institute Bahadurnagar, Okara. pp. 115-122.

Naylor R. Steinfeld H, Falcon W, Galloway J, Smil V, Bradford E, Alder J, oone H (2005). " osing the lin s between livestoc and land." J. ivest. Res. Rural Dev. 3(1):1621-1622.

PBIT (2011). Sector for investment. Available on line on http:// pbit.gop.mpk/ Sectors/ Livestock. Aspx. Accessed on 22-02-2012.

Ramsey R, Doye D, Ward C, McGrann J. Falconer L, Bevers S (2005). "Factors Affecting Beef Cow-Herd Costs, Production, and Prof-it s." J. Agric. Appl. Econ. 37(1):91-99.

Raza SH, Riaz MV, Iqbal A (2006). Milk productivity: A changing scenario for future investment. In: Proceed. Int. Conf. Productivity and Growth in Agriculture: Strategies andInterventions, 6-7 December, Univ. Agri. Faisalabad.

Raza S, Rabbani M (2012). Buffalo farming in Pakistan. Pakistan Today, Dated 23 June.

Riley DG, Chase CC, Olson TA, Coleman SW, Hammond AC (2004). Genetic and nongenetic influences on vigor at birth and preweaning mortality of purebred and high percentage Brahman calves. J. Anim. Sci. 82:1581-1588.

Tomaszewski MA (1993). Record-keeping systems and control of data flow and information retrieval to manage large high producing herds. J. Dairy Sci. 76:3188-3194.

6

Studies on the productive performance of jersey x sahiwal cows in Chittoor district of Andhra Pradesh

A. Reddy Varaprasad[1] , T. Raghunandan[2], M. Kishan Kumar[3] and M. Gnana Prakash[3]

[1]Sri Venkateswara Veterinary University, Tirupati, India.
[2]Livestock Research Institute, College of Veterinary Science, Rajendranagar, Hyderabad, Andhra Pradesh, India.
[3]Department of Animal Genetics and Breeding, College of Veterinary Science, Rajendranagar, Hyderabad, Andhra Pradesh, India.

Season of birth and age at sexual maturity had a non-significant effect on first lactation milk yield. The overall least-squares mean lactation milk yield was 2154.07 ± 16.88 L. Significant effect of batch was found on first lactation milk yield in the present study. Cows belonging to 4th batch recorded significantly (P<0.0) higher lactation milk yield of 2475.93 ± 19.65 L. The effects of season of birth and age at sexual maturity were found to have no-significant effect on lactation length. The overall least-squares mean value for lactation length was 300.16 ± 0.06 days.

Key words: Progeny Testing Programme, lactation milk yield, peak yield, age at sexual Maturity.

INTRODUCTION

Chittoor District of Andhra Pradesh has 1.10 million cattle, out of which 0.56 million are Jersey X Sahiwal crosses (18th Livestock census - 2007). Progeny Testing Programme was started in the year 1987 in Chittoor district and at present most of the cattle are stabilized at 50% Jersey X Sahiwal level. This breed is considered to be highly adoptable to hot and humid conditions, average milk yielder and well adapted to management conditions of Chittoor district. However, there is a dearth of information on the performance of these animals. Hence, in the present study the productive performance of Jersey X Sahiwal cattle in Chittoor district was studied.

MATERIALS AND METHODS

Data obtained from progeny testing programme

A total number of 1,411 records from 1994 to till date were collected From the Deputy Director (AH) , Progeny Testing Programme

Chittoor. The data is categorized into six batches according to the mode of progeny Testing Programme running at Chittoor:

I. Batch- AI was started in July 94 and milk recording for the calves born out of AI was started during 1997-99;
II. Batch- AI was started in May 96 and milk recording for the calves born out of AI was started during 1999-01;
III. Batch- AI was started in July 98 and milk recording for the calves born out of AI was started during 2001-04;
IV. Batch- AI was started in July 2001 and milk recording for the calves born out of AI was started during 2004-07;
V. Batch- AI was started in July 2003 and milk recording for the calves born out of AI was started during 2006-08
VI. Batch- AI was started in July 2005 and milk recording for the calves born out of AI was started during 2008 to till to date.

The data pertaining to lactation milk yield and lactation length were collected from the available records.

Data collection

Twenty three villages are selected for the study according to the

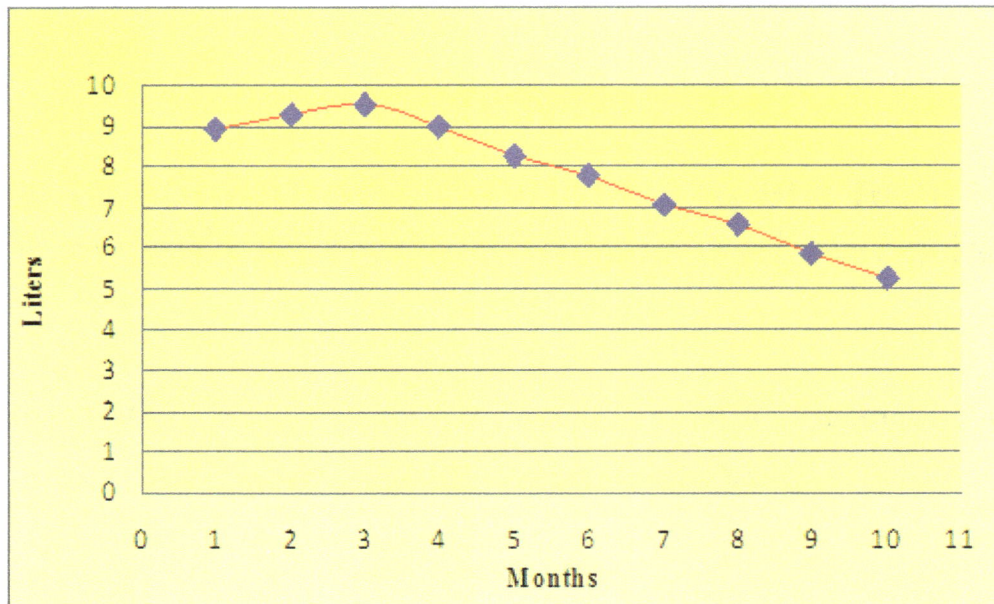

Figure 1. Lactation curve.

Fig 1. Lactation Curve

services offered by the Animal Husbandry department; 63.16, 26.32 and 10.52% farmers in the selected mandals are covered by Veterinary Dispensary, Rural Livestock Unit and gopalamitra, respectively. Data on the productive performance of animals maintained by the farmers were collected by interviewing them to arrive at values by memory recall method. A total of 190 farmers from 8 mandals in and around Chittoor where the Progeny Testing Programme is going on since two decades were interviewed. The family members of the farmers were also involved in collection of the data for more accuracy. The productive parameters include lactation length (LL), lactation milk yield (LMY), days to attain peak yield and peak yield. Data recorded on 1411 J X S cows under the progeny testing program were subjected to least-squares analysis through SPSS (10.0) software package.

RESULTS AND DISCUSSION

A non-significant influence of season of birth was observed on first lactation milk yield which ranged from 2132.10 to 2194.37 L. Similarly, non-significant effect of season of birth was reported by Thakur and Singh (2000, 2001) in Jersey X Red Sindhi and Jersey X Tharparkar cows respectively. The average peak yield calculated in the study is 9.51 L and the lactation curve is presented in the Figure 1. The highest mean FLMY was found in cows born during rainy season (2194.37 ± 16.66 L) followed by cows born during winter (2135.75 ± 19.57 L) and summer (2132.10 ± 14.19 L). The highest milk yield recorded in cows born in rainy season could be because of the availability of abundant green fodder and favorable climatic conditions. In contrast, summer born cows recorded less milk yield, which might be due to exposure to hot, dry climate and non availability of green fodder

during summer season.

Significant (P≥0.01) effect of batch was found on lactation milk yield with the cows of fourth batch yielding the highest milk yield (2475.93 ±19.65 L) followed by cows in sixth (2295.51 ± 16.85 L) and fifth (2272.17 ± 18.51 L) batch, which could be due to genetic differences among the cows. Cows of 1st and 2nd batches recorded significantly less milk among cows of all the batches (1952.27 ± 16.15 and 1953.78 ± 12.72 L, respectively), which might be due to severe drought conditions prevailing during that period. In general, the range of milk yield recorded in this investigation was on par with the mean lactation milk yield reported by Thakur and Singh (2000, 2001).

The effect of age at sexual maturity was non-significant on first lactation milk yield. The cows of ASM group-1 (400-600) were found with higher first lactation milk yield of 2198.04 L, followed by those of ASM group-2 where in the milk yield was 2132 kg.

The overall least-squares mean first lactation milk yield (2154.07 L) (Table 1) obtained in this study was on par with the means reported by Bhadauria and Katpatal (2003) in Holstein Friesian X Sahiwal cows (2097.27). However, higher mean lactation milk yields of 2796.33 and 2697 ± 25.8 L were reported by Tomar et al. (1996) in Holstein Friesian X Sahiwal cows and Reshmi and Stephan (2010) in Sunandini cows. Shubha Lakshmi et al. (2010) reported that the overall least squares means for lactation yield and 300 days lactation milk yield were 2864.30 and 2593.84 kg, respectively in Sahiwal Freisian crossbreds. The differences in lactation milk yield across the literature and in the present study could be due to the

Table 1. Least squares analysis of lactation milk yield (LMY).

Source of variation	d.f	Sum of squares	Mean sum of squares	F value
Batches	5	50806699.05	10161339.81	38.23**
Season	2	1031712.60	515856.30	1.94^{ns}
ASM group	3	148677.33	49559.11	0.18^{ns}
Error	1400	372041691.80	265744.06	
Total	1410	424028780.80		

**Significant at $P<0.01$; ns = non-significant.

Table 2. Least squares analysis of lactation length (LL).

Source of variation	d.f	Sum of squares	Mean sum of squares	F value
Season	2	1.439	0.720	0.55^{ns}
ASM group	3	0.720	0.240	0.18^{ns}
Error	425	552.97	1.30	
Total	430	555.13		

ns = Non-significant.

differences in the inheritance levels of exotic germplasm, environment and also due to the differences in the management especially pertaining to the feeding level.

From these findings it was found that the cows with early sexual maturity (Batch-4), early age at first calving and optimum gestation period recorded higher lactation milk yield, which is as per the natural trend observed in dairy cattle. This could also be due to the inheritance from new sires that were inducted in to the progeny testing programme. The cows in batches 1 to 3 recorded lower milk yield comparatively. The variation in milk yield observed among cows reared under different batches indicate the significant effect of genetic grades and genotype and environmental interactions which effect the expression of milk yield of different animals. The average peak yield calculated in the study are is 9.51 liters in Jersey X Sahiwal cows in First lactation milk yield and it is comparable with the observations of Deshmukh et al. (1995) in Jersey X Sahiwal and Zaman et al. (1998) in Jersey.

The effects of season of birth and age at sexual maturity were found to have non-significant effect on lactation length. The longest lactation length of 300.26 ± 0.14 days was observed in cows born during winter season followed by those born during summer (300.15 ± 0.07) and rainy (300.08 ± 0.10 days) seasons. The overall least-squares mean value for lactation length was 300.16 ± 0.06 days (Table 2). Among the four age groups of age at sexual maturity, the mean lactation length ranged from 300.07 ± 0.12 to 300.20 ± 0.11 days. The present finding is within the range of values reported in literature by Singh and Sharma (1984) and Thakur and Singh (2005) in Holstein Friesian X Sahiwal cows. ASM group had no significant effect on lactation length.

Conclusion

Season of birth and age at sexual maturity had a non-significant effect on first lactation milk yield. Significant effect of batch was found on first lactation milk yield in the present study. However, the farmers are advised for proper planning to breed their cows results in calves born which are well acclimatized to the existing environment. Cows belonging to 4th batch recorded significantly higher lactation milk yield of 2475.93 ± 19.65 L. The effects of season of birth and age at sexual maturity were found to have non-significant effect on lactation length.

REFERENCES

Bhadauria SS, Katpatal BG (2003). Effect of genetic and non- genetic factors on 300 day milk yield of first lactation in Fresian X Sahiwal crosses. Indian Vet. J. 80:1251-1254.

Deshmukh DP, Choudari KB, Deshpande KS (1995). Non genetic and genetic factors affecting production efficiency traits in Jersey, Sahiwal and Jersey X Sahiwal cows. Indian J. Diary Sci. 48:85-88.

Reshmi RC, Stephen C (2010). Evaluation of Lactation milk Yield in Crossbred cattle. Indian Vet. J. 87:363-334

Shubha Lakshmi B, Ramesh Gupta B, Gnana Prakash M, Sudhakar K, Lt. Col. Susheel Sharma (2010) Genetic Analysis of the Productuon Performance of Frieswal Cattle. Tamil Nadu J. Vet. Anim. Sci. 6(5)215-222.

Singh CSP, Sharma DB (1984). Comparative studies on Friesian X Sahiwal cows and buffaloes lactation length, milk yield, and milk producing efficiencies. Indian Med. Vet. J. 9:115-117.

Thakur YP, Singh BP (2000). Performance evaluation of Jersey X Zebu crossbreds involving different indigenous breed performance of Jersey X Tharparkar crossnbreds. Indian Vet. J. 77:169-171.

Thakur YP, Singh BP (2001). Performance evaluation of Jersey X Sindhi crossbreds. Indian Vet. J. 78:62-63.

Thakur YP, Singh BP (2005). Factors affecting first lactation milk yield traits in Jersey cows. Indian J. Anim. Res. 39:115-118.

Tomar AKS, Prasad RB, Bhadula SK (1996). First lactation

performance of Holstein X Sahiwal and their halfbreds in Tarai region of Northrern India. Indian J. Anim. Res. 30:129-133.

Zaman G, Das D, Roy TC, Aziz A (1998). Genetic studies on peak yield and days to attain peak yield in Jersey cattle in Assam. Indian J. Dairy Sci. 51:268-271.

7

Economic analysis of sheep farming activities in Turkey

Mehmet Arif Şahinli[1] and Ahmet Özçelik[2]

[1]Economics Department, Karamanoğlu Mehmetbey University, Karaman, Turkey.
[2]Ankara Üniversitesi, Ziraat Fakültesi Tarım Ekonomisi Bölümü, Dıkkapı Ankara, Turkey.

In this study, economic analysis of agricultural farming that also involves sheep farming in Konya Province in Turkey was done and then the effective factors in sheep farming activities were determined. The average household size of farms was 3.97 people and the average size of farms was 137.95 ha. The value of the total assets was composed of 57.44% fixed capital and 42.56% operating capital in farms. The average gross production account was 44.71% of crop production value and 55.29% of animal production. 36.77% of animal production value belonged to the sheep farming. The biggest share, 63.47% of feed costs and second share, with 24.24% of labour costs contained in the variable costs belonged to sheep farming activities. According to the factor analysis, 27 of variables affecting sheep farming are gathered by 4 factors. These factors are: Income, volumetric, costs and labour factors.

Key words: Sheep farming activities, economic analysis, factor analysis, Konya.

INTRODUCTION

Adequate and balanced nutrition of the growing population of Turkey which is used as raw material for the livestock industry in many areas has an important place. In addition, the animal husbandry sector, due to a lot of activities included in Turkey's economy, can bring solution to social problems. This sector helps to decrease unemployment in rural areas and urban migration, thereby avoiding unplanned urbanization and social functions like reducing the pressure of population. Livestock contribute to the balanced development of a country, increase national income and provide raw materials to other sectors. Sheep farming requires low capital and not much specialized machinery compared with most of the other agricultural production alternatives (Nix, 1988). The most important factor affecting the gross margin of sheep farms in Tonk, Rajastan region of India was labor expenses. The labor used was below the optimum level in small farms while it was around the optimal level in big farms (Sirohi and Rawat, 2000). A large number of the sheep farmers in Karnata Region of India had no land and more than half of the farmers were involved in blanket weaving activity and the yearly activities resulted in an Rs 13.000 net profit (Geeta et.al., 1999). According to the Turkish Statistical Institute (TurkStat) data, Konya Province has 31 districts. 15 districts train indigenous breed of sheep; 16 districts train indigenous sheep breeds and varieties and breeds of merino sheep. Konya Province has the most intensive sheep farming enterprises and agricultural technique, in geographic and economic situation (Anonymous, 2009).

MATERIALS AND METHODS

An important part of the material used in the study includes the area of agricultural holdings engaged in sheep breeding from where the survey is done. Sample businesses were selected and questionnaires were filled by making personal interview. Information was collected from the agricultural enterprises in this survey from October to November 2009. In addition to the primary data obtained, which are related to the subject of previous research

Table 1. Enterprises surveyed land nevis (also) and distribution (%).

Business groups	Arable land								Kitchen garden		Fruit garden		Total land	
	Irrigated land		Dry land		Fallow		Total arable land (irrigated land + dry land − fallow)							
	(da)	(%)	(da)	(%)	(da)	(%)	(da)	(%)	(da)	(%)	(da)	(%)	(da)	(%)
1	10.33	12.21	73.61	87.03	1.23	1.46	82.71	97.78	0.29	0.34	1.58	1.87	84.58	100.00
2	35.73	21.39	130.53	78.16	0.00	0.00	166.25	99.55	0.48	0.28	0.28	0.16	167.00	100.00
3	125.18	28.50	364.91	83.07	51.09	11.63	439.00	99.94	0.00	0.00	0.27	0.06	439.27	100.00
Mean of businesses	27.36	19.83	115.37	83.63	6.27	4.54	136.46	98.92	0.30	0.21	1.19	0.86	137.95	100.00

findings, published records and secondary data were used by various organizations. Methods applied in the study are given as follows.

Under the preliminary study, the characteristics that could represent the province as purposeful districts respectively were chosen. There are three districts within the scope of this research. In these districts in 2009, there was a total number of 334.795 heads of sheep in 28.58% of the province of Konya. There was careful selection of sample districts and villages so as to represent natural factors in terms of farming and sheep farming area.

This is the main material for the study of Cihanbeyli, Karatay and Karapinar districts of Konya province engaged in raising sheep. If sheep farming enterprises have at least 25 and more ewes mated, these enterprises were recorded. The criterion is based on the number of at least 25 and more ewes mated. Total population size is 392. The sample population was drawn by simple random sampling (SRS) method. Proportional method was the formula used for finding the value of n (Yamane, 1967). N value is founded by formula in the proportional method.

$$ n = \frac{N \sum N_h S_h^2}{N^2 D^2 + \sum N_h S_h^2} $$

Businesses dealing with sheep are divided into 3 groups: 100 and below heads in the first group, enterprises with 101 to 200 heads in the second group and enterprises with 201 and over in the third group. According to the method of SRS, population, as a result of the withdrawal of the sample size, is 104. As a result of the sample based on the method of proportionate distribution of the first layer, n_1 = 73; second layer, n_2 = 20 and the third layer, n_3 = 11. In

addition, reserve up to 25% of the sample volume of the business has been identified. Villages to do the survey sample survey were chosen by the operators in the absence of reserves.

Economic profitability was calculated by dividing the net return with the total assets. In order to calculate the financial profitability, interest on debt was subtracted from net return and the result was divided with net worth (ErkuK et al., 1995).

Family labour potential was calculated by using man-days (Aras and Çakir, 1975). The assets and liabilities in the balance sheet were organized according to the functional structure of agricultural enterprise (Hopkins and Heady, 1955).

The completed survey forms and data entry of information were made in a spreadsheet environment. The analysis was carried out using the SAS Enterprise Guide 3.0 program (Anonymous, 2004).

Factor analysis of the 27 variables is included in the high element of partnership, formed by the 4 factors obtained from these variables (Anonymous, 2004). Stages of factor analysis are as follows: Factor extraction methods, principal component analysis (PCA) method and Varimax method.

RESULTS

At the first stage, factor analysis of the correlation matrix examined the correlation between variables. Variables that contain the common factors are expected to have high correlations. The common

factor variance (CFV) for factor analysis of the feasibility test was considered.

Economic analysis findings

As a result of the economic analysis of agricultural holdings, evidence obtained from the average farm size groups and businesses are summarized as follows. The average farm enterprises had 1 to 100 width and 84.58 acres of land; 101 to 200 groups had farm size of 167.00 acres and 201+ groups had 439.27 hectare farm size; while the average businesses had 137.95 acres. Mean of businesses is 83.63% dry land, 19.83% of irrigated agricultural land and 4.54% fallow land (Table 1).

In relation to average businesses, 85.21% of the total labour force presence in the family labor, and foreign labor force is calculated as a ratio of 14.79%. Business groups and the total labor force based on the presence of Male Labor Unit (MLU) ranged from 3.51 to 4.58. According to the size of enterprises, total labor force increases smoothly. Business average, calculated as the total labor force in the presence of MLU is 3.73. The family labor's numbers are increasing for business (Table 2). In enterprises surveyed, farm capital with business means of 57.44% and 42.56%

Table 2. Family labor and the presence of foreign workers in enterprises surveyed (MLU, %).

Business groups	Family labor	%	Foreign labor	%	Total labor	%
1	3.09	88.03	0.42	11.97	3.51	100.00
2	3.35	81.91	0.74	18.09	4.09	100.00
3	3.49	76.20	1.09	23.80	4.58	100.00
Mean of businesses	3.18	85.21	0.55	14.79	3.73	100.00

Table 3. Examined in the presence of capital in enterprises.

Class of capital	Business groups						Mean of businesses	
	1		2		3			
	Turkish Lira	%	Turkish Lira	%	Turkish Lira	%	Turkish Lira	%
I. Capital asset	127,337.60	100.00	274,195.53	100.00	556,715.65	100.00	200,994.49	100.00
1. Farm capital	79,255.12	62.24	143,393.33	52.30	304,815.61	54.75	115,446.75	57.44
a. Territorial capital	37,023.52	29.08	75,505.47	27.54	209,854.95	37.70	62,704.14	31.20
b. Building capital	20,827.88	16.36	27,538.49	10.04	37,877.47	6.80	23,921.71	11.90
c. Plant capital	21,022.90	16.51	38,277.94	13.96	42,488.34	7.63	26,611.56	13.24
d. Reclamation of land capital	380.82	0.30	2,071.43	0.76	14,594.85	2.62	2,209.35	1.10
2. Business capital	48,082.47	37.76	130,802.19	47.70	251,900.05	45.25	85,547.74	42.56
a. Animals capital	23,748.28	18.65	87,477.20	31.90	182,859.48	32.85	52,832.91	26.29
b. Tool-machine capital	16,728.58	13.14	23,635.43	8.62	32,653.04	5.87	19,741.14	9.82
c. Material ammunition capital	1,532.76	1.20	6,162.75	2.25	11,707.49	2.10	3,499.32	1.74
d.Money capital	6,072.85	4.77	13,526.81	4.93	24,680.03	4.43	9,474.37	4.71
II. Liabilities capital	127,337.60	100.00	274,195.53	100.00	556,715.65	100.00	200,994.49	100.00
1. Foreign capital	6,694.14	5.26	11,774.42	4.29	11,271.81	2.02	8,155.30	4.06
2. Shareholder's equity	120,643.45	94.74	262,421.11	95.71	545,443.84	97.98	192,839.20	95.94

Table 4. Examined distribution of enterprises, financial and economic profitability.

Business groups	Financial profitability (%)	Economic profitability (%)
1	3.64	4.65
2	4.15	4.72
3	5.06	5.29
Mean of businesses	4.20	4.73

constitutes the working capital. As seen, 31.20% is land capital; 26.29%, animal capital; 13.24%, plant capital; 11.90%, building capital and 9.82% consists of tool-machine capital.

The remaining portion is composed of the other elements of capital (Table 3). Yıldırım (1993) specified that the active capital in territorial capital ratio is 57.30%, while Arısoy (2004) found that it is 63.48%. Building an active share in the ratio of capital, Dellal et al. (2002) found it to be 9.91%; Bayaner (1995), 12.00%; Erkan et al. (1989), 10.42%; and Oğuz (1991), 12.48%. In the ratio of capital to capital assets in animals in Turkey, Iklı et al. (1994) found this rate to be between 0.20 and 73.26%, while Oğuz and Mülayim (1997) found it as 4.64%. In the rate of active capital in the capital of the tool-machine,

Oğuz and Mülayim (1997) found it to be12.72%, while Dellal et al. (2002) found it to 4.62%.

Businesses surveyed in relation average, 95.94% and 4.06% of the shareholder's equity consists of foreign capital. Businesses are made on the basis of shareholder's equity production (Table 3). Oğuz and Mülayim (1997) found 98.89% shareholder's equity capital to be passive. Business average rates of economic and financial profitability (4.73 and 4.20%) were identified (Table 4).

Factor analysis findings

Total unit, average per unit values and proportion were

Table 5. Used variables in factor analysis and common factor variance values.

Variable numbers	Variable names	Common factor variance
1	Working Capital (Turkish Lira)	0.94341913
2	Active Capital (Turkish Lira)	0.98105887
3	Gross Domestic Product (Turkish Lira)	0.98949828
4	Total Operating Costs (Turkish Lira)	0.98051855
5	Pure Product (Turkish Lira)	0.85793021
6	Gross Profit (Turkish Lira)	0.98258853
7	Agricultural Income (Turkish Lira)	0.98703055
8	Existence of Sheep (BAU)	0.94129765
9	Feed Cost (Turkish Lira)	0.95896263
10	Gross Profit / BAU	0.99379614
11	Agricultural Income / BAU	0.99527129
12	Gross Profit / BAU	0.96427262
13	Gross Domestic Product / BAU	0.99506889
14	Business Land (Da)	0.89182875
15	Total Arable Land (Da)	0.92375122
16	Property Land (Da)	0.94840909
17	Irrigated Land (Da)	0.61705438
18	Total Operating Costs / BAU	0.99413989
19	Total Variable Costs / BAU	0.99590038
20	Fixed Charges / BAU	0.98350249
21	Cost of feed / BAU	0.86271240
22	Labor Costs / BAU	0.84780237
23	Family Labor Wage Provisions (Turkish Lira)	0.90149066
24	Labor Used (MLU)	0.94013131
25	BAU / MLU	0.85187833
26	Gross Profit / MLU	0.91308359
27	Pure Product / MLU	0.78936204
Mean		**0.927102**

used. The common factor variance and variables in determining the ability of the variables are represented. The common factor variance (CFV) that shows the amount of variance explained by each variable is included in the analysis. CFV of all variables with values is quite large. It was concluded that the variables showed a good fit with the factor solution.

Factor analysis of the applicability test used in making an element of partnership and the partnership element of the variables were obtained as the average of 0.927102. This average is applied to the variables for indicating factor analysis (Table 5). Eigenvalue of the factor analysis, difference, ratio and the cumulative values are given in Table 6. The eigenvalues of these factors are going down. After the properties of the other explanatory factors, factor 4 is reduced. According to Kaiser, Eigen values ≥ 1 criteria were applied to a large variable (Kaiser, 1960).

The sum of the eigenvalues of factors and the number of variables were found to be equal to 27. Factors and the value of the difference are the difference between two eigenvalues. Ratio value of each factor indicates the ratio

of the eigenvalues of the total eigenvalue. The value of this ratio also gives the percentage of variance of that factor. Cumulative value and the cumulative sum of the rates can be obtained, which shows the cumulative variance. In general, the cumulative distribution is expected to be above 70%. The resulting 4-factor variance is 0.9271, that is, the sum of the percentages of 92.71 is 92.71 of the variation in percent value that is quite high. This can be explained by the 4 factors (Table 6).

Factor in determining the number of factors can be given by the graphics. Factor is used to determine the number of the graph and the horizontal axis factor number, while the eigenvalues are on the vertical axis. This is the first break point on the graph by determining the factors seen in that point (Figure 1). Designations and associated factors of the eigenvalues, variance and cumulative percentage of variance values are given in Table 7. Accordingly, 49.99% of the first factor, 26.55% of the second factor, 10.06% of the third factor and 6.11% of the fourth factor of the total variance are described. The 4 factors determined the percentage of the total

Table 6. Eigenvalue of the factor analysis, difference, ratio, and the cumulative values.

Factors	Eigenvalue	Difference	Percentage of variance	Cumulative percentage
1	13.4963264	6.3274906	0.4999	0.4999
2	7.1688359	4.4521997	0.2655	0.7654
3	2.7166362	1.0666743	0.1006	0.8660
4	1.6499618	1.0342142	0.0611	0.9271
5	0.6157476	0.1103278	0.0228	0.9499
6	0.5054199	0.1556322	0.0187	0.9686
7	0.3497876	0.1493598	0.0130	0.9816
8	0.2004278	0.1193226	0.0074	0.9890
9	0.0811052	0.0187263	0.0030	0.9920
10	0.0623789	0.0097576	0.0023	0.9943
11	0.0526213	0.0198991	0.0019	0.9963
12	0.0327223	0.0074745	0.0012	0.9975
13	0.0252478	0.0066142	0.0009	0.9984
14	0.0186336	0.0054969	0.0007	0.9991
15	0.0131367	0.0078915	0.0005	0.9996
16	0.0052452	0.0025090	0.0002	0.9998
17	0.0027362	0.0006049	0.0001	0.9999
18	0.0021313	0.0017987	0.0001	1.0000
19	0.0003327	0.0000511	0.0000	1.0000
20	0.0002815	0.0000897	0.0000	1.0000
21	0.0001919	0.0001143	0.0000	1.0000
22	0.0000775	0.0000650	0.0000	1.0000
23	0.0000125	0.0000104	0.0000	1.0000
24	0.0000021	0.0000021	0.0000	1.0000
25	0.0000000	0.0000000	0.0000	1.0000
26	0.0000000	0.0000000	0.0000	1.0000
27	0.0000000		0.0000	1.0000

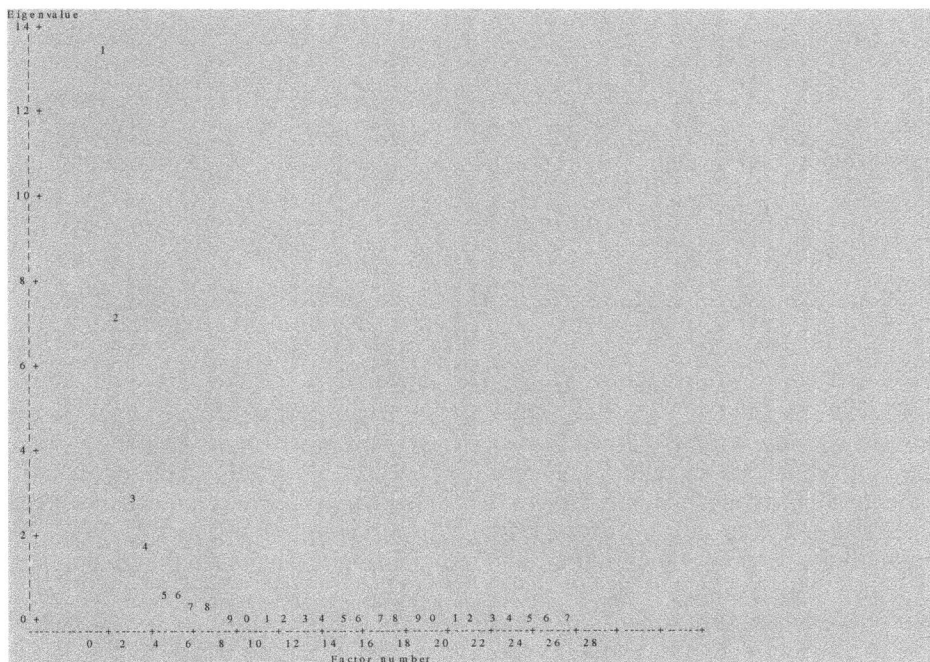

Figure 1. Factors and eigenvalue distribution.

Table 7. According to the analysis of factors related to these factors, the naming and the Eigenvalues, variance and cumulative variance values.

Factor No.	Factor names	Eigenvalue	Percentage of variance	Cumulative percentage
1	Income factor	13.4963264	0.4999	0.4999
2	Volumetric factor	7.1688359	0.2655	0.7654
3	Costs factor	2.7166362	0.1006	0.8660
4	Labor factor	1.6499618	0.0611	0.9271

Table 8. Factor loading matrix.

Variable	Factors			
	1	2	3	4
Working capital (Turkish Lira)	0.69612	-0.06567	0.44392	0.19972
Active capital (Turkish Lira)	0.89339	0.03688	0.38616	0.18012
Gross domestic product (Turkish Lira)	0.85261	0.01634	0.48357	0.16866
Total operating costs (Turkish Lira)	0.83927	0.00696	0.47603	0.22249
Pure product (Turkish Lira)	0.80461	0.03595	0.45629	0.03219
Gross profit (Turkish Lira)	0.89686	0.04404	0.39256	0.14896
Agricultural income (Turkish Lira)	0.86204	0.01292	0.46460	0.16699
Existence of sheep (BAU)	0.48840	-0.16315	0.81088	0.13645
Feed cost (Turkish Lira)	0.47236	-0.13151	0.83385	0.15247
Gross profit / BAU	0.06342	0.99438	0.02624	-0.0173
Agricultural income / BAU	0.06939	0.99476	0.02904	-0.0087
Gross profit / BAU	0.05386	0.97486	0.09179	-0.0509
Gross domestic product / BAU	0.06762	0.99445	0.03792	-0.0116
Business land (Da)	0.91132	0.17086	-0.03328	0.17611
Total arable land (Da)	0.92301	0.17893	-0.00619	0.19937
Property land (Da)	0.94204	0.16818	0.02927	0.17838
Irrigated land (Da)	0.77749	-0.03658	-0.03039	0.10151
Total operating costs / BAU	0.07338	0.99426	0.01257	0.00678
Total variable costs / BAU	0.07226	0.99253	0.07453	-0.0024
Fixed charges / BAU	0.07374	0.98810	-0.03894	0.01434
Cost of feed / BAU	0.10957	0.40888	0.81321	0.14904
Labor costs / BAU	0.08475	0.42070	0.80033	0.15201
Family labor wage provisions (Turkish Lira)	0.26392	-0.00356	0.08359	0.90821
Labor used (MLU)	0.41856	-0.07069	0.11492	0.86414
BAU / MLU	0.23451	-0.16106	0.82261	-0.307
Gross profit / MLU	0.77269	0.16428	0.46105	-0.2765
Pure product / MLU	0.70291	0.15982	0.49304	-0.1633

variance as 92.71%. As we can see, 92.71% of the total variance can be explained by these factors (Table 7).

Factors obtained by factor analysis and factor loadings are given in Table 8. According to the significance level of 5% for 200 observations, 0.180 and larger values were used for the determination of factor loadings (Joseph et al., 1992). Factors were taken into account when interpreting the vertical and horizontal values. First, each factor is evaluated in itself, followed by the variables that described the dependence of factors. Then, the factors and relationships of each variable were evaluated separately. Factor loadings are perpendicular to the original variables; dependent and independent variables are factors in which the multiple regression equation represents the standardized regression coefficients. According to Table 8, Agricultural Income/BAU variable has the highest loading factor in these variables. For this variable, the first factor $(0.06939)^2$, the second factor $(0.99438)^2$, the third factor $(0.02624)^2$ and the fourth factor $(-0.0173)^2$ describe a variance. Explanation rate of

the total variance of this variable is found to be,
$(0.06939)^2+(0.99438)^2+(0.02624)^2+(-0.0173)^2 = 0.99459$.

DISCUSSION

In the average of businesses surveyed, the total gross production value of 44.72% is for crop production value, while 55.28% consists of the value of animal production. Animal production value consists of 36.77% active sheep, 17.59% of active cattle and 0.92% consists of other livestock operations.

Emphasis on crop production enterprises, as well as the emphasis on farming and animal husbandry enterprises through development will significantly increase their revenues. Average farm size group of enterprises surveyed both in the gross production value of animal production and sheep farming has an important place.

Economic and financial profitability ratios of 4.20 and 4.73% respectively, compared to the average of businesses have been identified. Annual real returns on financial investment instruments, Producer Price Index (PPI) and Consumer Price Index (CPI) were calculated from the purified. Profitability ratios are evaluated according to the CPI, the economic profitability rate of the European Currency Unit (Euro) and the United States of America Currency (Dollars). Deposit Interest, Stock Index and gold bullion have low rate. Calculating these profitability ratios using Euro and the Dollar, business owners and sheep farming enterprises did not have loss in their labor and efforts.

According to the results of the feasibility analysis for the whole enterprises, enterprise size increased, increasing rates of both fiscal and economic profitability. In this situation, more efficient use of capital is in large enterprises. In addition, businesses that are working effectively are revealed. Small businesses cannot be used quite efficiently and economically in factors of production.

Factors are interpreted, evaluated and variables within each independent factor are described. In addition to the factors set relations were evaluated separately for each variable. There are four factors: income (factor 1), volumetric (factor 2), cost (factor 3) and labor factors (factor 4).

REFERENCES

Anonymous (2004). Sas enterprise guide. Sas Institute Inc., Cary, NC, USA.
Anonymous (2009). http://www.turkstat.gov.tr, Accessed 03.08.2009.
Aras A, Çakır C (1975). Gediz sulama projesi kapsamına giren tarım iĸletmelerinin ekonomik etüdü, E.Ü. Ziraat Fakültesi Yayınları 211:69.

Arısoy H (2004). The usage level of wheat varieties newly improved by agricultural research institutes and their economical analysis with conventional varieties -example of Konya province. Master Thesis. Selçuk University Institute of Science, Konya.
Bayaner A (1995). Economic analysis of wheat producing enterprises and investigating the functional form of fertilizer use in wheat production in Konya Province. PhD. Thesis. Ankara University- Graduate School of Natural and Applied Sciences Department of Agricultural Economics, Ankara.
Dellal G, Eliçin A, Tekel N, Dellal İ (2002). Small ruminants of structural properties in the the GAP. Ministry of Agriculture and Rural Affairs, Agricultural Economics Research Institute Publications: Ankara. p. 82.
Erkan O, Orhan ME, Budak F, ᵞengül H, Karlı B, Hartoka İ (1989). Economic analysis of agricultural holdings in Mardin-down Ceylanpinar plains and the forward-looking planning. Scientific Technical and Research Council of Turkey Agriculture and Forestry Research Group, Project No: TOAG-613, Adana.
ErkuĶ A, Bülbül M, Kıral T, Açıl F, Demirci R (1995). Tarım ekonomisi. Ankara Üniversitesi, Ziraat Fakültesi, Eğitim, AraĶtırma ve GeliĶtirme Vakfı Yayları, Ankara 5:298.
Geeta M, Sunanda K, Bhavani K (1999). Karnataka sheep farmers, Indian J. Small Rumin. 5(2):82-84.
Hopkins J, Heady EA (1955). Farm records and accounting. Fourth Edition, The Iowa State College Press, Ames, Iowa, U.S.A. p. 346.
IĶıklı E, Üzmez Y, Atlı F, Pekince Ö (1994). Capital issues in agriculture of Turkey. Agriculture Week Symposium, Chamber of Agricultural Engineers, Ankara.
Joseph F, Hair J, Rolph EA, Ronald LT, William CB (1992). Multivariate data analysis, Macmillian Publishing Company, Third Edition, New York, USA. p. 239,
Kaiser HF (1960). The application of electronic computers to factor analysis. Educational Psychological Measurement 20:141–151.
Nix J (1988). The economics of sheep production, British Vet. J. 144(5):246.
Oğuz C, Mülayim Ü (1997). Grow sugar beets under contract of economic situation of agricultural enterprises in Konya, L. L. Konya Sugar Beet Growers Foundation for Education and Health Publications, Konya.
Oğuz C (1991). The economic results of cereal + lentil and cereal growing farms under dry farming conditions in the Konya province of Turkey. PhD. Thesis. Çukurova University- Graduate School of Natural and Applied Sciences Department of Agricultural Economics, Adana.
Sirohi S, Rawat PS (2000). Resource use efficiency in sheep farming: a case study of district tonk, rajasthan, Indian J. Small Rumin. 6(1):42–47.
Yamane T (1967). Elementary sampling theory. Prentice-Hall., Englewood Cliffs, N.J.
Yıldırım İ (1993). Production economy of sheep enterprises and marketing in the Çatak town of Van province. PhD thesis. Ege University Science and Technology Institute, İzmir.

8

Effects of natural building ventilation on live weights performance of does during reproduction

Lamidi W. A.[1], Osunade J. A.[2] and Ogunjimi L. A. O.[2]

[1]Department of Agricultural Education, Osun State College of Education, Ila-Orangun, Osun State, Nigeria.
[2]Department of Agricultural Engineering, Obafemi Awolowo University, Ile-Ife, Nigeria.

Four identical adjoining buildings, each with eight pens (number of does = 32) of dimensions 1.2 × 0.6 × 1.2 m were constructed with floor area per pen (1.2 × 0.6 m). Three factors were considered in the study; building orientations (Or.45° and Or.90° to wind directions) and ventilation openings (Op.20, Op.40, Op.60 and Op.80% side openings) and seasons (dry and rainy). Ventilation rates inside each pen were estimated using opening effectiveness and magnitudes of prevailing winds' speed measured. Rabbits were mated to produce litters at two consecutively times per season. Significant ($p< 0.05$) differences due to season were observed for the weight gains during and after conception and also for size at birth. Or.90° with Op.60% had mean weight of 14.0 g day^{-1}, highest recorded in does weight gain before conception but conception rate (CR) of 32.81% and gestation period (GP) of 31.19 days. CR was highest for Op.80% (42.18) with lower GP of 31.38 days. Litter size (Ls) was high for higher openings Op.60, Op.80 and Op.100. Increasing side building ventilations increases weight gain and CR with 30 to 31 days GP and high Ls in does before, during and after conception.

Key words: Conception, gestation, litters, orientation, openings.

INTRODUCTION

Ventilation systems in building may be natural or artificial (mechanical) and are installed to maintain a comfortable indoor environment for the animals occupying it. Natural ventilation can be generated by wind pressure and/or thermal buoyancy due to stack effect. The natural ventilation allows a chimney effect due to buoyant forces to be created, as well as creating a suction force while wind blows across the building (Nicol, 2004). The natural ventilation opening acts as an exhaust to allow warm, moist air to exit the building. This is very important for hot climate when outdoor air temperature is extremely high (Humphreys and Nicol, 1998).

In rabbits' building, ventilation systems continuously remove the heat, moisture and odours created by rabbits,

solar heat gain through the building envelope and casual heat from equipment and artificial lighting (Patial et al., 1991; Bassuny, 1999). All these replenish the oxygen supply by bringing in drier and clear outside air (Gugliermetti et al., 2004). Adequate air exchange also removes gases such as ammonia (NH_3), hydrogen sulphide (H_2S) and methane (CH_4) which are generated in the pen via their wastes and which can be harmful to the rabbits (Shove et al., 2008). The merit of ventilation is how such polluted proportion of air within the enclosed space should be continuously withdrawn and replaced by fresh air. This must be drawn in from a clear external source, as high as elevation and as practical (Lamidi, 2011). Thus, ventilation is needed to maintain air purity.

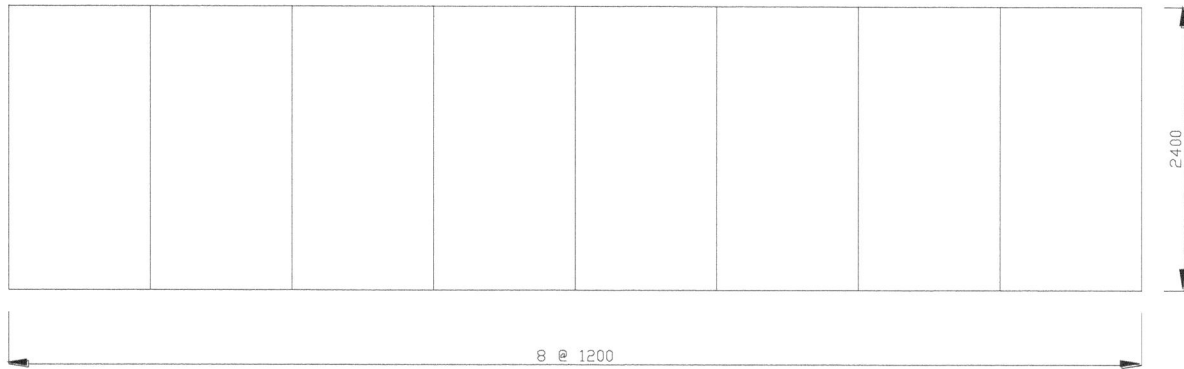

Figure 1. Plan view of the experimental building for 45 and 90° orientations.

Building shape or geometry may greatly influence air movement into and around the building and it accounts for increase or decrease in heat and its accompanying stress. The geometry of the building determines to a large extent the size and shape of the eddy, it should be realized that building height, width and depth determine the size of the eddy, or the calm area for the airflow. However, a small change in both building form and shape or feature arrangement may create large changes in air movement. As ventilation can never be compromised in building, building configuration assists in cross-ventilation, if, however, due to poor design and therefore compromised, air movement may not reach deep down into building's space depending on the level of the design and the distance (or height) between the floor and the ceiling (Sherman and Matson, 2008; Steemers, 2003; Chappells and Shovel, 2005). The objectives were to determine the effects of building ventilation opening resulting from variations in temperature and humidity inside the building on the weights and carcass development in does at different stages (before, during and after conception) of their reproductive lives.

MATERIALS AND METHODS

Experimental design

The study was carried out at the rabbitry section of the Teaching and Research Farm, Obafemi Awolowo University, Ile-Ife. Ile-Ife is in the rain forest zone of Nigeria, situated at longitude 14° 33'E and latitude 07° 28'N. Four identical adjoining buildings with dimensions 1.2 × 4.8 × 1.2 m (width × length × height, respectively) were constructed with floor area per pen (1.2 × 0.6 m). Each of the building housed eight pens for 32 does altogether. Another building, 100% open at all sides housed four pens of 1 doe each and with dimension 1.2 × 0.6 m each was used as control experiment.

Three factors were considered in the study; building orientations (Or.45° and Or.90° to the directions of the wind), ventilation openings (Op.20, Op.40, Op.60 and Op.80% side openings) and seasons of the year (dry and rainy seasons). The Op.100% opening with four rabbits served as control. The experimental set up was a 2 × 2 × 4 design with the effects of orientation and ventilation

openings investigated during the dry and rainy seasons; each treatment was replicated four times. The does were mated to produce two sets of offspring consecutively per season. All the experimental animals were obtained from the same population and therefore the influence of genetic factors should be minimal. In addition pure rabbits breeds are not available in Nigeria on demand because there are no pure breed parent materials in the country. The available rabbit stocks in Nigeria are made of several generations of crosses of about 4 different breeds imported into the country several years back and therefore the influence of genetic factors should be minimal.

Figures 1 to 5 show the experimental plans of the building and some rabbits in their pens. Thirty-two rabbits with age range 32 weeks to 34 weeks were allocated to each of the 4 pens at Op.20%; each of 4 pens at Op.40%, each of 4 pens at Op.60%, each of 4 pens at Op.80% in Or.45° and these were all repeated in pens in Or.90° orientations. The Op.100% opening had only Or.90° orientation with 1 rabbit doe in each of its four pens. Their individual concentrate feeds and water were equally weighed and given the same time in individual pot daily throughout the period of the experiment. The parameters measured for the determination of growth performance in the gestating rabbits were weight of the doe before conception, during gestation period and weight of does during breastfeeding (before weaning kits). Litter size was also measured. Triple beam balance MB with a capacity of 2610 ± 0.01 g was used to measure the weekly weight gain of does. Ventilation rates (the quantity of air) inflow into each constructed pen were estimated using approximate results based on empirical data as given by ASHRAE (1997) and ASHRAE/HVAC&R (2004) written as

$$Q = E\,Av \qquad (1)$$

where Q= wind or airflow rate, $m^3/$ s, A = area of inlet opening; m^2 (the outlet openings are assumed to be equal in area, which is true in the case of the experiment), v = wind or air velocity, m/s; and E = opening effectiveness and it is a function of ρ, v, μ, l, z, ϕ and θ, ρ = air density / (kg/m^3) = 1.65 kg/m^3 at 30°C, v = air or (prevailing) wind velocity, m/s, μ = absolute air viscosity Ns/m^2 = 0.0000248 Ns/m^2 at 30°C, ϕ = wind flow angle of incidence, z / l = ratio of opening height to opening length, θ = roof slope.

Then, according to Naas et al. (1998), effectiveness of the building openings (E) varies according to direction of the prevailing wind relative to the opening and is given as:

$$E = 16.33 \left[\frac{0.21\rho v}{\mu} \right]^{-0.3515} \times \text{Sin}(\phi)^{1.201} \times \left(\frac{4z}{l} \right)^{-0.1213} \times (\text{Sin } \theta)^{-0.153} \qquad (2)$$

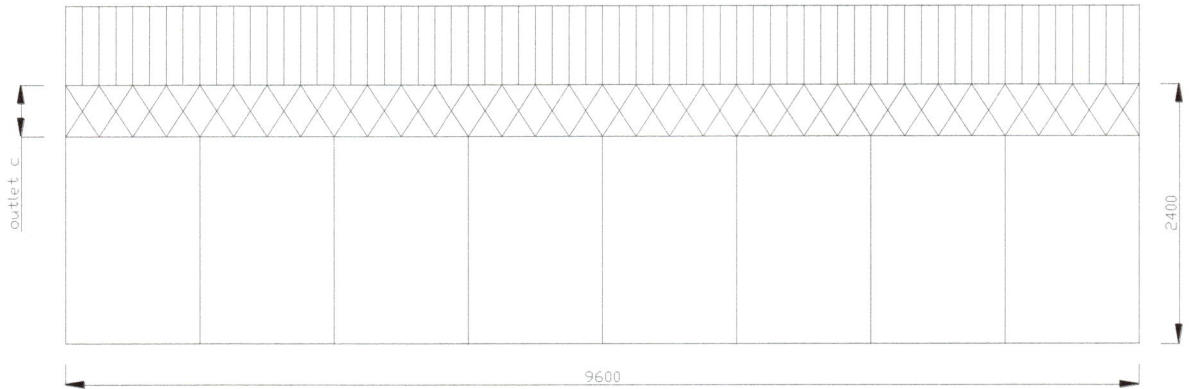

Figure 2. Side view of the experimental building for 45, 90° orientation and at leeward openings c = 20%: At 45° and 90°, whether inlet a = 20% or inlet b = 40% or inlet a = 60% or inlet b = 80%, outlet c = 20% constant throughout.

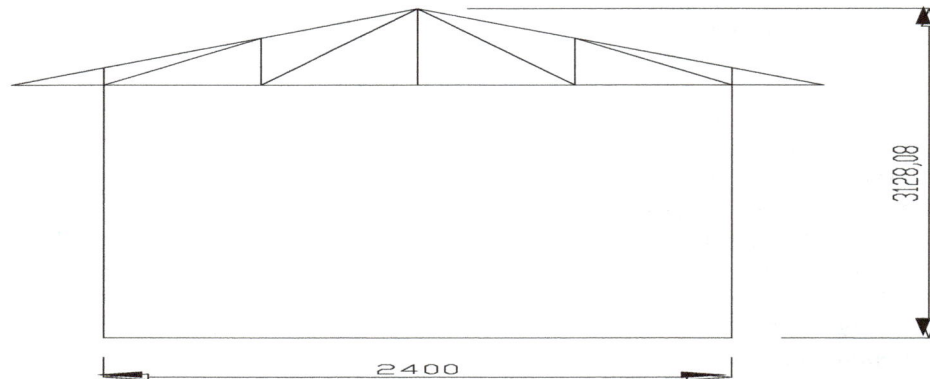

Figure 3. End view of the experimental building for 45 and 90° orientation and at different openings; roof slope = 30°.

Figure 4. Two of the four experimental housing built with wood showing pens in the 45° (background) and 90° (foreground) orientations buildings.

Figure 5. Experimental housing built with wood showing rabbits in their individual pens in 100% side opening building used as control, present conventional building.

Table 1. p-values showing interactions of seasons, building orientation and opening on the weekly weight gain by does at various stages in dry and rainy seasons.

Parameter (kg)	Source	p-values	
		Dry season	Rainy season
Wt.g.bCR	Or.	0.523	0.123
	Op.	0.356	0.236
	Or. * Op.	0.0001**	0.012*
	Season * Or. * Op.	0.278	0.059
Wt.g.dCR	Or.	0.034*	0.043*
	Op.	0.207	0.005*
	Or. * Op.	0.0001**	0.013*
	Season * Or. * Op.	0.00204*	0.007*
Wt.g.p	Or.	0.007*	0.021*
	Op.	0.021*	0.015*
	Or. * Op.	0.021*	0.015*
	Season * Or. * Op.	0.002*	0.002*

Or.*Op.; Season*Or.*Op = interactions; Wt.g. bCR – weight gain by doe before conception; Wt.g. dCR – weight gain by does during conception; Wt.g. p – weight gain by does during post-natal.

Statistical analysis

Data concerning the doe's liveweights before, during and after conception were statistically analysed by 2 × 2 × 4 factorial design according to the following experimental mixed model:

$$Y_{ijk} = \mu + S_i + Or_j + Op_k + S_i. Or_j + S_i. Op_k + Or_j. Op_k + S_i. Or_j. Op_k + e_{ijk}$$

where, Y_{ijk} = observation on ijk^{th} weights of doe due to all factors; μ = overall mean effect; S_i = the fixed effect of ith season (i = dry, rainy); Or_j = the fixed effect of jth orientation (j = 45^0, 90^0); Op_k = the fixed effect of kth opening (k = 20, 40, 60 and 80%); $S_i. Or_j.$ = interaction effect between the ith season and j^{th} orientation; $S_i. Op_k$ = interaction effect between the ith season and k^{th} opening; $Or_j. Op_k$ = interaction effect between the jth orientation and k^{th} opening; $S_i. Or_j. Op_k$ = interaction effect between the ith season, j^{th} orientation and k^{th} opening and e_{ijk} = random error.

RESULTS AND DISCUSSION

Weight gain by doe before conception

Significant differences (p < 0.05) due to the interactions between building orientations and openings were observed on the weight gain by the does before conception in the dry and rainy seasons (Table 1). In contrast, the interaction

Table 2. Means depicting body weights (± SE) of Does as affected by seasons, orientations and openings before conception (n = 30 days).

Items	Body wt (g)		Daily wt mean (g)	Weekly wt gain/loss (g)	Carcass wt (g)
	Initial	Final			
Effect of season					
Dry	1540 ± 13[a]	1656 ± 6[a]	7.2 ± 0.1[a]	11.6[a] ± 0.4[a]	1480 ± 6[a]
Rain	1550 ± 12[a]	1690 ± 6[a]	7.5 ± 0.2[a]	14.0[a] ± 0.6[a]	1500 ± 10[a]
Significance	NS	NS	NS	NS	NS
Effect of orientation					
Or.45°	1535 ±11[a]	1623± 10[b]	5.8± 0.8[a]	8.8 ± 0.4[a]	1500 ± 6[a]
Or.90°	1540 ± 7[a]	1696± 6[a]	6.5± 0.6[a]	15.6 ± 0.4[a]	1540 ± 16[a]
Significance	NS	NS	NS	NS	NS
Effect of opening (%)					
Op.20	1487 ± 2[b]	1647± 10[a]	6.3 ± 0.5[a]	6.0 ± 0.4[a]	1400 ± 4[a]
Op.40	1480 ±11[b]	1580± 6[a]	4.0 ± 0.6[a]	10.0 ± 0.2[a]	1480 ± 16[a]
Op.60	1600 ± 2[a]	1935 ± 10[b]	14.0 ± 0.5[b]	33.5 ± 1.4[b]	1400 ± 4[a]
Op.80	1580 ± 10[a]	1728 ± 16[a]	5.8 ± 1.2[a]	14.8 ± 1.3[a]	1480 ± 16[a]
Op.100	1700 ±10[a]	1731± 11[a]	8.8 ± 1.2[a]	12.8 ±1.3[a]	1420± 3.1[a]
Significance	NS	*	NS	*	NS

*$p < 0.05$; NS- not significant, Means bearing different supercripts in the same classification differ significantly.

among the season, the orientation and the openings as at then was not significant ($p>0.05$). Table 2 showed that in the rainy season, Or.90°and Op.60% were with mean weights respectively of values 7.5, 6.5 and 14.0 g day^{-1}, highest recorded in does' weight gain before conception. There were statistical differences among the weights (daily and weekly) in their respective classifications. Differences between initial and final weights was 140 g in rainy season compare with 116 g in dry season, 156 g in Or.90° and 335 g in Op.60 when compared with other openings, it was the highest (Table 2). In both dry and rainy seasons, the mean values from multiple range test showed building opening Op.40% had the lowest daily weight gain of 0.004 kg day^{-1}, whereas, Op.60% had 0.014 kg day^{-1}.

Weight gain by doe during conception

Significant differences ($p < 0.05$) due to (i) building orientation, (ii) the interaction between building orientation and opening and (iii) interactions between the seasons, orientations and openings were observed on the weight gains by the does at the conception period during the dry and rainy seasons (Table 1). Effect of opening was significant in the rainy season on the weight gain by doe. The season, orientation and the openings were statistically different from each other according to their classifications. Differences between initial and final weights of does during conception show higher weight in

Op.60% (220 g) and Op.100 (141 g), (Table 3). The higher weight and weight gain recorded at rainy season in Or.90° at Op.80% (16.2, 10.0 and 4.8 g week^{-1}); Table 3, were due to lower Temperature-Humidity Index (THI) recorded (23.00 ± 0.05) in the pen, lower THI value implied lesser heat stress on the animal (Figure 7a), this was in agreement with Finzi et al. (1995) and Ogunjimi et al. (2008) that rabbits have increased weight at lesser heat load. The study showed that mean weight's increment was highest for Or.90° (10.6 g week^{-1}). Conception rate (CR) was highest for Op.80% (42.18) with lower gestation period (GP) of 31.38 days compare to 33.38 days in Op.60%.

Weight gain by doe during pre-weaning (post-natal) period

The significant effects ($p < 0.05$) due to building orientation, opening and their interactions and interactions between the season, orientation and opening were observed in the does' weight gain at post-natal (pre-weaning) period during the dry and rainy seasons (Table 1). Statistical means showed 60% opening with highest litter size of 2.92, the 90° orientation with litter size of 2.45 (45° orientation, 2.15) and in the rainy season, 2.53 (dry season, 1.60). There were statistical differences among the same parameters in the same classifications. There was significant ($p< 0.05$) effect due to season on litter size at birth (Table 4). Randi (1982) and Patial et al.

Table 3. Means (± SE) depicting some Does' parameters as affected by seasons, orientations and openings during conception (n = 35 days).

Items	Body wt (g)		Daily wt mean (g)	Weekly wt gain/loss (g)	Conception rate (%)	Gestation period (days)
	Initial	Final				
Effect of season						
Dry	1550 ± 2^a	1666 ± 6^a	4.2 ± 0.1^a	15.4 ± 4^a	34.60 ± 0.6^a	34.30 ± 0.2^b
Rain	1570 ± 10^a	1685 ± 16^a	5.1 ± 0.2^a	16.2 ± 6^a	35.40 ± 1.0^a	32.13 ± 0.6^a
Significance	NS	*	*	*	*	**
Effect of orientation						
Or.45^0	1520 ± 14^a	1628 ± 1^a	1.1 ± 0.15^a	$9.0^b \pm 0.6^a$	35.41 ± 1.0^a	32.31 ± 0.6^a
Or.90^0	1582 ± 6^b	1688 ± 1.5^b	1.3 ± 0.1^a	$10.0^b \pm 0.6^a$	35.26 ± 2.3^a	32.13 ± 0.6
Significance	NS	*	*	*	*	**
Effect of opening, %						
Op.20	1540 ± 2^a	1646 ± 10^a	2.3 ± 0.5^a	6.0 ± 0.2^a	33.91 ± 1.3^a	33.07 ± 0.2^a
Op.40	1580 ± 16^a	1684 ± 6^a	4.0 ± 0.6^b	5.5 ± 0.3^a	32.27 ± 2.4^a	33.38 ± 0.7^a
Op.60	1690 ± 2^c	1910 ± 8^b	1.4 ± 0.5^a	3.6 ± 1.4^a	32.81 ± 4.2^a	31.19 ± 0.2^a
Op.80	1620 ± 10^c	1731 ± 11^a	1.8 ± 1.2^a	4.8 ± 1.3^a	42.18 ± 3.1^b	31.38 ± 0.1^a
Op.100	1580 ± 10^c	1721 ± 11^a	2.8 ± 1.2^a	5.8 ± 1.3^a	32.00 ± 2.1^a	31.00 ± 0.1^a
Significance	NS	*	*	*	*	**

*$p < 0.05$; **$p < 0.001$; NS- not significant, Means bearing different supercripts in the same classification differ significantly.

(1991) also found that seasons significantly affect litter size in Himachal Pradesh (India). This result however contrasted Nasr (1994) findings in Egypt (a sub-saharan tropical country) that effect of season was not significant on the litter size at birth. Litter size was high for higher openings Op.60, Op.80 and Op.100% respectively with mean values 2.92, 2.29 and 2.92. Figures 6 and 7 showed how the weights of doe before their conception varied in their different pens; there were gains and losses in their weights during the lactation (or pre-weaning) period. It was generally observed that as the building opening size increases, the weight gain by doe after conception increases as shown by openings Op.20, Op.40 and Op.60% in the rainy seasons (Table 2). No mortality was recorded from the does during the seasons at both orientations and openings.

Maximum weight gain of 5.8 and 0.45 kg week^{-1} was recorded for Op.80% ventilation opening and 7.8g week^{-1} for Or.90° orientation in the rainy season. The differences between initial and final doe's weight after conception period were higher at Or.90 (120 g), Op.60 (170 g) and Op.100 (180 g) (Table 4). Tables 5 and 6 revealed that there was linear relationship between the THI values and the weight gains in does as shown by the linear model equations respectively for the dry and rainy seasons. In all, the high R^2 values shown that there was positive correlation between the THI, weights and weight gains in the does during the post natal period (Tables 5 and 6). At Op.60, Op.80 and Op.100% openings, the correlations were stronger. The combined effects of

different orientations and openings, Figures 8 and 9 show that the does' weights were not increasing but fluctuate non-linearly throughout the periods. This may be due to the fluctuations in the THI values calculated from measured temperatures and humidities (Marai and El-Kelawy, 1999). It may also be as a result of the combination of different openings and different orientations [these led to different ventilation rate (m^3/s) in the individual pens] that had significant effect ($p < 0.05$) on the weights of the does and by extension on their metabolic activities and reproductive performances.

Airflow rates in the pens

The ventilation rates in m^3/s calculated from Equations 1 and 2, 100% opening had 0.00577 m^3/s; Op 80. 90° Or. had 0.00417 m^3/s, Op 80. 45° Or. had 0.00303 m^3/s (37.62% increase of 90° Or. over 45° Or. at 80% opening); Op 60. 90° Or. had 0.00279 m^3/s, Op 60. 45° Or. had 0.00330 m^3/s (18.27% increase of 45° Or. over 90° Or. at 60% opening) during the dry season. During the rainy season, Op 80. 90° Or. had 0.00492 m^3/s, Op 80. 45^0 Or. had 0.00307 m^3/s (60.26% increase of 90° Or. over 45^0 Or. at 80% opening); Op 60. 90° Or. had 0.00427 m^3/s, Op 60. 45° Or. had 0.00201 m^3/s (112.43% increase of 90° Or. over 45° Or. at 60% opening). It can be seen that for the same opening at different orientations, there were appreciable, significant differences in their magnitudes. These different airflows in different

Table 4. Means (± SE) depicting weight, litter size, does' mortality as affected by seasons, orientations, openings after conception (n = 30 days).

Items	Body wt (g)		Daily wt mean (g)	Weekly wt gain/loss (g)	Wt. of litter (g)	Litter size	Mortality of does (%)
	Initial	Final					
Effect of season							
Dry	1650 ± 2^a	1730 ± 16^b	0.8 ± 0.01^a	2.4 ± 0.03^a	114 ± 0.6^a	1.60 ± 0.02^a	Nil
Rain	1672 ± 10^a	1756 ± 16^b	0.5 ± 0.02^a	2.0 ± 0.06^a	134 ± 1.0^b	2.53 ± 0.06^a	Nil
Significance	NS	*	*	*	*	**	NS
Effect of orientation							
Or.45^0	1618 ± 1.4^a	1724 ± 1^b	2.4 ± 0.15^b	6.0 ± 0.6^b	126 ± 1.0^a	2.15 ± 0.02^a	Nil
Or.90^0	1680 ± 6^a	1800 ± 1.5^c	1.3 ± 0.1^a	7.8 ± 0.03^b	133 ± 2.3^b	2.45 ± 0.12^b	Nil
Significance	NS	*	*	*	*	*	NS
Effect of opening (%)							
Op.20	1635 ± 2.4^a	1742 ± 10^b	2.3 ± 0.5^b	6.0 ± 0.2^b	136 ± 1.3^b	2.10 ± 0.02^a	Nil
Op.40	1570 ± 16^a	1680 ± 6^a	2.0 ± 0.6^b	5.5 ± 0.3^b	117 ± 2.4^a	1.83 ± 0.17^a	Nil
Op.60	1710 ± 10^a	1890 ± 8.3^{ab}	1.3 ± 0.05^a	3.6 ± 1.4^a	127 ± 4.2^a	2.92 ± 0.2^a	Nil
Op.80	1700 ± 6^a	1806 ± 10.5^c	1.8 ± 0.2^b	5.8 ± 1.3^b	140 ± 3.1^b	2.29 ± 0.11^a	Nil
Op.100	1600 ± 6^a	1780 ± 12.5^c	2.8 ± 0.2^b	7.8 ± 1.3^b	144 ± 3.1^b	2.92 ± 0.11^a	Nil
Significance	NS	*	*	*	*	*	NS

*$p < 0.05$; **$p < 0.001$; NS- not significant, Means bearing different superscripts in the same classification differ significantly.

pens provided different thermo-comfort conditions for the rabbits and thereby may likely contribute to their reproductive performance.

The different resulted weight gains and losses in does at various periods of the experiment as shown in Figures 6a and b to 11a and b led to the different cumulative weights of does at the end of each period. Though all the rabbits were reared in their different pens at the weight bracket of 1.54 ± 0.20 kg, their weights at the end of the first week of rearing were no longer within the same weight brackets and at the subsequent weeks, they continued to be different and far from same weight brackets (Figure 6a and b). This shows that their weights were affected by the prevailing conditions inside their individual pens as a result of the different orientations and openings. The Op.60, Op.80 and Op.100% openings respectively had higher cumulative weights' values of 2.22 ± 0.08, 2.20 ± 0.17 and 2.04 ± 0.25 kg at Or.90° orientation. At Or.45° orientation, the Op.60 and Op.80% openings respectively had 1.93 ± 0.20 and 1.97 ± 0.32 kg cumulative weights' values for the does in the rainy season. The lowest final weight of doe recorded was 1.11 ± 0.26 kg for Op.60% opening in the Or.45° orientation in the dry season, Figure 3a. The fluctuations in the does' weights were as a result of the fluctuations in both temperature and humidity that have effects on the animal physiological and metabolic activities.

Figures 6 to 11 show weight gains/losses in does before, during and after conception periods. However, there were more weight gains than losses at openings Op.60 and Op.80% and especially, at the Or.90° orientation than other openings.

Conclusions

Seasons affect weight and weight gain in does at different ventilation openings and different orientations. The interaction between season, building orientation and opening significantly affected ($p < 0.05$) the weight gain by the does

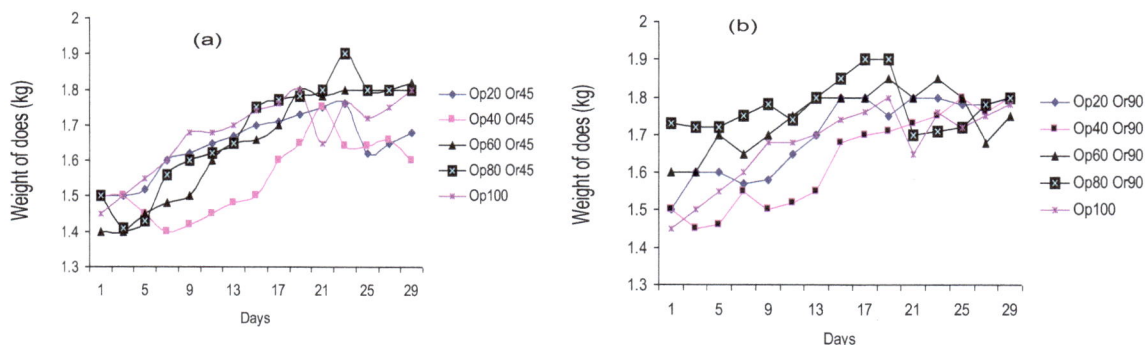

Figure 6. Weight of doe in every other day before conception during the dry season (a) Or.45°(b) Or.90°.

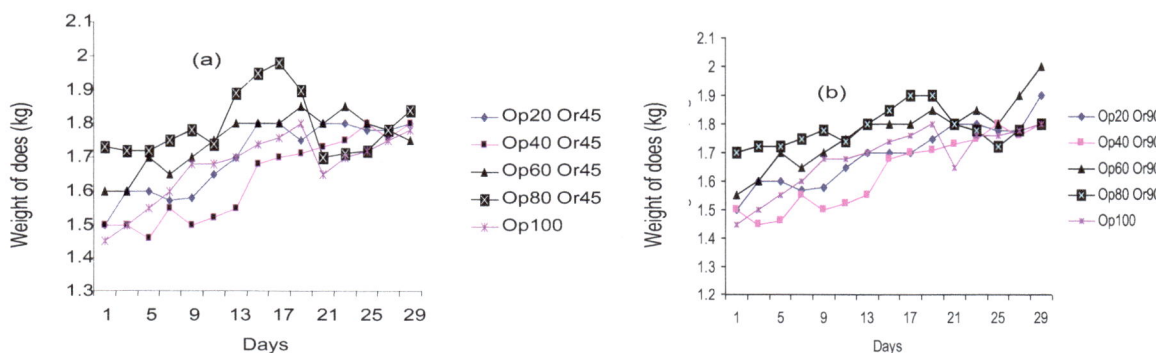

Figure 7. Weight of doe in every other day before conception during the rainy season (a) Or.45° (b) Or.90°.

Table 5. Regression analysis between building openings, THI values and weights of doe after gestation (during lactation) at both orientation (dry season), n = 35 days.

Orientation- Or.45°

Op.%	Linear model		Quadratic model	
	Regression equation	R^2 values	Regression equation	R^2 values
Op.20	w = -0.224 t + 25.55	0.42	w = 0.0203 t^2 - 0.548 t + 26.48	0.72
Op.40	w = -0.194 t + 25.41	0.32	w = 0.0103 t^2 - 0.359 t + 25.88	0.42
Op.60	w = -0.163 t + 25.53	0.51	w = 0.0241 t^2 - 0.549 t + 26.63	0.61
Op.80	w = -0.191 t + 25.40	0.62	w = 0.0137 t^2 – 0.411 t + 26.022	0.72
Op.100	w = -0.213 t +25.004	0.42	w = 0.0223 t^2 - 0.148 t + 26.48	0.72

Orientation-Or.90°

Op.%	Linear model		Quadratic model	
	Regression equation	R^2 values	Regression equation	R^2 values
Op.20	w = -0.218 t + 25.54	0.32	w = 0.012 t^2 - 0.537 t + 26.43	0.62
Op.40	w = -0.141 t + 25.209	0.71	w = 0.011 t^2 - 0.314 t + 25.70	0.91
Op.60	w = -0.213 t + 25.61	0.62	w = 0.012 t^2 - 0.410 t + 26.17	0.82
Op.80	w = -0.158 t + 25.48	0.82	w = 0.025 t^2 - 0.552 t + 26.59	0.72
Op.100	w = -0.213 t + 25.004	0.42	w = 0.0223 t^2 - 0.148 t + 26.48	0.73

w = weight of doe, t = THI.

Table 6. Regression analysis between building openings, THI values and weights of doe after gestation (during lactation) at both orientation (rainy season).

Orientation- Or.45°

Op.%	Linear model		Quadratic model	
	Regression equation	R^2 values	Regression equation	R^2 values
Op.20	$w = -0.182\,t + 25.71$	0.92	$w = -0.0079\,t^2 - 0.056\,t + 25.35$	0.92
Op.40	$w = -0.091\,t + 24.99$	0.21	$w = 0.0079\,t^2 + 0.0029\,t + 24.65$	0.31
Op.60	$w = 0.0920\,t - 23.66$	0.33	$w = 0.0027\,t^2 + 0.049\,t + 23.78$	0.43
Op.80	$w = -0.214\,t + 25.16$	0.93	$w = 0.0157\,t^2 - 0.464\,t + 25.87$	0.82
Op.100	$w = -0.113\,t + 24.004$	0.42	$w = 0.0023\,t^2 - 0.448\,t + 25.41$	0.72

Orientation- Or.90°

Op.%	Linear model		Quadratic model	
	Regression equation	R^2 values	Regression equation	R^2 values
Op.20	$w = -0.191\,t + 25.84$	0.41	$w = -0.020\,t^2 + 0.127\,t + 24.94$	0.64
Op.40	$w = -0.078\,t + 24.94$	0.70	$w = 0.0210\,t^2 - 0.411\,t + 25.89$	0.41
Op.60	$w = 0.140\,t + 25.57$	0.72	$w = -0.010\,t^2 + 0.011\,t + 25.15$	0.92
Op.80	$w = -0.110\,t + 25.52$	0.61	$w = 0.0027\,t^2 - 0.152\,t + 25.65$	0.61
Op.100	$w = -0.113\,t + 24.004$	0.42	$w = 0.0023\,t^2 - 0.448\,t + 25.41$	0.72

w = weight of doe, t = THI.

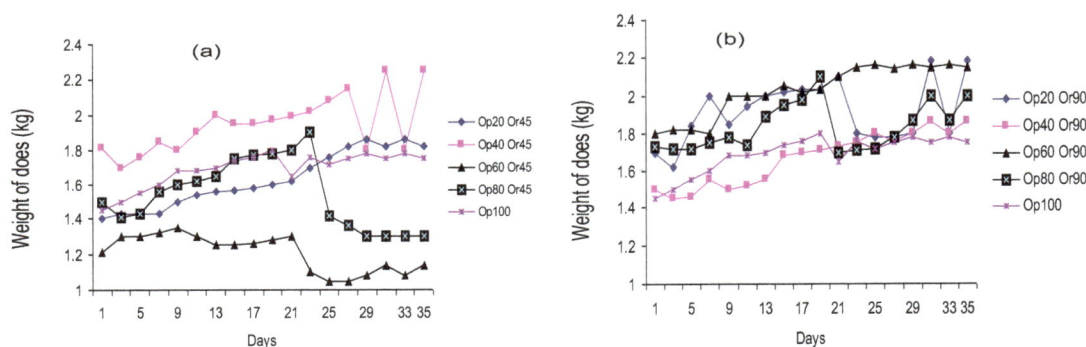

Figure 8. Weight gain by Doe (kg) in every other day during conception in dry season (a) Or.45° (b) Or.90°.

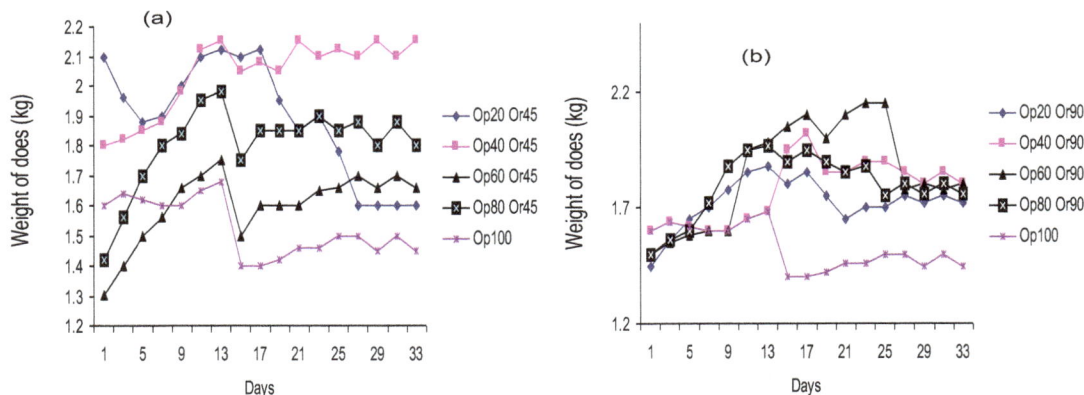

Figure 9. Weight gain by Doe in kg in every other day during conception in rainy season (a) Or.45° (b) Or.90°.

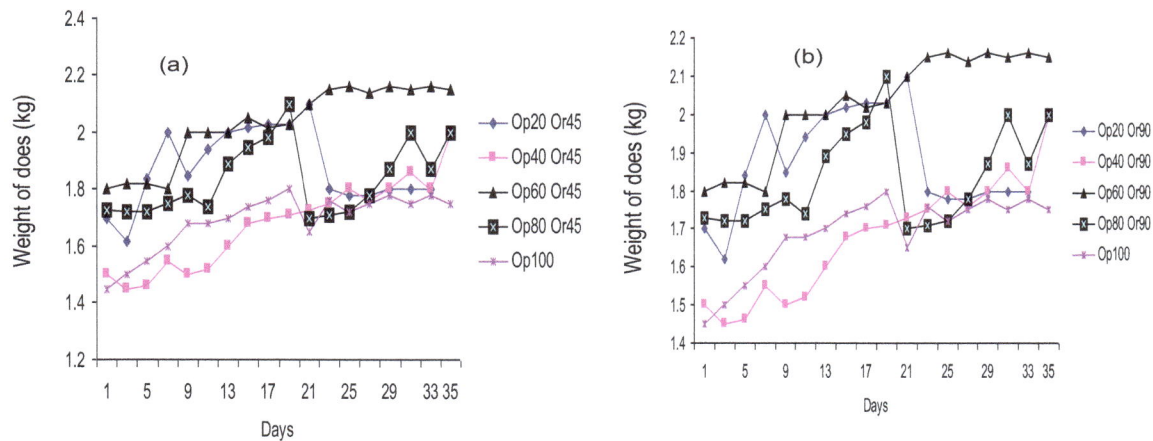

Figure 10. Weight gain by Doe in kg in every other day at post-natal in dry season (a) Or.45° (b) Or.90°.

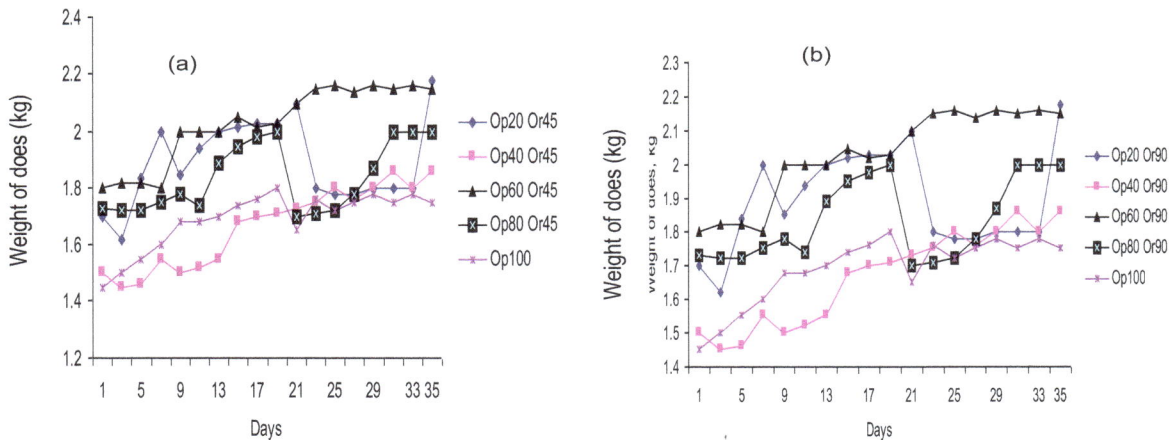

Figure 11. Weight gain by Doe in kg in every other day at post-natal in rainy season (a) Or.45° (b) Or.90°.

during and after conception. Building opening Op.40% had the lowest daily weight gain of 0.004 kg day^{-1}, whereas, Op.60% had 0.014 kg day^{-1}. The study also revealed that mean weight's increment was highest for Or.90° (10.6 g week^{-1}). It was generally observed that as the building opening size increases, ventilation rate increases and the weight gains by doe after conception increases as shown by openings Op.40, Op.60 and Op.80 in the rainy seasons. CR was highest for Op.80% (42.18) with lower GP of 31.38 days in Op.60%. There was linear relationship between the THI values and the weight gains in does. In all, the high R^2 values shown that there was positive correlation between the THI and weights and weight gains in the does during the post natal period. At Op.60, Op.80 and Op.100% openings, the correlations were stronger. Seasons significantly (p< 0.05) affected the litter size at birth. Litter size was high for higher openings Op.60, Op.80 and Op.100%. No

mortality was recorded for the does at both seasons.

ACKNOWLEDGMENT

Author appreciates the permission of the Teaching and Research Farm of the Obafemi Awolowo University, Ile-Ife in Nigeria to use their farm for the research.

REFERENCES

ASHRAE/HVAC&R (2004). Modeling of Hermetic Scroll Compressors: Model Validation and Aplication. American Society of Heating, Refrigerating and Air-conditioning Engineers. Atlanta, Georgia. Int. J. Heating, Ventil., Air-condition. Refrigerat. Res. 10(3):307-329

ASHRAE (1997). Environment for Animals. ASHRAE Handbook, Fundamentals, SI Edition. American Society of Heating, refrigerating and Air-conditioning Engineers. Atlanta, Georgia.

Bassuny SA (1999). Performance of Doe Rabbits and their weaning as affected by Heat Stress and their Alleviation by Nutritional Means under Egyptian conditions. J. World Rabbit Sci. 9:73-86. http://www. elsevier.com/locate/livsci. Accessed 22 / 2 / 10

Chappells H, Shovel E (2005). Debating the future of comfort: Environmental Sustainability, Energy Consumption and Indoor Environment', Buildi. Res. Inform. 33 (1):32-40

Finzi A, Morera P, Kuzminsky G (1995). 'Sperm abnormalities as possible indicators of rabbit's chronic heat stress'. J. World Rabbit Sci. 3(4):157-161. http:// www.sciencedirect.com Accessed 20 / 3 /10.

Gugliermetti F, Passerini G. Bisegna F (2004). 'Climate Models for Assessment of Office Building Energy Performance' Build. Environment 39:39-50

Humphreys MA, Nicol JF (1998). 'Outdoor Temperature and Indoor Thermal Comfort: Raising the Precision of the relationship for the 1998' ASHRAE Database of Fields Studies

Lamidi WA (2011). Effects of Building Ventilation on the Reproductive Performance of Female Rabbits in Humid Tropics. Ph.D Dissertation, Department of Agricultural Engineering, Faculty of Technology, Obafemi Awolowo University, Ile-Ife, Nigeria.

Marai IFM, El- Kelawy HM (1999). Effect of Heat Stress on the Reproduction in Female Rabbits. In: Proceedings of 1st International conference on Indigenous versus Acclimatized Rabbits, El-Arish North Sinai, Egypt.

Naas IA, Moura DJ, Buckin RA, Fialho FB (1998). An algorithm for determining opening effectiveness in natural ventilation by wind. Trans. ASAE 41(3):767-772.

Nasr AS (1994). Milk Yield and some associated traits affected by season of kindling, Parity and Kindling intervals in NZW doe rabbits under Egyptian conditions. Egyptian J. Rabbit Sci. 4(2):149-159.

Nicol F (2004). Adaptive Thermal Comfort Standards in the Hot Humid Tropics'. Energy. Build. 40:87-97

Ogunjimi LAO, Oseni SO, Lasisi F (2008). Influence of Temperature Humidity Interaction on Heat and Moisture Production in Rabbits. 9th World Rabbit Congress, June 10-13, 2008, Verone, Italy. Manage. Economics. pp. 1579-1583

Patial KK, Monuja N, Gupta K, Sanjeet K (1991). The effect of season on litter size of broiler rabbits in Himachal Pradesh (India). J. Appl. Rabbit Res. 14(4):257-259

Randi E (1982). Productivity traits in two rabbits breeds: New Zealand White and Californian. Revista di Zootecnia e Veterinaria 10(2):81-86.

Sherman M, Matson N (2008). Residential Ventilation and Energy Characteristics', Energy Efficiency and Renewable Energy, Office of Building Technology of the U. S. Department of Energy DE-ACO3-765F00098.

Shove E, Chappels H, Lutzenheiser L, Hackett B (2008). 'Comfort in a Lower Carbon Society.' Building Res. Inf. 36(4):307-311

Steemers SK (2003). 'Towards a Research Agenda for Adapting Climate Change' Build. Res. Inform. 31:291-301(a).

9

Intraspecies variation in nutritive potentials of eggs from two ectotypes of giant African land snail (*Archachatina marginata* var. *saturalis*) in Calabar, Nigeria

Okon, B.[1], Ibom, L. A. [1], Ifut, O. J.[2] and Bassey, A. E.[2]

[1]Department of Animal Science, University of Calabar, Calabar, Cross River State, Nigeria.
[2]Department of Animal Science, University of Uyo, Uyo, Akwa Ibom State, Nigeria.

This pilot study provides information on the nutritive potentials of eggs laid by two ectotypes of *Archachatina marginata* var. *saturalis* (P) snails. One hundred (100) adult *A. marginata* snails, 50 each of the black-skinned ectotype and white-skinned ectotype used for the study were selected from the snail sanctuary of the Department of Animal Science, University of Calabar, Nigeria. The snails were raised 5 per hutch and fed fresh paw-paw (*Carica papaya*) leaves and concentrate. Freshly laid eggs from snails' matings were collected from the two ectotypes for chemical analyses (proximate and mineral compositions). Results of the proximate composition of the snail eggs revealed that there were no significant differences (P > 0.05) between the white-skinned and black-skinned *A. marginata* snails for all the fractions (moisture, crude protein, crude fibre, lipid/fat, carbohydrate and ash) analyzed. Results of the mineral composition of the snail eggs also revealed non-significant differences (P > 0.05) between the white-skinned and black-skinned *A. marginata* snails. The results of mineral composition further showed that snail eggs constitute good sources of calcium, iron, sodium, potassium and magnesium, and compared favourably in these minerals with snail meat, as well as meat of lean domestic livestock. These eggs are therefore recommended for both young and old as they will constitute alternative sources of essential nutritional elements (protein and basic minerals) at a lower cost for Nigerians.

Key words: Nutritive potentials, eggs, snail, ectotypes, Nigeria.

INTRODUCTION

Archachatina marginata is the largest known snail kept and reared in Nigeria (Okon et al., 2012). It is an herbivorous non-selective scavenger which dwells naturally in the forest litters of the tropical rainforest of the south and the fringing riparian derived guinea savannah zones of Nigeria (Adedire et al., 1999; Ibom et al., 2012).

There are two known ectotypes of the snail based on the foot (skin) colour, namely; the black-skinned and the white-skinned (Figures 1 and 2). The white-skinned (Albino) ectotype is highly discriminated against by many ethnic consumers in Nigeria (Ebenso, 2003; Ibom et al., 2008; Okon et al., 2009). The reason for this discrimination

Figure 1. Black-skinned *A. marginata*.

Figure 2. White-skinned *A. marginata*.

according to the authors is that the consumers belief that the white-skinned *A. marginata* are used by "witch doctors" in their activities. Thus, Okon et al. (2009) noted that this has led to the species being found in abundance in any environment it thrives.

Omole and Kehinde (2005) and Ibom et al. (2008) opined that snails lay 4 to 18 eggs in 1 to 2 min unlike chickens that lay one egg per day. Akinnusi (1979) noted that *A. marginata* lays 5 to 11 eggs within the same period (1 to 2 min). Ogogo (2004) described the egg as being spherical and cream yellow in colour; whereas Raut and Barker (2002) stated that the colour of eggs from *A. marginata* is chalky white, Okon and Ibom (2012) opined that the eggs of *A. marginata* are spherical, translucent and yellowish in colour.

The size of giant African land snail (GALS) varies from species to species. It is therefore interesting to know that a small egg size is characteristic of snail. However, the consumption of snail egg is not very common or popular in Nigeria, partly because of its small size and partly because of lack of information on the nutritional qualities of the eggs. Thus, the study was carried out to assess the nutritive values of eggs from both the black-skinned and white-skinned ectotypes of *A. marginata* snail; thereby recommending it as another source of animal protein for the teeming population of the Calabarians in Nigeria.

MATERIALS AND METHODS

The research was carried out in the Department of Animal Science Snailery and the Analytical Laboratory of the Department, University of Calabar, Calabar, Nigeria. The location and climate of Calabar is as prescribed in Okon and Ibom (2011). One hundred (100) adult *A. marginata* var. *saturalis* (P) snails, 50 each of the black-skinned ectotype and white-skinned ectotype were selected from the snail sanctuary of the Departmental Snailery. The snails were raised 5 per hutch and fed fresh paw-paw (*Carica papaya*) leaves and concentrate as recommended by Okon and Ibom (2012). The snails were mated and freshly laid eggs (Figures 3 and 4) were collected from the two ectotypes for chemical analyses.

Proximate composition analyses of the freshly laid eggs were determined by the methods of the Association of Official Analytical Chemists (AOAC, 1990). Nitrogen was determined by the Micro-Kjeldahl method as described by Pearson (1976) and the percentage nitrogen was converted into crude protein by multiplying by 6.25. Lipid content was extracted and determined by methods outlined by Bligh and Dyer (1959) and Holman and Hayes (1958), respectively. Carbohydrate was estimated by the difference between the sum of the values of the previous nutritional components (protein, moisture, fibre, fat and ash) and 100% (accepted overall value of nutritional components).

The mineral components of the snail eggs were analyzed from solutions obtained by first dry-ashing the egg samples at 550°C and dissolving the ash in standard flasks with distilled, de-ionized water containing a few drops of concentrated hydrochloric acid.

Phosphorous was determined calorimetrically from sample using Spectronic-20 (Gallenkamp, U.K.) described by Pearson (1976) with KH_2PO_4, as a standard. Sodium and potassium were analyzed by means of flame photometer (Model 405, Corn, U.K.), using NaCl and KCl to prepare the standards. Calcium, magnesium, iron and zinc were analyzed by means of atomic absorption spectrophotometry (Modak SP 9, Pye Unicam, Uk).

The data obtained were analyzed using T-test of GENSTAT (2004) software package for comparison between the two ectotypes.

RESULTS AND DISCUSSION

The chemical composition of the snail eggs (Table 1) revealed that the crude protein content of the white-skinned and black-skinned *A. marginata* snail eggs were 64.01 ± 2.40% and 63.89 ± 2.35%, respectively. There was no significant difference (P > 0.05) in crude protein between the eggs of these two ectotypes. These values compared favourably with the value of 63.46 ± 2.56% for flesh (foot) obtained by Uboh et al. (2010), but lower than

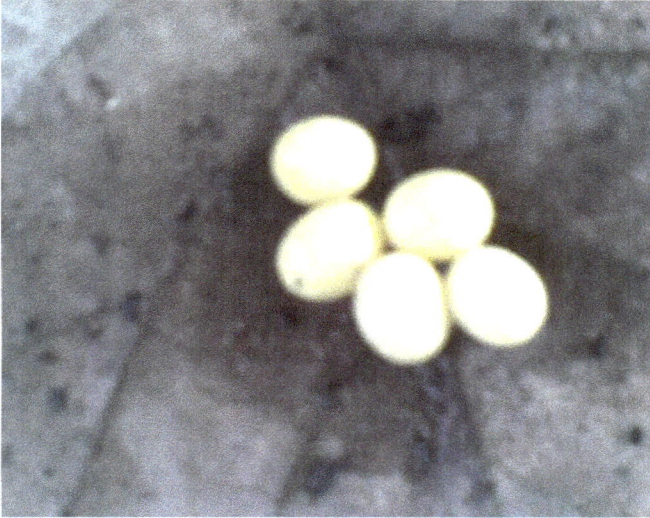

Figure 3. Eggs from white-skinned *A. marginata.*

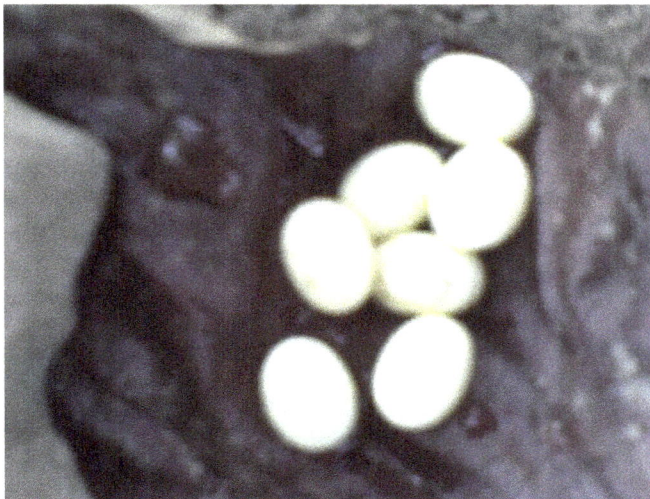

Figure 4. Eggs from black-skinned *A. marginata.*

the 80.95 ± 0.01% for flesh (foot) reported by Eneji et al. (2008). The protein values of the snail eggs were also higher than the values of 9.7 to 10.6% and 15.7 to 16.6% reported by Li-Chan et al. (1995) as protein values for chicken egg albumen and yolk, respectively. The high value for protein content of the snail eggs implies that the snail eggs, though small in size is a good source of protein. Hence, due to high cost of poultry products (meat and eggs) and beef, snail eggs that can be obtained at minimal cost may be used as a suitable source of protein.

The percentage moisture contents of the snail eggs obtained were 60.00 ± 2.18% and 59.94 ± 2.00% for white-skinned and black-skinned *A. marginata* snails, respectively. There was no significant difference (P > 0.05) in moisture content between the eggs of these two

ectotypes. These values were quite lower than the 80.30 ± 0.03% and 73.69% obtained by Eneji et al. (2008) and Babaloa and Akinsoyinu (2009) for snail meat, respectively. Similarly, the moisture content values of these snail eggs were also lower than the value of 84.3 to 88.8% reported by Li-Chan et al. (1995) for chicken egg albumen. However, the snail egg values obtained in this study were almost similar to the 63.1% moisture content of snail meat reported by Malik et al. (2011). The variation in moisture content obtained may be due to either the method of analysis, strain of snail or snail flesh (meat) used. This analysis was for snail eggs.

The percentage crude fibre contents of *A. marginata* snail eggs were negligible, 0.01 ± 0.01 and 0.00 ± 0.00 for the white-skinned and black-skinned snail eggs, respectively (Table 1). These agreed with the reports of Fagbuaro et al. (2006) and Malik et al. (2011) for snail meat. Uboh et al. (2010) reported a higher crude fibre value of 3.01 ± 0.01% for the snail meat of the same species, making the snail eggs with lower crude fibre level more suitable for consumption than the meat. There was no significant difference (P > 0.05) in crude fibre between the eggs of these two ectotypes studied.

The percentage lipid contents of white-skinned and black-skinned *A. marginata* snail eggs obtained were 1.03 ± 0.02% and 1.01 ± 0.01%. These values were higher than the value of 0.03% reported by Li-Chan et al. (1995) as lipid content value of chicken albumen. However, the snail egg values were quite related to the 1.3 to 1.5% and 1.23 ± 0.01% reported by Asibey and Eyeson (1995) and Fagbuaro et al. (2006) for snail meat. The lipid content values obtained in this study were lower than the value of 31.8 to 35.5% reported by Li-Chan et al. (1995) for chicken egg yolk. Lipid provides energy for the body. Thus, the low fat content makes the snail egg a good antidote for hypertensive patients and those that have fat related diseases. The percentage carbohydrate contents of the two ectotypes of snail eggs were 3.97 ± 0.05% and 3.85 ± 0.04% which was quite higher than the 1.70% obtained by Adeola et al. (2010) for black-skinned *A. marginata* snail meat. Similarly, Li-Chan et al. (1995) also reported lower values of 0.4 to 0.9% and 0.2 to 1.0% for chicken egg albumen and yolk, respectively. There was no significance (P > 0.05) in percent carbohydrate between the eggs of white-skinned and black-skinned *A. marginata.* The results were however lower than the 22.53 ± 1.08% reported by Uboh et al. (2010) for black-skinned *A. marginata* snail meat. The disparity in carbohydrate contents here might be due to the strain or species of snail used as well as the method of analysis either on wet basis or dry basis. Besides, this analysis was done on snail eggs and not the flesh or meat.

This pilot study provides information on mineral elements of snail eggs. The study revealed that snail eggs compared favourably in mineral contents with snail meat, as well as those of lean domestic livestock. Eggs from the two ectotypes of *A. marginata* snail studied constitute good sources of calcium, iron and sodium (Table 2),

Table 2. Mineral composition of GALS (*A. marginata* var. *saturalis*) eggs and meat (DW basis).

Mineral (mg/100 g)	White-skinned eggs	Black-skinned eggs	Meat**
Potassium (K$^+$)	4.65 ± 0.05	4.62 ± 0.05*	98.47 ± 2.87
Sodium (Na$^+$)	17.16 ± 1.00	17.28 ± 1.15*	30.89 ± 3.25
Calcium (Ca^{++})	184.20 ± 9.38	185.10 ± 10.15*	199.26 ± 15.32
Iron (Fe^{++})	5.10 ± 0.07	5.00 ± 0.06*	0.64 ± 0.01
Magnesium (Mg^{++})	4.70 ± 0.05	4.72 ± 0.06*	31.00 ± 3.02
Zinc (Zn^{++})	0.08 ± 0.01	0.09 ± 0.01*	2.01 ± 1.01

Values are presented as means ± SD of five determinations. *Not significantly different from each other. **Uboh et al. (2010).

Table 1. Proximate composition of GALS (*A. marginata* var. *saturalis*) eggs and meat (DW basis).

Proximate composition (%)	White-skinned eggs	Black-skinned eggs	Meat**
Moisture content	60.00 ± 2.18	59.94 ± 2.00*	
Crude protein	64.01 ± 2.40	63.89 ± 2.35*	63.46 ± 2.56
Ash	30.98 ± 0.03	31.25 ± 1.80*	2.08 ± 0.01
Crude fibre	0.01 ± 0.01	0.00 ± 0.00*	3.01 ± 0.01
Lipid/Fat	1.03 ± 0.02	1.01 ± 0.01*	2.40 ± 0.02
Carbohydrate	3.97 ± 0.05	3.85 ± 0.04*	22.53 ± 1.08

Values are presented as means ± SD of five determinations. *Not significantly different from each other. **Uboh et al. (2010).

indicating that the consumption of these eggs could increase the levels of these minerals in the human body. Minerals play a vital role in the maintenance of various biochemical activities. For instance, calcium present in high concentrations, 184.20 ± 9.38 mg/100 g and 185.10 ± 10.15 mg/100 g for white-skinned and black-skinned *A. marginata* snails, respectively is known to play a crucial role in blood clotting and bone development in human. No wonder snail meat is used to stop bleeding during child birth (Akinnusi, 2002; Okon and Ibom, 2012). The calcium values of this study were higher than the values of 0.008 to 0.02% and 0.121 to 0.262% reported by Li-Chan et al. (1995) for chicken egg albumen and yolk, respectively. According to Pearson and Gillel (1999), calcium is the most abundant mineral element in the animal body and considered as an important constituent of the skeleton and teeth, in which about 99% of total calcium in the body is found. Besides, calcium is also important for the activity of a number of enzyme systems, including those necessary for the transmission of nerve impulses.

The eggs from the two ectotypes of *A. marginata* snails studied also constitute good sources of iron. The iron contents of 5.10 ± 0.07 mg/100 g and 5.00 ± 0.06 mg/100 g obtained for white-skinned and black-skinned *A. marginata* snails, respectively, though not significantly different (P > 0.05) from each other were quite higher than the 2.29 mg/100 g and 0.64 ± 0.01 mg/100 g values reported by Babalola and Akinsoyinu (2009), and Uboh et al. (2010), respectively for snail meat of the same species. Li-Chan et al. (1995) reported 0.0009% and

0.00053 to 0.011% as iron content of albumen and yolk, respectively of chicken egg, which is also lower than the iron values obtained in this study for the eggs of the two snail ectotypes evaluated. The disparity in the levels of iron is attributed to variations in iron contents from one locality to another, depending on the mineral content of the soils in which the snails were raised (Wosu, 2003). Iron facilitates the oxidation of carbohydrate, protein and fats. The iron in the eggs will not only enhance the absorption of iron from other sources such as cereals but will also increase considerably the level of iron absorption in the blood and prevent anemia (Andrew, 2011). Iron also helps in bone and teeth formation as well as for haemoglobin of the red blood cells in human.

Magnesium contents of the two ectotypes snail eggs obtained were 4.70 ± 0.05 mg/100 g and 4.72 ± 0.06 mg/100 g for white-skinned and black-skinned *A. marginata* snail eggs, respectively. These values though quite lower than the value of 31.00 ± 3.02 mg/100 g reported by Uboh et al. (2010) for *A. marginata* snail meat, were higher than the values of 0.009% and 0.032 to 0.128% reported by Li-Chan et al. (1995) for chicken egg albumen and yolk, respectively. The low magnesium content of snail eggs obtained here might be due to differences in the level of magnesium in the soil used for the rearing and the inability of the snails used to transfer a reasonable amount of this magnesium to the eggs.

Magnesium is a key element in cellular biochemistry and functions. Magnesium is closely associated with calcium and phosphorous and about 70% of the total magnesium is found in the skeleton. Magnesium is an enzyme activator, for example, in systems with thiamine pyro-phosphate as a co-factor and oxidative phosphorylation is reduced in magnesium deficiency. It is an essential activatior of phosphate transferase, activates pyruvate carboxylase, pyruvate oxidase and the reactions of the tricarboxylic acid cycle (Lehninger, 1984; Uboh et al., 2010).

Potassium values (4.62 ± 0.05 mg/100 g and 4.65 ± 0.05 mg/100 g for black-skinned and white-skinned ectotypes, respectively) obtained for these snail eggs though low will help to play an important role in osmotic regulation of the body fluid and in acid-base balance in the animal. The potassium values obtained for the snail eggs in this study were higher than the values of 0.145 to 0.167% and 0.112 to 0.360% reported by Li-Chan et al. (1995) for chicken egg albumen and yolk, respectively. The potassium content of these eggs will also participate in nerve and muscle excitability as well as carbohydrate metabolism (Aganga et al., 2003; Uboh et al., 2010).

Zinc contents obtained from the study were 0.08 ± 0.01 mg/100 g and 0.09 ± 0.01 mg/100 g for white-skinned and black-skinned A. marginata snail eggs, respectively. These values were not significantly different ($P > 0.05$) from each other but compare with the value of 0.5 mg/100 g reported by Andrew (2011) for chicken egg. The zinc values obtained in this study were lower than the 1.69 mg/100 g and 1.16 ± 0.01 mg/100 g values reported by Fagbuaro et al. (2006) and Eneji et al. (2008), respectively for snail meat. The low zinc content in snail eggs implies that the eggs are non-toxic to health, thus, recommended for human consumption.

Conclusion

The results of this pilot study showed that the eggs of the two ectotypes of A. marginata (white-skinned and black-skinned) snail are not significantly different ($P > 0.05$), thus, they are good sources of protein and basic minerals. These eggs are highly recommended for both young and old. The snail eggs constitute an alternative source of essential nutritional elements at a lower cost for Nigerians.

REFERENCES

Adedire CO, Imevbore EA, Eyide EO, Ayodele WI (1999). Aspects of digestive physiology and complementary roles of the microbial enzymes in the intestinal tract of giant land snail Archachatina marginata (Swainson). J. Technosci. P. 3.

Adeola AI, Adeyemi AJ, Ogunjobi JA, Alaye SA, Adelakun KM (2010). Effect of Natural and Concentrate diets on proximate composition and sensory properties of giant land snail (Archachatina marginata) meat. J. Appl. Sci. Environ. Sanit. 5(2):185-189.

Aganga AA, Aganga AO, Therna T, Obocheleng KO (2003). Carcass Analysis and meat composition of the donkey. Pak. J. Nutr. 2(3):138-147.

Akinnusi O (1979). Introduction to Snails and Snail Farming. 1st ed. Abeokuta, Nigeria. Triolas Exquisite Ventures. P. 98.

Akinnusi O (2002). Introduction to Snails and Snail Farming. 2nd ed. Abeokuta, Nigeria. Triolas Exquisite Ventures. P. 98.

Andrew T (2011). Vitamins and minerals in eggs. Livestrong Foundation/Livestrong.com.

AOAC (1990). Association of Official Analytical Chemists. Official Methods of Analysis, 15th ed. Washington DC.

Asibey EOA, Eyeson KK (1995). Additional Information on the Importance of Wild Animals as Food Source in South Africa, Bungo. J. Ghana Wildl. Soc. 1:13-17.

Babalola OO, Akinsoyinu AO (2009). Proximate Composition and Mineral profile of Snail meat from different breeds of land snail in Nigeria. J. Nutr. 8:1842-1844.

Bligh EG, Dyer WJ (1959). A rapid method of total lipid extraction and purification. Can. J. Biochem. Physiol. 3:911.

Ebenso IE (2003). Nutritive Potentials of White Snails (Archachatina marginata) in Nigeria. Discov. Innov. 15(3/4):156-158.

Eneji CA, Ogogo AU, Emmanuel Ikpeme CA, Okon OE (2008). Nutritional Assessment of Some Nigerian Land Snail Species. Ethiop. J. Environ. Stud. Manag. 1(2):56-60.

Fagbuaro O, Oso JA. Edward JB, Ogunleye RF (2006). Nutritional Status of Four Species of giant land snails in Nigeria. J. Zhejinng. Univer. Sci. B. 7(9):686-689.

GENSTAT (2004). Genstat release 12.1 (PC/Windows XP) Lowes Agricultural trut - (Rothmted experimental Station, VSN International Ltd.).

Holman RT, Hayes H (1958). Determination of polyunsaturated acids in lipids of plasma and tissue. Analyt. Chem. 30:1422.

Ibom LA, Okon B, Essien A (2008). Morphometric analysis of eggs laid by two ectotypes of Snail Archachatina marginata (Swainson) raised in captivity. Global J. Agric. Sci. 7(2):119-121.

Lehninger SA (1984). Principles of Biochemistry. U.S.A., Worth Publishers, Inc.

Li-Chan ECY, Powrie WD, Nakai S (1995). The chemistry of eggs and egg products. In: Egg Science and Technology, W. J. Stadelman and Cotterill (Eds.), The Haworth Press Inc., New York.

Malik AA, Aremu A, Bayoda GB, Ibrahim BA (2011). A Nutritional and Organoleptic Assessment of the meat of the giant African land snail Archachatina marginata (Swainson) Compared to the meat of ither livestock. Livest. Res. Rural Dev. 23(3).

Ogogo AU (2004). Wildlife Management in Nigeria. Objectives, Principles and Procedures. Calabar. Median Communications. pp. 134-154.

Okon B, Ibom LA (2011). Phenotypic Correlation and Body weights Prediction using Morphometric Traits of Snails in Calabar, Nigeria. Ethiop.J. Environ. Stud. Manage. 4(3):7-11.

Okon B, Ibom LA (2012). Snail Breeding and Snailery Management. Freshdew Productions, Calabar, Nigeria.

Okon B, Ibom LA, Ettah HE, Ukpuho IE (2012). Effects of Genotype, Dietary Protein and Energy on the Reproductive and Growth Traits of Parents and F₁ Hatchlings of Achatina achatina (L) Snails in Nigeria. Int. J. Appl. Sci. Technol. 2(1):179-185.

Okon B, Ibom LA, Ewa EC (2009). Evaluation of Reproductive and Some Egg Quality Parameters of Albino Snails [Archachatina marginata saturalis (Swainson)]. J. Appl. Sci. 12(1):8234-8240.

Omole AJ, Kehinde AS (2005). Backyard snail farming at a glance. Back to Agricultural series (1). Ibadan, Nigeria. Technovisor Agricultural Publications.

Pearson D (1976). Chemical Analysis of Feeds. 7th ed. London J, Churchill A, Pearson AM, Gillel TA (1999). Processed Meats. Galthers-burg. M.C. Aspen.

Raut SK, Barker GM (2002). Achatina fulica Rowdies and other Achatiidae as Pests in Tropical Agriculture. In: G. M. Barker ed. Molluscs as Crop Pests, Hamiton, New Zealand. CABI Publishing. pp. 55-114.

Uboh FE, Ebong PE, Mbi E (2010). Cultural Discrimination in theConsumption of black snail (Archachatina marginata) and white snail (Achatina achatina); any Scientific Justification? Int. Res. J. Microbiol. 1(1):013-017.

Wosu LO (2003). Commercial Snail Farming in West Africa. A Guide, AP Express Publishers, Nsukka, Nigeria.

10

Assessment of feed resource availability and livestock production constraints in selected Kebeles of Adami Tullu Jiddo Kombolcha District, Ethiopia

Dawit Assefa[1], Ajebu Nurfeta[2], Sandip Banerjee[2]

[1]Adami Tullu Agricultural Research Center, P. O. Box 35, Ziway, Ethiopia.
[2]School of Animal and Range Sciences, Collage of Agriculture, Hawassa University, P. O. Box 5, Hawassa, Ethiopia.

This study was conducted in Adami Tullu Jiddo Kombolcha district, Oromia Regional State to assess the major available feed resources in the area and to identify and rank feeding problems and possible improvement options for livestock feeding in the district. Multi-stage sampling techniques were used to select the study sites. Sixty respondents were selected from rural and peri-urban Kebeles. The total annual feed DM available was higher (P < 0.05) in rural (13.98 tons) than in the peri-urban (9.45 tons) kebeles. An average of 11.72 tons of feed dry matter (DM) was produced per household from the major available feed resources, in which 74.57% was obtained from crop residues. A total of 419.4 and 283.5 tones DM/year vs 423.6 and 394.3 tones DM/year was the requirements in rural and peri-urban areas, respectively. Hence, the study indicates that the available feed DM satisfies 99 and 71.9% of DM requirements of rural and peri-urban sites, respectively. The estimated annual DM requirements for maintenance were 13.63 tons with a deficit of 1.91 tons. Feed shortage, water scarcity, disease and low productivity of animals were assessed to be the major livestock production constraints.

Key words: Crop residues, feed availability, feed balance, feed requirement, urban-peri urban.

INTRODUCTION

Livestock industry is an important and integral part of the agricultural sector in Ethiopia. Livestock farming is vital for the supply of meat and milk; it also serves as a source of additional income both for smallholder farmers and livestock owners' (Ehui et al., 2002). Livestock production constraints can be grouped into socio-economic and technical limitations (Mengistu, 2003). Inadequate feed, widespread diseases, poor breeding stock, and inadequate livestock policies with respect to credit, extension, marketing and infrastructure are the major constraints affecting livestock performance in Ethiopia (Desta et al., 2000). Feed resources as reported by

Tolera et al. (2012) can be classified as natural pasture, crop residue, improved pasture and forage and agro industrial by-products of which the first two contribute the largest share. The fibrous agricultural residues contributes a major parts of livestock feed especially in the populated countries where land is prioritized for crop cultivation. Tolera et al. (2012) reported that crop residues contribute to about 50% the total feed supply in Ethiopia.

Under smallholder livestock production system, animals are dependent on a variety of feed resources which vary both in quantity and quality. For optimum livestock productivity, the available feed resource should match with the number of animals in a given area. However, there is scanty of information regarding the assessment of feed resources in Adami Tullu area. Few literatures at hand mainly focuses on available feed resources without quantifying the amount obtained from each feed types without indicating their values on the bases of dry matter available which could satisfy the DM requirement of the livestock. For example, the study by Alemu et al. (2006) evaluated the utilization of crop residues in selected agro-ecological zones of Eastern Shoa which mainly focused on rural households. Moreover, the land which used to be allocated for grazing and crop production is being converted to other businesses which require regular assessment. Livestock production constraints could vary not only across agro-ecology but also among production systems. For example different classes of animals are kept by the urban and peri-urban farmers which are dictated by the demand for the products such as milk and availability of the supplemental feeds. The peri-urban and urban farmers usually purchase basal feeds (grasses and crop residues) from the rural area. However, the supply of feeds to the urban farmers depends on the availability of feed resources in the rural area. Shortage of feeds in the rural area affects the management and productivity of livestock in the urban and peri-urban areas. Therefore, it is necessary to study the feed production status and production constraints in the rural and urban/peri-urban livestock production systems. Such information is necessary for policy makers and farmers in order to alleviate the problems. There is very little information which asses the availability and utilization of feed resources in rural and peri-urban areas. Feed resource assessment is important to diagnose the problems and suggest intervention measures to be taken by farmers and policy makers. Therefore, it is important to assess the available feed resources in relation to the requirements of livestock on annual basis in a given area. Therefore, this study was initiated with the objective to assess the major available feed resources and investigate feed related problems in Adami Tullu Jiddo Kombolcha district.

MATERIALS AND METHODS

Description of the study area

The study was carried out in four Kebeles of Adami Tullu Jido Kombolcha (ATJK) district of East Showa Zone, Oromia, Ethiopia. The district was selected due to the severity of feed shortage and the rapid expansion of flower investment which limits feed production which necessitates the assessment of feed resources. The district is located between 38°20' and 38.5°5' E and 7°35'and 8°05' N. It lies at altitudinal range from 1500 to 2000 masl. The district consists of 43 Kebeles and 4 urban towns. The livestock population of the district was estimated to be 211559 cattle, 116585 goats, 25114 sheep, 23720 donkeys, 1441 horses, 423 mules and 13059 Chicken (ATJK district Livestock Development and Health

bureau, 2008). The agro-ecological zone of the district is semi-arid and sub-humid in which 90% of the area is lowland while the remaining 10% is intermediate. The average annual temperature ranges from 22 to 28°C. The area receives average annual rainfall ranging from 760 to 1000 mm in which the distribution is uneven and erratic in nature (Kebede, 2010).

Adami Tullu Jido Kombolcha district is characterized by mixed crop-livestock farming system. Cattle, goat, sheep, donkey and poultry are important livestock species reared in the area. Cereals (maize, barley, and sorghum), haricot bean are the major crops produced under rain fed condition. The vegetation cover of the area is generally characterized by open woodland that consists of acacia and Balanite species. The major soil type of the district is sandy, loam and black (ATJK district livestock Development and Health, 2008).

Sampling technique

Information was obtained from Adami Tullu Jiddo Kombolcha District Office of Livestock Development and Health (LDH) on locally developed organizational structure of the Kebeles (lowest administrative unit), livestock population and distribution of cross breed animals in the district. First, rural and peri-urban potential Kebeles were purposively selected. The criteria for selection of kebeles and farmers were multifold vs livestock population, accessibility and experience of farmers keeping livestock for not less than two years. Accordingly, two Kebeles each from rural and peri-urban sites were selected for the study. Based on the aforementioned criteria, the selected rural kebeles were Elica Calamo and Arba while Edo-Gojola and Garbi-Gilgile kebeles were selected from the peri-urban areas. Sixty respondents (fifteen from each kebeles) were randomly selected for the study.

Types of data and methods of data collection

Data was collected both from primary and secondary sources. Secondary sources of data on climate, soil, topography, agro-ecology, human population, livestock population and livestock production constraints were collected by reviewing different documents and from Adami Tullu Jido Kombolcha LDH office. Primary data (household size, land utilization pattern, major feed resource, production of grain and crop residues, household herd size; seasonality of feed resources) were obtained from the questionnaire survey during the course of the study. The primary data was collected using semi-structured questionnaire between November and December 2010. The questionnaire was first pre tested before the commencement of the survey. Focused group discussions were made at each Kebele to clarify issues not well addressed thought survey and to validate some information collected by individual interview. A total of 28 individuals, 7 from each Kebele were involved in the group discussion. The discussion focused on identifying constraints related to livestock feed and identifying the major livestock production constraints.

Estimation of annual feed resources and livestock feed requirement

The quantity of feed dry matter (DM) obtained from crop residues per household farm were estimated from crop yield to crop residue ratio using conversion factors of FAO (1987). Accordingly, for a ton of maize stover conversion value of 2.0 was used, for a ton of wheat, barley and teff (*Eragrostis abyssinica*) straw, the conversion value of 1.5 was used, while conversion Figures of 1.2 and 2.5 were used for the haricot bean and sorghum, respectively. The quantity of crop residue on the basis of DM available and those

Table 1. Average land use patterns and holding size (ha) per house hold in rural (N=30) and peri-urban (N=30) Kebeles.

Land use type	Location of Kebeles	
	Rural	Peri-urban
Homestead	0.43 ± 0.04	0.27 ± 0.02
Cultivated land	2.44 ± 0.22	2.04 ± 0.13
Private grazing land	0.92 ± 0.24	0.18 ± 0.07
Plantation/wood land	0.15 ± 0.05	0.02 ± 0.02
Total land holding	3.81 ± 0.46	2.65 ± 0.17
Mean land holding of the area	3.23 ± 0.25	

N= Number of respondent.

actually available for livestock consumption was estimated by deducting 10% of the same as wastage (Tolera and Said, 1994).

The quantity of feed DM obtained annually from different land use type was determined by multiplying the hectare under each land use type (FAO, 1987). Conversion factor of 2.0, 0.5, 3.0, 1.8 and 0.7 tDM/ha/year were used for natural pasture, aftermath, private grazing land, fallow land and forest/wood land, respectively. The quantity of DM obtained from irrigation practices was estimated by multiplying the irrigated land size by 0.3 tDM/ha/seasons (FAO, 1987).

The livestock population per household was converted to tropical livestock unit (TLU) as recommended by Gryseels (1988) and Shiferaw (1991) for local and cross breed animals, respectively. The DM requirement was calculated based on daily DM requirement of 250 kg dual purpose tropical cattle (an equivalent of one TLU) for maintenances according to Kearl (1982).

Statistical analysis

Primary data from surveyed households was organized and analyzed using Statistical Package for Social Science (SPSS version 13). Leven's test was used to check homogeneity of variances in the data analysis. Mean and percentage values of various parameters were compared between the two study locations.

RESULTS AND DISCUSION

Analysis of *household survey*

General household characteristics of the sampled household

The average family size of the respondents was 9.92 ± 0.52 heads; however the range was quite broad and spanned between 3 to 20 heads per household. This result is similar to the average family size reported by Kebede (2010) and Wondatir (2010). This large family size may be attributed to lack of awareness towards family planning measures and having many family members is considered as an asset for extensive farm activity. The majority (85%) of the respondents were male household heads. The study further indicated that 21.7% of the respondents were illiterate, while the rest (23.3, 46.7, 5 and 3.3%) had educational background for basic education, primary education, junior secondary education and high school level, respectively.

Land holding and land use pattern

Land is one of the most important resources required for successful implementation of any agricultural farming activities. The results indicate that about 69% of the land was allocated for cultivation while the rest was allocated for private grazing land, homestead land and enclosed plantation/wood land, respectively (Table 1). Rural household farmers allocated more (p<0.05) land for homestead, private grazing land and closed plantations. As expected, the ownership pattern indicated that the land ownership was higher in the rural areas when compared with peri-urban areas. This may be ascribed to diversion of land for commercial purposes and/or for developmental activities in the area.

Livestock holding and population trend

The total numbers of livestock, cattle and cross bred cattle holding per household in the study area are presented in Table 2. The results indicate that the average total and cross bred cattle populations were 8.27 and 2.67 TLU, respectively. The average holdings of local breed type were higher (p<0.05) in rural areas than that urban Kebeles, while for the crossbred type the reverse was true. This may be attributed to the easy access opportunity for the necessary input such as veterinary services, agro-industrial by-products available in the peri urban areas thus ensuring the availability of nutrients in the lean periods and better access to market and thereby ensuring product off take.

Constraints of livestock production

The results from Table 3 indicate the constraints influencing the overall productivity of the livestock in the studied area.

Table 2. Total livestock and cattle, local and cross bred cattle holding per household in rural and peri-urban Kebeles.

Livestock type	Rural (N=30)	Peri urban (N=30)	Overall (N=60)
Total livestock (TLU)	9.76±1.31	9.10±0.66	9.43±0.73
Total cattle (TLU)	8.50±1.21	8.04±0.54	8.27±0.65
Local cattle (TLU)	6.77±1.06[a]	4.43±0.34[b]	5.60±0.57
Cross bred cattle (TLU)	1.73±0.25[b]	3.61±0.37[a]	2.67±0.25
Goat (TLU)	0.33±0.08	0.18±0.08	0.26±0.05
Sheep (TLU)	0.07 ±0.03	0.21± 0.07	0.14±0.04
Donkey (TLU)	0.6±0.12	0.62±0.13	0.61±0.09
Horses and mules	0.21±0.11	0.03±0.26	0.12±0.19

N = number of respondent; means with different superscript in a row are significantly different at P<0.05.

Table 3. Major Livestock production constraints in Adami Tullu Jiddo Kombolcha district.

Major constraints	Constraint level (N= 60)					Rank
	1	2	3	4	5	
Feed shortage	40 (67.7%)	13 (21.7%)	3 (5.0%)	3 (5.0%)	1 (1.7%)	1
Water scarcity	19 (31.7%)	34 (56.7%)	4 (6.7%)	3 (5.0%)	-	2
Disease	-	11 (18.3%)	35 (58.3%)	12 (20.0%)	2 (3.3%)	3
Low productivity of animals	1 (1.7%)	1 (1.7%)	11 (18.3%)	31 (51.7%)	16 (26.7%)	4
Others	-	-	7 (11.7%)	10 (16.7%)	43 (71.7%)	5

N = number of respondent; Others = change of crop farm to flower and winery farm, expansion of crop land at expenses of grazing land.

The results indicate that feed shortage is the major constraint identified by most of the respondents. Farmers indicated that increment in crop land at the expense of grazing land, shortage of land for forage production, renting and allocation of open grazing lands around Lake Zeway for investors has resulted in a decrease grazing land. The observations are in agreement with that of Keftasa (1996) who also indicated that shifting of grazing land into crop cultivation has dwindled the potential of the livestock in the area and also put immense pressure on the existing land. Recurrent drought, prolonged dry period and uneven distribution of rainfall which affects crop production and re-growth potential of grasses were also the factors which cause feed shortage in the study area. As most of the water bodies are recharged annually during the rains, erratic rainfall is hampering their proper recharge thereby affecting the livestock in the region as poor quality and inadequate quantity of water is affecting both livestock health and production. Farmers participated on group discussion pointed out that Lake Zeway and River Bulbula are the main sources of water for most of livestock in the district. They claimed that due to use of water from Lake Zeway, the volume of these water bodies is shrinking rapidly. Therefore, the maintenance of the existing water bodies and also identifying new water bodies to satisfy the water needs for both man and livestock alike need to be carried out on a priority basis in the area.

The result of the discussion also indicated that the prevalence of disease in the area is common. The most economically important disease are black leg, foot and mouth disease, anthrax, mastitis, pastreulosis and pox. The participated farmers ranked disease outbreak to be the third constraint. The disease generally occurs during the short rainy season spanning between March and May, which may be because the livestock are immune compromised due to lack of fodder in the preceding dry season. The observations in the present study are in accordance with the observations of Burmby and Scholtens (1986) who indicated that the animals health problem are closely linked to the kind of environment in which the herd is kept and the management methods used in production system.

The other challenge reported was the emerging investment growth in the area as the land available for grazing purpose are being taken away for industries adding to further woes to the already scarce feed resources prevailing in the area. Setting of land use policy for every type of activities and production system was the recommendation given by farmers during group discussion. The livestock production constraints as reported in the present study are in consonance with the observations of Desta et al. (2000) who indicated that the inadequate feed and nutrition, poor health , low productivity of local breeding stock are the main livestock production constraints in Ethiopia.

Table 4. Local name and scientific names of naturally grown grass and legume types identified by farmers.

| Grass species | | Legume species | |
Local name	Scientific name	Local name	Scientific name
Korto	Cynodon dactylon	Ejisisa	Crotalaria incana
Marga hillo (grass)	Chrysopogon plumulosus	Calcabbi	Indigofera spinosa
Hufe	Aristida adoensis	Qore Hare	Crotalaria spinosa
O'aa	Hyparrhenia cymbaria	Galee	Psydrax schimpefiana
Sumaro	Bothriochloa radicans		
Egee sare	Cenchrus ciliaris		

Table 5. Estimated annual feed dry matter obtained per household farm from different crop residues, land use type and utilizable DM yield in rural and peri-urban kebeles.

| Sources of feeds | Feed production (t DM) | | | % |
	Rural	Peri urban	Average DM (t)	
Maize	8.7 ± 0.9	5.6 ± 0.5	7.2 ± 0.5	73.63
Teff	0.22 ± 0.1	0.34 ± 0.1	0.28 ± 0.04	2.86
Wheat	0.32 ± 0.18	0.31 ± 0.10	0.31 ± 1.0	3.23
Barley	0.6 ± 0.21	0.94 ± 0.20	0.77 ± 0.15	7.92
Sorghum	0.10 ± 0.08	0.07 ± 0.04	0.08 ± 0.05	0.84
Haricot bean	0.9 ± 0.23	1.3 ± 0.2	1.1 ± 0.2	11.21
Vegetables	0.04 ± 0.01	0.02 ± 0.01	0.03 ± 0.01	0.31
Total crop residues	10.9 ± 1.1	8.5 ± 0.5	9.7 ± 0.6	100
Utilizable crop residue	9.79	7.69	8.74	74.57
Aftermath	1.22 ± 0.11	1.02 ± 0.07	1.12 ± 0.06	9.56
Grazing land	2.76 ± 0.74	0.54 ± 0.20	1.65 ± 0.41	14.08
Fallow land	0.2 ± 0.06	0.07 ± 0.03	0.14 ± 0.03	1.19
Wheat bran	0.01 ± 0.01	0.11 ± 0.08	0.06 ± 0.04	0.51
Noug seed cake	-	0.02 ± 0.02	0.01 ± 0.01	0.09
Total DM available	13.98	9.45	11.72	100

Feed resource and feed availability

During group discussion farmers were asked to provide the types feed resources available in the area. Accordingly, natural pasture, aftermath grazing, crop residues, and maize thinning were the major feed resources during the wet season. However, crop residues, natural pasture and aftermath grazing were the major feed resources for dry season, in their descending order. In general crop residues and natural pasture are the major feed resources of the area which agree with the report of Tolera et al. (2012) who reported natural pasture and crop residue to be the major feed resources for highlands of Ethiopia. The major grasses species grown in the area is presented in Table 4. *Cynodon dactylon, Bothriochloa radicans, Aristida adoensis, Cenchrus ciliaris, Psydrax schimpefiana, Chrysopogon plumulosus* and *Hyparrhenia cymbaria* were the abundant grass species indentified by the local community while *Crotlaria Incana, Indigofera Spinosa, Crotaria Spinosa* and *Psydrax schimpefiaana* were legumes grown in which grass species

were dominant in the natural pasture. The studies further indicated that *Psydrax schimpefiana* and *Chrysopogon plumulosus* are fast disappearing and are rarely seen now days. As reported by Chadhokar (1984) such incidences may be related to overgrazing, as is evident in the area and this has resulted in the domination of unpalatable species of grasses in the study area resulting the pasture to be dominated by unpalatable grass species. This may in turn affect the photosynthetic area of a plant as well as species composition in the area (Stoddart et al., 1975). The presence of many grass species in natural pasture as indicated by the farmers are in agreement with the findings of Abate (2007) who also observed high proportion of grass (78.6%) in the total biomass in arid zones of Bale.

Estimation of feed dry matter production

It is perceived from the results presented in Table 5 that the average utilizable feed DM yield per household

Table 6. Estimated annual utilizable feed DM supply, DM requirement and feed balance per household in rural and peri-urban kebeles.

Kebeles	Annual feed DM supply (t)	Estimated annual DM requirement for maintenance (t)	Balance of supply versus requirement (t)
Rural	13.98	14.12	-0.14 (99%)
Peri-urban	9.45	13.14	-3.69 (71.9%)
Total	11.72	13.63	-1.91 (86%)
Total/district	703.2	817.8	-114.6 (86%)

farm from crop residues was 9.79 and 7.69 tons per annum by using 10% loss for rural and peri-urban kebeles, respectively. McDonald et al. (2002) observed that in spite of the importance and availability of the crop residues, there are several constraints to their utilization which may be primarily attributed to the higher structural carbohydrates which limits its digestibility. The proportion of leaf to stem ratio is the major factor affecting the nutritional differences among crop varieties (McDowell, 1988). The higher production of DM in the rural areas may be related to the larger land holdings by individual household which is larger for rural households. The differences in varieties and types of the crop besides agronomic management viz. usage of quality and quantity of fertilizers; plant protection measures which may also lead to the difference in vegetative growth and thereby affecting the yield (Reddy et al., 1998). The relationship between grain yield and crop residues suggest that grain yield and vegetative growth is positively correlated (Keftasa, 1988). The present study indicates that 8.74 tons feed DM was annually produced per household farm from crop residues in the study district. The results are higher than the values (6.7 tons DM) reported by Bogale et al. (2008) in the Bale high land. From the annual dry matter obtained, a sizable amount of feed DM was obtained from crop residue in which maize stover contributes the highest proportion followed by the dry matter obtained from the residues of haricot bean (Table 5). From the overall crop residues on average, 73.63% of the total DM was obtained from maize stover alone. Following the harvesting of crops, aftermath grazing is used as potential source of feed for the livestock in the area which usually lasts for three to four months.

The mean annual utilizable dry matter production from different types of feed resources was estimated to be 13.98 and 9.45 tDM for rural and peri-urban Kebeles, respectively. The average value of feed dry matter produced per household farm in rural areas was higher (p < 0.05) than that of feed DM produced in peri-urban kebeles. In general a total of 11.72 tons of feed dry matter (DM) per household were obtained from the major available feed resources in which 74.57% was obtained from various crop residues and the rest (14.84, 9.56, 1.19, and 0.6%) were obtained from grazing land, aftermath, fallow land and industrial by-product. The

value of crop residues dry matter contribution obtained in the present study was comparable to the findings of Admasu (2008) and Wondatir (2010) who found that crop residues contributed to 78.72 and 86.38% of total feed DM production in Alaba Wereda and Central Rift Valley of Ethiopia, respectively. The present study indicated that crop residues and aftermath grazing contributed to 84.67% of the total annual feed dry matter supply per household.

Estimated annual feed balance

The average annual utilizable feed DM supply was estimated to be 13.98 and 9.45 tDM per household farm for rural and peri-urban kebeles, respectively (Table 2). Based on the suggested estimation by Kearl (1982) the annual feed DM requirements for maintenances for rural and peri-urban areas was 14.12 and 13.14 tDM, respectively (Table 6).

According to the result obtained in this study, about 419.4 and 283.5 tDM/year feed was produced from all feed resources in rural and peri-urban kebeles, respectively, whereas about 423.6 and 394.2 tDM/year feed was the actual requirement for both locations, respectively. Hence the annual utilizable feed dry matter satisfied about 99 and 71.9% of the livestock maintenance requirement for rural and peri-urban kebeles, respectively. This result further indicates that the feed shortage was more aggravated in peri-urban than rural Kebeles. The negative feed balance observed in this study was comparable with the observations of Tolera and Said (1994), Admasu (2008) and Wondatir (2010) in Wolayita Sodo, Alaba Wereda and central Rift Valley of Ethiopia, respectively. However, Amare (2006) and Mulu (2009) reported a positive balance for DM matter requirements conducted at north Gonder and Bure wereda. These positive values reported may be related to the small livestock population in the area and also the fertility of the land favoring feed production. Another reason for the positive values may be related to the average moisture content in the area as these study cover different agro-ecology like the highland which have got surplus moisture compared with the current study area which is moisture deficit.

In general caution should be taken when analyzing feed

balance. For example, home made by-products and tree leaves are often utilized by smallholder farmers which is not easy to quantify. Also alternative use of crop residues is very common. For example, Alemu et al. (2006) indicated that farmers use crop residues as a source of cash, fire wood, construction material, mulching and mattress. Therefore, there could be overestimation or under estimation of the values obtained.

CONCLUSION AND RECOMMENDATION

From this study it can be concluded that, the availability of feed DM did not satisfy the maintenance requirements of total livestock reared in the study area. The study further highlights that the scarcity of feed was more serious in the peri-urban Kebeles. The feed deficit observed in the study area could be one of the contributing factors affecting livestock productivity. There should be land use policy regulation in the area which can secure area for livestock feed production to make the livestock sector contributes to poverty eradication and encourage smallholder farmer food secured household. Finally, with regards to feed resource assessment detailed study should be conducted on chemical composition especially crude protein and energy content and digestibility of available feed resources in the area to recommend concrete development strategies.

ACKNOWLEDGEMENTS

The authors would like to thank Oromia Agricultural Research Institute for granting financial support as well as Adami Tullu Agricultural Research Center for giving all necessary facilities. Also they particularly acknowledge Ato Abebe Mabirate and Ato Taha Mume for their support to facilitate material and financial needs, and Ato Aman Abdulkadir for his assistance in feed preparation and data collection of feeding trial. They appreciate farmers for their involvement during data collection.

REFERENCES

Abate T (2007). Traditional utilization practices and condition assessment of rangeland in Rayitu district of Bale zone, Ethiopia. M.Sc. Thesis. Haramaya University, Ethiopia.   P. 128.

Admasu Y (2008). Assessment of livestock feed resources utilization in Alaba wereda, southern Ethiopia. M.Sc. Thesis. Haramaya University. Dire Dewa, Ethiopia. P. 58.

Amare S (2006). Livestock production systems and availability of feed resources in different agro-ecologies of north Gonder Zone, Ethiopia. M.Sc. Thesis. Haramaya University. Dire Dewa, Ethiopia. P. 64.

Bogale S, Melaku S, Yami A (2008). Potential use of crop residues as livestock feed resource   under smallholder farmer's condition in Bale Highlands of Ethiopia. Trop. Subtrop. Agroecosyst. 8:107-114.

Burmby PJ, Scholtens RG (1986). Management and health constraints for small-scale dairy production in Africa. pp. 9-12. In: ILCA Bulletin No. 25. Addis Ababa, Ethiopia, August 1986, International Livestock Centre for Africa (ILCA).

Chadhokar PA (1984). Multipurpose plant species for soil and water conservation, field document No. 14. FAO, Addis Ababa, Ethiopia. P. 94.

Desta L, Kassie M, Benins S, Pender J (2000). Land degradation and strategies for sustainable development in Ethiopian highlands: Amhara region. ILRI socioeconomics and policy research working paper 32. International Livestock Research Institute (ILRI), Nairobi, Kenya. P. 122.

Ehui S, Benin S, Williams T and Meijer S (2002). Food security in Sub-Saharan Africa to 2002, socio-economic and policy research working paper 49, ILRI (International Livestock Research Institute), Nairobi, Kenya. P. 60.

FAO (Food and Agriculture Organization of the United Nations) (1987). Land use, production regions, and farming systems inventory. Technical Report, 3 vol. 1. FAO Project ETH/78/003, Addis Abeba, Ethiopia.

Gryseels G (1988). Role of livestock on a mixed smallholder farmers in Debre Berhan. PhD Dissertation. Agricultural University of Wageningen,   The   Netherlands.   P.   149. http://www.fao.org/ag/agp/agpc/doc/counprof/ethiopia/ethiopia.htm.

Kearl LC (1982). Nutrient requirement of ruminants in developing countries. International   Feedstuff Institute, Utah Agricultural Experiment Station. Utah State University, London, USA. P. 381.

Kebede T (2010). Assessment of on-farm breeding practices and estimation of genetic and phenotypic parameters for reproductive and survival traits in indigenous Arsi Bale goats. M.Sc. Thesis. Haramaya University. Dire Dewa, Ethiopia. P. 8.

Keftasa D (1988). Role of Crop Residues as Livestock Feed in Ethiopian Highlands. In: B. H. Dzowela (Ed.). African Forage Genetic Resources, Evaluation of Forage Germplasm and extensive Livestock Production Systems. Proceedings of a Workshop Held in Arusha, Tanzania, 27-30 April 1987. pp. 430-439.

Keftasa D (1996). Research on the integration of forage legumes in wheat-based cropping systems in Ethiopia: A review. In: J., Ndikumana and P., De Leeuw (eds.). Proceedings of the second African Feed Resources Network (AFRNET) held at Harare, Zimbabwe. Nairobi, Kenya, 6–10 Dec. 1993, African Feed Resources Network (AFRNET). P. 201.

McDonald P, Edwards AR, Greenhalgh DFJ, Morgan AC (2002). Animal nutrition. 6th ed. Longman Group, Harlow (UK) P. 607.

McDowell RE (1988). Improvement of crop residues for feeding livestock in small-holder   farming systems. In: J.D. Reed, B.S. Capper and P.J.H. Neate (eds.). Plant Breeding and Nutritive Value of Crop Residues. Proceedings of a Workshop Held at ILCA, Addis Ababa, Ethiopia, 7-10 December 1987. ILCA, Addis Ababa. pp. 3-27.

Mengistu A (2003). Country pasture/forage resources profiles: Ethiopia. Food and Agriculture Organization of the United Nations (FAO).

Mulu S (2009). Feed resource availability, cattle fattening practices and marketing system in Bure Wereda, Amahara Region, Ethiopia. M.Sc. Thesis. Mekele University. P. 51.

Reddy RMG, Rao MM, Murthy SKK (1998). A multiple regression model for predicting groundnut yields in arid zones using weather parameters. Trop. Agric. 75(4):321-331.

Shiferaw B (1991). Crop livestock production in the Ethiopian highlands and effects on sustainability of mixed farming: a case study of Ada district, Debrezeit. M.Sc. Thesis. Agricultural University of Norway, Oslo, Norway. P. 163.

Stoddart LA, Smith AD, and Box TW (1975). Range management, 3rd edition, McGraw Hill Book, INC. New York. P. 532.

Tolera A, Said AN (1994). Assessment of feed resources in Wolayita Sodo. Ethiopian. J. Agric. Sci. 14:69-87.

Tolera A, Yami A, Alemu D (2012). Livestock feed resources in Ethiopia: Challenges, Opportunities and the need for transformation. Ethiopia Animal Feed Industry Association, Addis Ababa, Ethiopia.

Wondatir Z (2010). Livestock production system in relation to feed availability in the highlands and central rift valley of Ethiopia. M.Sc. Thesis. Haramaya Uninversity. Dire Dewa, Ethiopia. P. 31.

11

Effects of Guar meal, Guar gum and saponin rich Guar meal extract on productive performance of starter broiler chicks

S. M. Hassan

Animal and Fish Production Department, King Faisal University, Al-Ahsa, Kingdom of Saudi Arabia.

The study was set up to evaluate whether Saponin rich guar meal extract (GS) or residual Guar gum (GG) is the main anti-nutritional compound contributing to Guar meal (GM) relatively poor feeding value for poultry. Two hundred forty 1-days-old chicks were randomly distributed among 4 treatments with 4 replicates of 15 chicks each from 1 to 21 days. Chicks were fed one of four treatments: Control broiler diet, control diet containing 5.0% GM, control diet containing 0.90% GG, and control diet containing 0.250% GS. Feed intake was the highest in chicks fed 5.0% GM from 1 to 7 days, but was the lowest in chicks fed 0.90% GG from 8 to 14 days. Over the entire course of the study from 1 to 21 days, feed conversion ratio was very poor; the highest was for chicks fed 0.250% GS as compared to other groups. The final body weight at 21 days was lower in chicks fed 0.250% GS than chicks fed 0.90% GG and control. Total body weight gain from 1 to 21 days was lower in chicks fed 0.250% GS than chicks fed 0.90% GG and control. We conclude that that there are more negative effects associated with adding 0.250% GS than 0.90% GG suggesting saponins may play a prominent role in the growth inhibition effects on feeding GM to broiler chicks.

Key words: Starter broiler chicks, Guar meal, Guar gum, saponin, performance.

INTRODUCTION

Using feed ingredient alternatives to supplement or replace traditional feed ingredients in poultry diets is an economical interest used by poultry nutritionists worldwide. Guar, *Cyamopsis tetragonoloba* L. (syn. *C. psoraloides*) or cluster bean is a drought-tolerant summer annual legume native to India and Pakistan (Rahman and Shafivr, 1967; Patel and McGinnis, 1985).

Guar meal (GM) is a by-product of the isolation of guar gum (GG) from guar bean. GM consists of a mixture of germ and hull fractions (Rahman and Shafivr, 1967). GM contains about 33 to 47.5% crude protein on a dry matter basis (Ambegaokar et al., 1969; Nagpal et al., 1971; Lee et al., 2004), about 18% residual GG (Anderson and Warnick, 1964; Nagpal et al., 1971; Lee et al., 2004)

and about 5.0% crude saponin by weight of the dry matter basis (Hassan et al., 2010).

GG is produced by further processing the endosperm fraction of the guar bean (Lee et al., 2004). Chemically, GG is a linear chain of D-mannose units connected by β-1-4 glycoside bonds. Every other D-mannose unit bonds a D-galactose unit by α-1-6 glycoside linkage. Commercial GG is composed of approximately 8.0 to 14% moisture, 75 to 85% galactomannan, 5 to 6% crude protein, 2 to 3% crude fiber and 0.5 to 1.0% ash (Maier et al., 1993).

GM is rarely used as a feed ingredient because of several anti-nutritional factors such as saponin (Thakur and Pradhan, 1975a, b; Yejuman et al., 1998) and

Table 1. Composition of isocaloric and isonitrogenous broiler starter diets[1] containing 5.0% guar meal (GM), 0.90% guar gum (GG), or 0.250% saponin rich guar meal extract (GS), respectively from 1 to 21 days of age.

Ingredients (%)	Dietary treatments			
	C	GM	GG	GS
Corn	60.38	58.5.00	58.88	59.78
Guar saponin extract	0.00	0.00	0.00	0.250
Guar gum	0.00	0.00	0.90	0.00
Guar meal[2]	0.00	5.0	0.00	0.00
Dehulled soybean meal	32.00	28.28	32.20	32.20
DL-Methionine	0.27	0.27	0.27	0.27
L-Lysine HCl	0.29	0.29	0.29	0.29
Corn oil	2.00	2.60	2.40	2.15
Limestone	1.63	1.63	1.63	1.63
Dicalcium phosphate	1.00	1.00	1.00	1.00
Mono-dicalcium PO_4	1.72	1.72	1.72	1.72
Salt	0.21	0.21	0.21	0.21
Trace minerals[3]	0.25	0.25	0.25	0.25
Vitamins[4]	0.25	0.25	0.25	0.25

[1]Average calculated analysis of of isocaloric and isonitrogenous broiler starter diets were as follows: CP, 22.06%; ME, 3,059 kcal/kg; Ca, 1.22%; non-phytin P, 0.66%; methionine, 0.57%; lysine, 1.30%; threonine, 0.77%; tryptophan, 0.28%. [2]The guar meal nutrient matrix used was CP, 39.75%; ME, 2,033 kcal/kg; Ca, 0.16%; non-phytin P, 0.16%; methionine, 0.45%; lysine, 1.64%; arginine, 4.90%; threonine, 1.04%; and tryptophan 0.43%. [3]Trace minerals premix added at this rate yields: 149.60 mg Mn, 16.50 mg Fe, 1.70 mg Cu, 125.40 mg Zn, 0.25 mg Se, 1.05 mg I per kg diet. [4]Vitamin premix added at this rate yields: 11,023 IU vitamin A, 46 IU vitamin E, 3,858 IU vitamin D_3, 1.47 mg minadione, 2.94 mg thiamine, 5.85 mg riboflavin, 20.21 mg pantothenic acid, 0.55 mg biotin, 1.75 mg folic acid, 478 mg choline, 16.50 μg vitamin B_{12}, 45.93 mg niacin, and 7.17 mg pyridoxine per kg diet.

residual GG (Vohra and Kratzer, 1964a, b, 1965; Katoch et al., 1971). However, Bakshi (1966) and Couch et al. (1967) recognized trypsin inhibitor as an important antinutritional factor in feed ingredients. These findings were contradicted by Conner (2002) and Lee et al. (2004) who noted that GM contained lower levels of trypsin inhibitor than processed soybean meal.

It is not yet clear whether residual GG or guar saponin contribute to a greater extent regarding the growth inhibitory effects of GM in broiler diets. No data is available in the scientific literature directly comparing the effects of GM, GG or saponin rich guar meal extract (GS) in a single broiler growth trial. Therefore, this study was conducted to investigate whether addition of either 0.90% GG or 0.250% GS when used at low concentrations roughly equivalent to feeding 5.0% GM in broiler diets would affect productive performance of broiler chicks.

MATERIALS AND METHODS

A commercial GG and GM powders was purchased from Rama Industries, Manufacturer and Exporter of Guar Gum Split and Powder, Government Recognized Export House, Gujarat, India. GS was isolated according to the procedure described by Hassan et al. (2010) from GM powder.

Experimental design

Two hundred forty one-day-old unsexed Ross broiler chicks were purchased from a local commercial hatchery, weighed and randomly distributed in battery cages among four treatments with four replicates of 15 chicks per replicate. Chicks were assigned to one of the following four treatment groups: (1) the control broiler starter diet, (2) the control broiler starter diet reformulated with 5.0% GM, (3) the control broiler starter diet supplemented with 0.90% GG, and (4) the control broiler starter diet supplemented with a 0.250% GS. The broiler starter diets used in this study were calculated to be iso caloric and iso nitrogenous (Table 1). Feed and water were provided *ad libitum* with a 22:2 h light: dark schedule throughout the entire 21 days course of the study. Weekly feed intake, feed conversion ratio, mortality rate, body weight, and body weight gain were recorded from 1 to 21 day of age.

Statistical analysis

Data obtained were subjected to one-way ANOVA using the GLM procedure of a statistical software package (SPSS 18.0, SPSS Inc., Chicago, IL). Experimental units were based on cage averages. Treatment means were expressed as mean ± standard error of means (SEM) and separated (P ≤ 0.05) using the Duncan's multiple range test (Duncan, 1955).

RESULTS

During the first wk of the study (1 to 7 days of age), feed intake was significantly higher in chicks fed 5.0% GM than all the other treatments. However, feed intake was significantly lower in chicks fed 0.90% GG than all the other treatments from 8 to 14 days of age. Although feed

Table 2. Weekly feed intake, feed conversion ratio, body weight and body weight gain for broiler chicks from 1 to 21 days of age.

Age (days)	Treatments			
	Control	5.0% GM	0.90% GG	0.250% GS
Feed Intake (g)				
1-7	116.99 ± 2.66 ^{b}	127.00 ± 0.66^a	115.67 ± 1.84^b	117.67 ± 0.84^b
8-14	345.33 ± 3.50^b	365.00 ± 5.34^a	323.00± 4.00^c	347.00 ± 8.19^{ab}
15-21	599.00 ± 10.67	559.00± 17.33	563.33± 6.84	559.00.00± 10.11
1-21	1061.32± 16.83	1051.00 ± 23.33	1002.00± 12.95	1023.67± 34.12
Feed conversion ratio (g feed intake/g body weight gain)				
1-7	1.19 ± 0.00^c	1.27 ± 0.00^a	1.23 ± 0.01^b	1.23 ± 0.00^b
8-14	1.24 ± 0.00^c	1.29 ± 0.00^b	1.18 ± 0.01^d	1.37 ± 0.01^a
15-21	1.53 ± 0.00^b	1.89 ± 0.07^a	1.64 ± 0.03^b	1.98 ± 0.07^a
1-21	1.38 ± 0.00^c	1.54 ± 0.02^b	1.41 ± 0.01^c	1.62± 0.03^a
Body weight (g)				
1	44.33 ± 0.19	44.67 ± 0.19	44.67 ± 0.19	44.63 ± 0.07
7	142.67 ± 2.34^{ab}	144.65 ± 0.84^a	139.00 ± 0.90^b	141.92± 1.17^{ab}
14	421.33 ± 5.50^a	428.33± 4.01^a	412.67±8.17^{ab}	395.67± 9.17^b
21	813.33± 13.17^a	726.33± 25.40^{bc}	755.54± 6.54^{ab}	697.67± 25.50^c
Body weight gain (g)				
1-7	98.34 ± 2.15^{ab}	99.98 ± 0.65^a	94.59 ± 0.71^b	97.31 ± 0.98^{ab}
8-14	278.66± 3.16^a	283.68 ± 3.17^a	273.67 ± 7.29^a	253.75± 8.00^b
15-21	392.00 ± 7.67^a	298.00 ± 21.40^{bc}	342.87 ± 4.12^b	284.00 ± 16.33^c
1-21	769.00 ± 12.98^a	681.66 ± 25.21^{bc}	710.87 ± 6.38^{ab}	634.34± 25.31^c

$^{a-d}$ Means ± standard errors of mean within a row that do not share a common superscript are significantly different (P ≤ 0.05).

intake was significantly higher in chicks fed 5.0% GM than control, there were no significant differences observed between chicks fed 0.250% GS and both chicks fed 5.0% GM and control diet during the same period (8 to 14 days) (Table 2).

Feed conversion ratio was poor and significantly higher for chicks fed 5.0% GM than all the other treatments from 1 to 7 days of age, but there were no significant differences in feed conversion ratio between chicks fed 0.250% GS and chicks fed 0.90% GG during the same period (1 to 7 days). From 8 to 14 days of age, feed conversion ratio was poor and significantly higher than all the other treatments for chicks fed with 0.250% GS. Feed conversion ratios for chicks fed 0.250% GS and 5.0% GM were significantly higher than chicks fed 0.90% GG and control diet from 15 to 21 days of age. Over the entire course of the study (1 to 21 days), feed conversion ratio was very poor and significantly higher than all the other treatments for chicks fed 0.250% GS (Table 2).

The initial body weight of broiler chicks distributed among the four dietary treatments was not significantly different at the start of the experiment. Body weight was significantly lower in chicks fed 0.90% GS/GG than chicks fed 5.0% GM at 7 days of age, but were not different from chicks fed 0.250% GS and control treatment. At 14 days of age, body weight was significantly lower in chicks fed 0.250% GS than both chicks fed 5.0% GM and control treatment, but were not different from chicks fed 0.90%

GG. The final body weight at 21 days of age was significantly lower in chicks fed 0.250% GS than chicks fed 0.90% GG and control treatment, but were not different from chicks fed 5.0% GM treatment (Table 2).

Body weight gain was significantly lower in chicks fed 0.90% GG than chicks fed 5.0% GM from 1 to 7 days of age, but were not different from chicks fed 0.250% GS and control treatment. From 8 to 14 days of age, body weight gain was significantly lower in chicks fed 0.250% GS than all the other treatments, but were no significant differences among chicks fed 0.90% GG, 5.0% GM and control treatment. Body weight gains from 15 to 21 days of age were significantly different among all treatment groups ranging from 284 g of body weight gain for chicks fed 0.250% GS to 392 g of body weight gain for the control treatment. Body weight gain was significantly lower in chicks fed 0.250% GS than chicks fed 0.90% GG and control treatment, but were not different from chicks fed 5.0% GM treatment. Total body weight gain from 1 to 21 days of age was significantly lower in chicks fed 0.250% GS than chicks fed 0.90% GG and control treatment, but were not different from chicks fed 5.0% GM treatment (Table 2).

DISCUSSION

Results obtained were in disagreement with Miah et al.

(2004) who noted that adding 75 mg steroid saponin per kg feed for broiler chicks increased feed intake. On the other hand, Cheeke (1996), Ueda and Ohshima (1987), and Makkar and Becker (1996) noted that adding saponin to the broiler diet decreased feed intake. These differences may be attributed to the specific chemical structures and concentrations of the saponins fed. From 1 to 14 days of age, feed intake for chicks fed the 5.0% GM was significantly higher than those fed the control diet. Thakur and Pradhan (1975a, b) reported that GM use in poultry diets historically was limited by its adverse effects on feed intake.

Results obtained in this study were in agreement with several studies reported that adding GM in broiler and laying hen diets showed deleterious effects on feed conversion ratio (Saxena and Pradhan, 1974; Nagra et al., 1985; Patel and McGinnis, 1985; Nagra and Virk, 1986). Lee et al. (2003) found that GM contains GG whose impact on intestinal viscosity adversely affects feed conversion ratio.

While our results exhibited negative effect for adding GS in broiler chicks on feed conversion ratio, Yejuman et al. (1998) reported that at appropriate concentration saponins have potential as dietary additives to favor better feed conversion ratio. Our results were in disagreement with the findings of Johnston et al. (1982), Al-Bar et al. (1993), and Miah et al. (2004) who found that adding 75 mg steroid saponin per kg feed for broiler chicks improved feed conversion ratio.

In the current study, there was no significant effect for of all dietary treatments on mortality rate (unshown data). Hassan et al. (2008) found no significant effect of adding GM in broiler diet on mortality rate. On the other hand, Al-Bar et al. (1993) noted that adding saponin to broiler diets decreased mortality rate. Other reports mentioned that adding GM in broiler diets increased mortality rate (Sathe and Bose, 1962; Anderson and Warnick, 1964; Thakur and Pradhan, 1975b; Verma and McNab, 1982; Patel and McGinnis, 1985).

Our results obtained from the present study were in agreement with the observation of Anderson and Warnick (1964) and Conner (2002) who reported no significant negative impacts on the body weight of broiler chicks fed a diet supplemented with 5.0% GM. Previous studies reported that the negative effects of adding GM on body weight might be attributed to the presence of anti-nutrient compounds such as saponins (Thakur and Pradhan, 1975a, b). It was reported that GM contains about 5.0% crude Guar saponin (Hassan et al., 2010).

While Lee et al. (2005) reported GM could be safely fed to broilers at 2.5% of the diet without adversely affecting performance, the findings of Saxena and Pradhan (1974) that showed that GM has deleterious effects on body weight gain of broiler chicks. Saxena and Pradhan (1974), Cheeke (1996), Ueda and Ohshima (1987), and Thakur and Pradhan (1975a, b) reported that GM use in poultry diets historically was limited by its adverse effects on body

weight gain. On the other hand, Daskiran et al. (2004) and Vohra and Kratzer (1964a, 1965) demonstrated that 1% GG in broiler diets causes a 25 to 30% depression of body weight gain. When diets contained 2% GG, the relative growth of broiler chicks was 61 to 67% of controls (Kratzer et al., 1967; Rogel and Vohra, 1982, 1983). Lee et al. (2005) also supported the idea that residual GG in GM was at least partially responsible for the negative effects GM on body weight gain.

The growth reduction of broiler chicks fed GG is likely due to increased viscosity of the digesta (Blackburn and Johnson, 1981). Previous studies reported that the negative effects of adding GM on body weight gain might be attributed to the presence of anti-nutrient compounds such as residual GG (Vohra and Kratzer, 1964a; 1965; Katoch et al., 1971; Annison and Choct, 1991). It was reported that GM contains about 18% residual GG on a dry matter basis (Anderson and Warnick, 1964; Nagpal et al., 1971; Lee et al., 2004).

The growth depressing properties of GG supplementation in poultry diets may be overcome by treating the feed with enzymes capable of hydrolyzing it, namely pectinase, and cellulase, a preparation from sprouted guar beans (Vohra and Kratzer, 1964a) or endo-β-D- mannanase (Vohra and Kratzer, 1964a; 1965; Ray et al., 1982; Verma and McNab, 1982; Patel and McGinnis, 1985; Lee et al., 2003; Daskiran et al., 2004). These exogenous enzymes are thought to reduce intestinal viscosity and alleviated the deleterious effects associated with excessive GG.

These results were in agreement with the previous studies which reported that the negative effects of adding GM on body weight gain might be attributed to the presence of anti-nutrient compounds such as saponins (Thakur and Pradhan, 1975a, b). The results obtained from the present study showed that adding 0.25% GS to broiler diets decreased body weight gain over the entire course of the study from 1 to 21 days (Table 2). The results were in agreement with the observations of Ueda et al. (1996), Cheeke (1996), Makkar and Becker (1996) that reported that adding saponins to broiler diets decreased body weight gain. On the other hand, several reports surprisingly found that adding saponins to poultry diets increased body weight gain (Johnston et al., 1982; Al-Bar et al., 1993; Yejuman et al., 1998). Also, Miah et al. (2004) noted that adding 75 mg steroid saponin per kg feed for broiler chicks increased body weight gain. However, Ishaaya et al. (1969) reported no adverse effect on body weight gain after feeding fenugreek seeds containing steroid saponins and soybean triterpenoid saponins (Petit et al., 1995) at concentrations as much as five times the concentration in a normal soybean-supplemented diet of chicks, rats and mice. Miah et al. (2004) reported that, recently some medicine companies are marketing saponins as feed additive in poultry production. Yejuman et al. (1998) reported that at appropriate concentrations saponins have potential as dietary additives to favor higher body weight gain.

In conclusions, it appears that guar saponin is the main compound in GM affecting body weight, body weight gain, and feed conversion ratio, although no significant differences in feed intake were observed compared with the control group. Productive performance of broiler chicks in the present study was less negatively inhibited by the 0.90% GG treatment suggesting that GM triterpenoid saponins may be the most important anti-nutritional factor present in GM.

ACKNOWLEDGMENT

Author expresses his sincere thanks to Deanship of Scientific Research, King Faisal University for funding this study.

REFERENCES

Al-Bar A, Ismail A, Cheeke PR, Nakaue H (1993). Effect of dietary *Yucca shidegera* extract (Deodorage) on environment ammonia and growth performance of chickens and rabbits. J. Anim. Sci. 71:114.

Ambegaokar SD, Kamath JK, Shinde VP (1969). Nutritional studies in protein of 'gawar' (*Cyamopsis tetragonoloba*). J. Nutr. Diet. 6:323–328.

Anderson JO, Warnick RE (1964). Value of enzyme supplements in rations containing certain legume seed meals or gums. Poult. Sci. 43:1091-1097.

Annison G, Choct M (1991). Antinutritional activities of cereal non-starch polysaccharides in broiler diets and strategies minimizing their effects. World's Poult. Sci. J. 47:232–241.

Bakshi YK (1966) Studies on toxicity and processing of guar meal. PhD dissertation, Texas A&M University, College Station, Texas, United States of America.

Blackburn NA, Johnson IT (1981). The effect of guar gum on the viscosity of the gastrointestinal contents and on glucose uptake from the perfused jejunum in the rat. Br. J. Nutr. 46:2:239-46.

Cheeke PR (1996). Biological effects of feed and forage saponins and their impacts on animal production. In: Waller GR, Yamasaki K (eds) Saponins Used in Food and Agriculture, Plenum Press, New York. pp. 377-385.

Conner SR (2002). Characterization of guar meal for use in poultry rations. PhD dissertation, Texas A&M University, College Station, Texas, United States of America.

Couch JR, Lazano JA, Creger CR (1967). Soy protein guar meal and excess calcium in nutrition of commercial layers. Poult. Sci. 46:1248.

Daskiran M, Teeter RG, Fodge D, Hsiao HY (2004). An Evaluation of Endo-β-D-mannanase (Hemicell) Effects on Broiler Performance and Energy Use in Diets Varying in β-Mannan Content. Poult. Sci. 83:662–668.

Duncan DB (1955). Multiple ranges and multiple F test. Biometrics 11:1-42.

Hassan SM, El-Gayar AK, Caldwell DJ, Bailey CA, Cartwright AL (2008). Guar meal ameliorates *Eimeria tenella* infection in broiler chicks, Vet. Parasitol. 157: 133-138.

Hassan SM, Haq AU, Byrd JA, Berhow MA, Cartwright AL, Bailey CA (2010). Haemolytic and antimicrobial activities of saponin-rich extracts from guar meal. Food Chem. 119:600-605.

Ishaaya I, Birk Y, Bondi A, Tencer Y (1969). Soybean saponins. IX. Studies of their effect on birds, mammals and cold-blooded organisms. J. Sci. Food Agric. 20: 433-436.

Johnston NL, Quarles CL, Fagerberg DJ (1982). Broiler performance with DSS40 yucca saponin in combination with monensin. Poult. Sci. 61:1052–1054.

Katoch BS, Chawla JS, Rekib A (1971). Absorption of amino acid (*in*

vitro) through intestinal wall of chicken in presence of guar gum. Ind. Vet. J. 4:142-146.

Kratzer FH, Rajaguru RWASB, Vohra P (1967). The effect of polysaccharides on energy utilization, nitrogen retention and fat absorption in chickens. Poult. Sci. 46:1489-1493.

Lee JT, Bailey CA, Cartwright AL (2003). Guar meal germ and hull fractions differently affect growth performance and intestinal viscosity of broiler chickens. Poult. Sci. 82:1589–1595.

Lee JT, Connor-Appleton S, Bailey CA, Cartwright AL (2005). Effects of guar meal by-product with and without beta-mannanase Hemicell® on broiler performance. Poult. Sci. 84: 1261–1267.

Lee JT, Connor-Appleton S, Haq AU, Bailey CA, Cartwright AL (2004). Quantitative measurement of negligible trypsin inhibitor activity and nutrient analysis of guar meal fractions. J. Agric. Food Chem. 20:6492-56495.

Maier H, Anderson M, Karl C, Magnuson K, Whistler RL (1993). Guar, locust bean, tara and fenugreek gums: Industrial Gums. In: Whistler RL, Be Miller JN (eds) Polysaccharides and Their Derivatives., Acadmy Press, London, United Kingdom. pp. 81–1221.

Makkar HPS, Becker K (1996). Effect of quillaja saponins on in vitro rumen fermentation: In: Waller GR, Yamasaki K (eds) Saponins Used in Food and Agriculture, Plenum Press, New York. pp. 387-394.

Miah MY, Rahman MS, Islam MK, Monir MM (2004). Effects of Saponin and L-Carnitine on the Performance and Reproductive Fitness of Male Broiler. Int. J. Poult. Sci. 3:530-533.

Nagpal ML, Agrawal OP, Bhatia IS (1971). Chemical and biological examination of guar –meal (*Cyampsis tetragonoloba* L.). Ind. J. Anim. Sci. 4:283-293.

Nagra SS, Shingari BK, Ichhponani JS (1985). Feeding of guar (*Cyampsis tetragonoloba*) meal to poultry.1. Growth of commercial broiler chicks. Ind. J. Poult. Sci. 20:188–193.

Nagra SS, Virk RS (1986). Growth and laying performance of White Leghorn pullets fed toasted guar meal alone or in combination with groundnut and mustard cakes as sources of protein. Ind. J. Poult. Sci. 21:16–20.

Patel MB, McGinnis J (1985). The effect of autoclaving and enzyme supplementation of guar meal on the performance of chicks and laying hens. Poult. Sci. 64:1148–1156.

Petit P, Sauvaire Y, Hillaire-Buys D, Leconte O, Baissac Y, Ponsin G, Ribes G (1995). Steroid saponins from fenugreek seeds: Extraction, purification, and pharmacological investigation on feeding behavior and plasma cholesterol. Steroids 60:674–680.

Rahman M, Shafivr M (1967). Guar meal in dairy cattle rations. PhD dissertation, Texas A&M University, College Station, Texas, United States of America.

Ray S, Pubols MH, McGinnis J (1982). The effect of a purified guar degrading enzyme on chick growth. Poult. Sci. 61:488–494.

Rogel AM, Vohra P (1982). The effects of complex polysaccharides on growth, digestibility and blood parameters in pair-fed chicks. Nutr. Res. 2:39-49.

Rogel AM, Vohra P (1983). Hypocholesterolemia and growth –depression in chicks fed guar gum and konjac mannan. J. Nutr. 113:873-879.

Sathe BS, Bose S (1962). Studies on the utilization of industrial and farm by-products in growing poultry rations. Ind. J. Vet. Sci. 32:74–84.

Saxena UC, Pradhan K (1974). Effect of high protein levels on the replacement value of guar meal (*Cyamopsis tetragonoloba*) in layer's ration. Ind. J. Anim. Sci. 44:190–193.

Thakur RS, Pradhan K (1975a). A note on inclusion of guar meal (*Gyamopsis tetragonoloba*) in broiler rations. Ind. J. Anim. Sci. 45:98-102.

Thakur RS, Pradhan K (1975b). A note on inclusion of guar meal (*Gyamopsis tetragonoloba*) in broiler rations. Effect on carcass yield and meat composition. Ind. J. Anim. Sci. 45:880-884.

Ueda H, Ohshima M (1987). Effects of alfalfa saponin on chick performance and plasma cholesterol level. Jpn. J. Zootech. Sci. 58:583-590.

Ueda H, Kakutou Y, Ohshima M (1996). Growth-depressing effect of alfalfa saponin in chicks. Anim. Sci. Technol. 67:772-779.

Verma SVS, McNab JM (1982). Guar meal in diets for broiler chickens. Br. Poult. Sci. 23:95–105.

Vohra P, Kratzer FH (1964a). Growth inhibitory effect of certain polysaccharides for chickens. Poult. Sci. 43:1164-1170.

Vohra P, Kratzer FH (1964b). The use of guar meal in chicken rations. Poult. Sci. 43:502-503.

Vohra P, Kratzer FH (1965). Improvement of guar meal by enzymes. Poult. Sci. 44:1201-1205.

Yejuman YH, Shiminghua, Niweiju Y. Hong, Chen WD, Ye MH, Shi, NIWJ (1998). Effect of herbal origin bioactive substances on growth rate and some biochemical parameters in blood of broilers. J. Zhejiang Agric. Univ. 24:405-408.

12

Traditional poultry production: The role of women in Kaura-Namoda local government area, Zamfara State, Nigeria

Garba J., A. Y. Yari, M. Haruna and S. Ibrahim

Department of Agricultural Education, Zamfara State College of Education, P. M. B. 1002 Maru, Zamfara State, Nigeria.

Poultry production has a high priority because poultry meat has better energy and protein conversion ratio than many other animals. Traditional poultry is a well-known livestock enterprise in northern Nigeria and women contribute immensely to agricultural production but still their contribution to the country's development remains largely not documented. This study therefore, aimed at assessing the role of rural women, their contribution and constraints to traditional poultry production in Kaura-Namoda local government area, Zamfara State-Nigeria. Sixty (60) women poultry keepers were randomly selected from twelve villages of four districts selected at random and administered with questionnaires to collect relevant data on socio-economic characteristics, management system adopted, output and constraints of poultry production. Data collected were analyzed using descriptive statistics. The result revealed that majority of the respondent were young (26-35 years) and married with no formal schooling whose kept between 11 to 20 bird with ratio of 1:2 of cock to hen. Extensive management system is practicing by the majority of respondent also; recorded more than 40 eggs laid by their hens with farm returns of N2000.00 to N5000.00 ($12.5 to $31.25) per week and sales their produce at market. Majority of the respondents provide local oral leaf and bark extract of mahogany and solanaceous plant for disease control while some uses local trap in controlling rodents, debeaked their birds to avoid cannivalism. Transportation, pest and diseases, weather changes, poor extension services and low capital are some of the constraints of poultry production in the study area.

Key words: Traditional poultry, women, Kaura-Namoda.

INTRODUCTION

Agricultural development is a complex process and a challenging one as well (Fabiyi et al., 2007). Poultry production has a high priority because poultry meat has better energy and protein conversion ratio than many other animals (Alabi and Aruna, 2009). They are also the most prolific of all farm animals being capable of producing up to 200 eggs or off-springs per year (Akinwumi and Ikpi, 1979) which give them a greater potential in contributing to increase in livestock output within short run. This high rate of return associated with poultry industry is however, coupled with capital/input, management and environment. Traditional or rural poultry production is a well-known livestock enterprise in Northern Nigeria where virtually every household kept small flock of between 5 to 20 birds. The term "traditional or rural poultry" is indicative of the low input husbandry of domestic poultry that is typically managed by rural subsistence farmers (Akinola and George, 2008). The enterprise has earned recognition in the rural socio-economy because it provides readily harvestable protein (meat and eggs) and revenue (Okitoi et al., 2007). Women in Nigeria contribute immensely to agricultural production as they play a vital role in food production for the household, farm labour, post-harvest activities, livestock husbandry as well as processing and marketing of farm produce (Yusuf et al., 2006).

According to Ironkwe and Ekwe (1998), more than 60% of the agricultural production is carried out by women in the Nigerian traditional setting. Mijindadi (1993) estimated that women are responsible for 70% of actual farm work and constitute up to 60% of the farming population. Recently, Yahaya (2002) reported that 76% of women from Oyo and Bauchi States are actively involved in farming activities. Women in North-western Nigeria despite the traditional structure of right and obligation within the rural Muslim families, still play a vital role in poultry management, animal fattening, processing and marketing of farm produce.

In spite of these roles played by women in agriculture and other economic activities, their contribution to the country's development still remains largely not documented. The present study therefore, aimed at assessing the role of rural women, their contribution and constraints to traditional poultry production in Kaura-Namoda local government area (LGA), Zamfara State, Nigeria.

MATERIALS AND METHODS

The study was carried out in 2009 in Kaura-Namoda LGA of Zamfara State, Nigeria. Kaura-Namoda LGA is situated between longitude 6° 38'East and latitude 12° 39' North. It is located in the northern part of Zamfara State. The climate of the area is variably hot, having November-April as dry season, March-April as hottest months, while May-October is the rainy season with August recording the highest amount of rainfall. Simple random sampling technique was used to select twelve villages from the four major districts (Kaura-Namoda, Kasuwar-Daji, Kuryar-Madaro and Yankaba) and 60 women poultry keepers from these twelve villages (Yardole, Matoya, Balankabe, Kanwa, Tudunwada, Magizawa, Katsaura, Kogi, Gundumi, Dandambo, Mallamawa and Kurnartullo). Questionnaires were administered to collect relevant data on socio-economic characteristics, management system adopted, output and constraints from the women selected. Data collected were analyzed using descriptive statistics, involving frequency and percentage.

RESULTS AND DISCUSSION

Socio-economic characteristics of respondents in the study area

Majority of the women poultry keepers in this area were young women. Table 1 shows that 38% of the respondents were within 26 to 35 years of age, 30% were 36 to 45

years of age, 23% were 15 to 25 years of age, 5% were 46 to 55 years of age and 4% were 56 and above years of age. The age of farmer is important because it determines the intensity of farm labour; age is therefore a factor that can significantly affect the productivity and profitability of the farm business (Falusi and Olayide, 1980; Yusuf et al., 2006); thus, women at younger age are expected to be more productive. Many of the respondents were married (80%) and few were widow (5%), while 15% were single.

In North-western Nigeria, women usually get married at 15 to 20 years; as they start a new family, rural women look for sources of additional income. Poultry rearing is considered as one of the most important way of generating additional income for rural women. Majority of the respondents had no formal education (45%); rather, they only attended Qur'anic Islamic schools. 28% attended primary schools, 18 and 5% had secondary and tertiary education, respectively, while 4% of the respondent do not attend any school.

Education makes human being great, wise and honourable (Paul and Saadullah, 1991). It is one of the most important factors for the improving socio-economic condition of the rural women, as it is very important for easy understanding, communication and adoption of new technology that helps in stimulating their production (Adams, 1982; Yusuf et al., 2006; Alabi and Aruna, 2009).

This study revealed that majority of the respondents had no formal education hence access to information and adoption of new technology will not be easy because it will be difficult for them to comprehend what they are thought or understand the reason(s) why some changes are necessary. 30% of the respondents kept 11 to 20 birds, 28% kept 21 to 30 birds, 22% kept 41 and above number of birds, while 17% kept 31 to 40 birds and 3% kept only 1 to 10 birds.

Flock size range of the respondents is an indication of the dominance of small-scale poultry production in the study area. It is low input husbandry of domestic poultry that is typically managed by rural subsistence farmers (Akinola and George, 2008).

The enterprise has earned recognition in the rural socio-economy because it provides readily harvestable protein (meat and eggs) and revenue (Okitoi et al., 2007). Majority of the respondent (38%) kept between 6 to 10 cocks, 35% kept between 1 to 5 cocks, 20% kept between 11 to 15 cocks, while 3.5% each kept between 16 to 20 and 21 and above cocks.

On the number of hens kept by the respondents, the table shows that, majority (35%) kept between 11 to 20 hens, 22% each kept between 1 to 10 and 21 to 30 hens, 15% kept between to 40 hens and 6% kept 41 and above number of hens, thus, the ratio of cocks to hens in the study area is 1:2.

This can be due to the fact that cocks were culled at an early age for either sale or slaughter.

Table 1. Socio-economic characteristics of the respondent in the study area.

Variable	Frequency	Percentage (%)
Age (years)		
15-25	14	23
26-35	23	38
36-45	18	30
46-55	3	5
56 and above	2	4
Marital status		
Single	9	15
Widow	3	5
Married	48	80
Educational status		
Never been to school	2	4
Qur'anic education	27	45
Primary education	17	28
Secondary education	11	18
Tertiary education	3	5
Flock size		
1-10	2	3
11-20	18	30
21-30	17	28
31-40	10	17
41 and above	13	22
Number of cocks in the flock		
1-5	21	35
6-10	23	38
11-15	12	20
16-20	2	3.5
21 and above	2	3.5
Number of hens in the flock		
1-10	13	22
11-20	21	35
21-30	13	22
31-40	9	15
41 and above	4	6

n = 60.

Management systems practiced and returns of the women in the study area

Majority of the respondents in the study area were practicing extensive system of poultry management. Figure 1 shows that 45% of the respondents rear their poultry in free range, 33% semi-extensive, while 22% kept their birds in doors mostly in local cages. This result revealed that majority of the birds in the study area is not provided with feed and housing instead they are allowed to scavenge for food and water during the day backyard, thus, the bird may be susceptible to high and in the night they perch on trees, nearby walls or mortality due to absence of any care and predation. Table 2 shows the percent distribution of quantity of eggs laid by the birds/week and the returns of selling the eggs and the birds to the women in the study area. Majority of the respondents (45%) reports that their birds can laid 41 eggs in a week, 20% of their birds laid between 11 to 20 eggs/week, 17% of their birds laid between 21 to 30 eggs/week, 10% of their birds

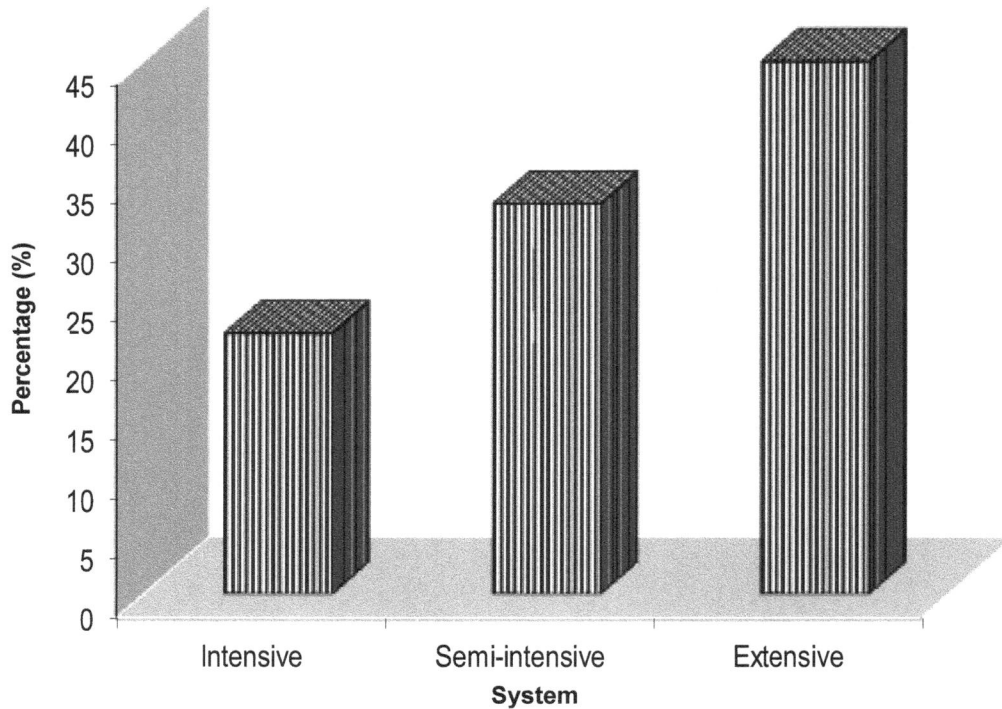

Figure 1. Management systems adopted in the study area

Table 2. Impact of traditional poultry rearing to the women of the study area.

Variable	Frequency	Percentage (%)
Number of eggs lay per week		
1-10	5	8
11-20	12	20
21-30	10	17
31-40	6	10
41 and above	27	45
Returns (N) per week		
500-2000	41	68
2000-4500	13	21
4500-6500	7	11

n = 60.

laid between 31 to 40 eggs/week and 8% of their birds laid between 1 to 10 eggs/week.

Majority of these women (68%) can realize between ₦500 to 2000 ($3.12 to 12.5) per week as returns, 21% can realize between ₦2000 to 4500 ($12.5 to 28.13) per week, while 11% realizes between ₦4500 to 6000 ($28.13 to 37.5) per week. These returns are usually spent for the off-keep of the family in buying cooking and cleaning utensils, payment of medical bills as well as donations for naming and wedding ceremonies, thus,

reducing or in most cases de-shouldering these responsibilities from the husbands. On the point of sell (Figure 2), majority of the respondents (47%) do sell their eggs and birds as well at market, 25% at home, while 28% sell both at home and market.

Majority of the women (85%) in the study area do not belong to any co-operatives society; thus, they neither gain any personal financial assistance nor getting loans from the financial institutions, while few of them (15%) belong to the co-operatives society, hence, they are

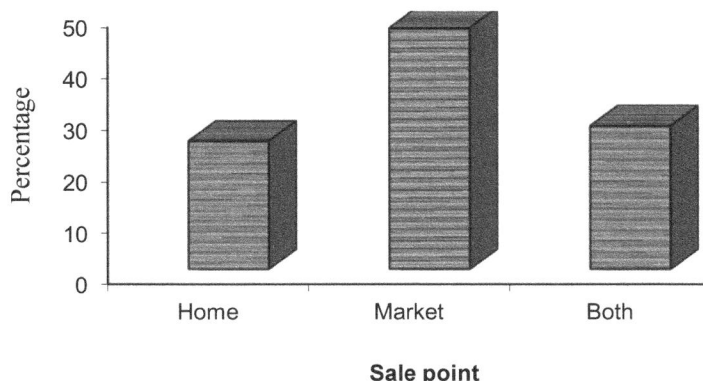

Figure 2. Sales points of poultry and their product in the study area

Table 3. Formation of association and access to loans/financial assistance by the respondents in the study area.

Variable	Frequency	Percentage (%)
Membership to co-operative society		
Yes	9	15
No	51	5
Access to loan /financial assistance		
Yes	10	17
No	50	83

n = 60.

Table 4. Health management techniques practiced by women in the study area.

Practice	Frequency	Percentage (%)
Separation	1	2
Deworming	40	67
Debeaking	3	5
Trapping	8	13
None	8	13

n = 60.

getting personal financial assistance and loans from financial institutions (Table 3).

Health management techniques practiced by women in the study area

Majority of the respondent in the study area (67%) provide their poultry with oral application of leafs and bark extract of mahogany and solanaceous plant, respectively in controlling Newcastle and Gomboro diseases, 13% of the respondent use local trap in controlling rodent and predators, 5% debeaked their poultry to avoid canivalism between them and 2% separates ills poultry from the disease-free birds to avoid infection, while the remaining

13% of the respondents are not giving any treatments to their poultry, they rather slaughter or sell them when infected (Table 4).

Women constraint in poultry production in the study area

Distribution of respondents according to the constraints militating against poultry production revealed that 50% of the respondents are having problems of transportation, due to the poor roads networks linking them to the main cities; they found it very difficult in transporting their birds and eggs especially in the rainy season and because of this, they normally encounter breakage eggs and mortality

Table 5. Distribution of respondents according to constraints of poultry production in the study area.

Problem	Frequency	Percentage (%)
Transportation	30	50
Pest and diseases	36	60
Modern vaccines	12	20
Cannivalism	6	10
Weather changes	9	15
Extension services	4	7
Low productivity	1	2
Feed availability	4	7
Capital	6	10

from the bird while taking them to/from market. 60% of the respondents also reported pest and diseases as one the major problems facing their birds especially during harmattan where most of their birds die due to Newcastle and Gomboro diseases (Table 5).

This led to mass sales of the birds at lower prizes implying lost/low farm return. Canivalism, weather changes and low productivity also are some of the constraint of poultry production as reported by 10, 15 and 2% of the respondents, respectively. Weather changes constitutes major problem in poultry industry as extremes of cold and heat causes chilling and excess hydration which led to the eventual death of bird. Cold weather (Hamattan) is also associated with outbreak of many poultry diseases like fowl cholera, Newcastle and Gomboro. Low productivity means low returns as such, these birds are local breeds without any genetic improvements in terms of growth and eggs productions thus, are smaller in size and produces small eggs of low quality. Finally, it was revealed from the present study that inadequate of extension services, modern vaccines and feed as well as low capital as some of the problems facing the poultry farmers as it represent 7, 20, and 10% respectively. Inadequate extension services led to less/no awareness of the farmers on improved breeds of these birds, proper management techniques and marketing channel. Inadequate modern vaccine and feed means continuous proliferation and attack of diseases because the birds lack feed of balanced ratio and immunity thus, susceptible to diseases which may not be controlled by the local herbs. Low capital constraint from the women of the study area led to possession of birds in smaller flocks and poor management, these will result into disease infection and eventual death of the birds hence, low farm returns.

CONCLUSION AND RECOMMENDATIONS

The present study revealed that women play a vital role in local poultry production as they posses flocks of the birds in various size, participate in the decision making of

these birds and their product and utilized the income derived from the industry for their sustained livelihood. This study also reports transportation, capital, pest and diseases, feed availability and inadequate extension services as some of the constraints facing poultry production in the area.

Based on the present study, the following recommendations were made:

(1) Our rural areas should be provided with various social amenities like roads, schools and hospitals for the improvement of their standard of living and overall agricultural productivity
(2) Women should be encourage on the formation of association and co-operatives societies so that their need will be known and to be in better position to pursue for the solution of their problems
(3) Government and other financial institution should provide a soft loan with minimum or no collateral requirement so as to encourage full participation and utilization of the loans by the women in improving poultry industry
(4) Modern vaccine and feed should be made available at affordable rate as well as development of appropriate vaccination protocols and feeding managements
(5) Agricultural extension services should be geared towards women and their concern which could be achieved through training and provision of more female extension workers

REFERENCES

Adams ME (1982). Agricultural Extension in Developing Countries. Longman Group ltd. Singapore P. 108.
Akinola LAF, George OS (2008). Small-scale Family Poultry Production as a Sustainable Source of Animal Protein in Selected Local Government Areas in Rivers State. In: Bawa, GS, Akpa GN, Jokthan GE, Kabir M, Abdu SB (eds). Repositioning Animal Agriculture for the Realization of National Vision 2020. Proceedings of 13th Annual Conference of Animal Science Association of Nigeria held at Ahmadu Bello University, Zaria. September 15-19, 2008.
Akinwumi JA, Ikpi AE (1979). Economic Analysis of the Nigeria Poultry Industry. Fed. Livest. Bull. pp. 3-54

Alabi RA, Aruna MB (2009). Technical Efficiency of Family Poultry production in Niger Delta,Nigeria. In: Alder RG, Spradbrow PB, Young MP (eds). Village Chickens, Poverty Alleviation and Sustainable Control of Newcastle Disease. Proceeding of an International Conference held in Daressalaam, Tanzania, 5-7th October, 2005. ACIAR Proc. pp. 131-235.

Fabiyi EF, Danladi BB, Akande KE, Mahmood Y (2007). Role of Women in Agricultural Development and their Constraints: A case study of Biliri local government area, Gombestate, Nigeria. Pak. J. Nutr. 6(6):676-680.

Falusi AO, Olayide SO (1980). Agricultural Inputs and Small Farmers in Nigeria. In: Olayide, SO, Eweka TA, Bello Osagie VF (eds). Nigerian Small Farmers: Problems and Prospects in Integrated Rural Development. CARD University of Ibadan, Nigeria, pp. 68-86

Ironkwe AG, Ekwe KC (1998). Rural of Women Participation in Agricultural Production in Abia State. Proceedings of the 33rd Annual Conference of the Agricultural Society of Nigeria, held at Cereal Research Institute, Badegi Niger State, 16-22:147-157

Mijindadi NB (1993). Agricultural Extension for Women: Experience for Nigeria. Paper present at 13th World Bank Agriculture Resource Management, Washington, D.C. pp. 6-7

Okitoi LO, Ondwasy HO, Obali MP, Murekefu F (2007). Gender issues in Poultry Production in Rural Households of Western Kenya. Livest. Res. Rural Dev. 19(2).

Paul DC, Saadullah M (1991). Role of Women in Homestead of Small Farm Category in an Area of Jessore, Bangladesh. Livest. Res. Rural Dev. 3(2).

Yahaya MK (2002). Gender and communication variables in Agricultural Information Dissemination in two Agro-ecological Zones of Nigeria. Research Monograph University of Ibadan, P. 68.

Yusuf HA, Abdulsalam A, Yusuf AA (2006).The Role of Women in Agricultural Development: A Case Study of Giwa Local Government Area of Kaduna State. Proceeding of the 31th Annual Conference of Soil Science Society of Nigeria held between 13th to 17th November, at Ahmadu Bello University, Zaria.

13

Endogenous methods for preservation of wagashi, a Beninese traditional cheese

Philippe Sessou[1,3], Souaïbou Farougou[1] Paulin Azokpota [2], Issaka Youssao [1], Boniface Yèhouenou[3], Serge Ahounou[1] and Dominique Codjo Koko Sohounhloué[3]

[1]Laboratory of Research in Applied Biology, Polytechnic School of Abomey-Calavi, University of Abomey-Calavi, 01 P. O. Box 2009 Cotonou, Benin.
[2]Laboratory of Food Microbiology and Biotechnology, Faculty of Agronomic Sciences, University of Abomey-Calavi, 01 P. O. Box 526 Cotonou, Benin.
[3]Laboratory of Study and Research in Applied Chemistry, Polytechnic School of Abomey-Calavi, University of Abomey-Calavi, 01 P. O. Box 2009 Cotonou, Benin.

Traditional cheese locally called wagashi is a good proteins source with high water content (60%) which is favorable for microorganism's growth that affects its quality. This work summarizes the endogenous methods used for wagashi preservation in Benin. Data have been collected during a survey toward 318 producers, 164 retailers and 464 consumers of Wagashi randomly selected from six agroecological zones. . Sun drying (62.26%), whey storage (21.07%) soaking in untreated water (8.49%) are the methods mostly used by producers to preserve wagashi. Daily cooking of wagashi at 80 to 100°C is the main method used by the sellers (86.58%) whereas traditional smoking was practiced by consumers (28.8%).Globally, these methods cannot preserve the product no more than 12 days of conservation, except for smoking method. In addition, cooking of spoiled wagashi with or without leaves of *Vitellaria paradoxa*, *Pennisetum polystachion* or *Piliostigma thonningii* by the producers, sellers or consumers for microbial decontamination and deodorization of the product could not avoid hazards inside the product. Probably, consumption of wagashi in Benin could be associated with microbial contamination.

Key words: Fulani, milk, whey, wagashi, pathogens, microbial decontamination.

INTRODUCTION

The economies of most African countries in the south of Sahara, including Benin, is mainly based on agriculture and livestock takes an important place (Diao et al., 2006). Among livestock products, cow's milk has a great socio-economic importance. Indeed, in Benin, milk contributes more than 50% of annual household income of Fulani ethnic group (Dossou et al., 2006). Because of its high water content and nutrients, fresh milk undergoes rapid degradation by pathogenic microorganisms. Due to the lack of a cold chain, many traditional methods of preservation of wagashi have been developed in Africa, particularly in Benin (Kèkè et al., 2008). In Benin, the milk is processed into various products such as yoghurt, curd and wagashi, a traditional cheese from an artisanal process developed by the Fulani ethnic group. The wagashi is the most popular and most consumed milk derivate products (Aïssi et al., 2009). Wagashi is an important source of animal protein, especially for people with low incomes and could efficaciously contribute to solving problems related to proteins deficiency in the

Table 1. Geographical distribution of investigation's areas.

Agro-ecological zones	Provinces	Municipality	Latitude of municipality	Longitude of Municipality
Zone A	Atacora	Pehunco	10° 13' 42.00" N	2° 00' 7.00" E
Zone B	Donga	Djougou	9° 42' 0.00" N	2° 18' 57.00" E
Zone C	Alibori	Gogounou	10° 47' 20.36" N	2° 37' 8.14" E
		Kandi	11° 07' 43.36" N	2° 56' 13.00" E
Zone D	Borgou	Parakou	9° 21' 00.00" N	2° 37' 0.00" E
	Collines	Savè	8° 01' 48.00" N	2° 29' 24.00" E
Zone E	Atlantique	Abomey-Calavi	06° 27' 0.00" N	2° 21' 0.00" E
Zone F	Littoral	Cotonou	6° 21' 45.00" N	2° 25' 31.80" E

Zone A = Foodstuff Zone of Southern Borgou; Zone B = Western Zone Atacora, Zone C = Cotton zone of northern Benin; Zone D = Cotton Zone of centre Benin, Zone E = Land area bar; Zone F = Area of fisheries.

diets in Africa (Kèkè et al., 2008). Unfortunately, wagashi creates a favorable environment for the growth of microorganisms, which may negatively affect its quality. Several research studies have essentially focused on wagashi process and its microbiological quality.

Moreover, Aworh and Egounlety (1985) and Kees (1996) reported the processing technology of wagashi and its stabilization by heat treatment and using chemical additives such as propionic acid and sorbates. In addition, Kèkè et al. (2008) reported a method of conservation of wagashi using strains of *Lactobacillus plantarum*. Furthermore, the conservation of wagashi by chemical method has a negative effect on the sensorial quality of the product (Kèkè et al., 2008). Traditional practices for preservation of wagashi, used on-farm persist over time and allow preserving the sensorial quality of the product for a relatively long time. The exhaustive inventories of all these practices so far are not well documented. The present work aims to inventory the traditional methods used by the producers, the sellers and the consumers to preserve wagashi in Benin, with a view to access these methods in order to improve the quality of the product.

MATERIALS AND METHODS

Study area

The study was conducted in eight municipalities located in six agro-ecological areas of Benin (Table 1 and Figure 1).

Selection of survey areas

Pehunco, Gogounou, Kandi and Djougou areas are known to be big regions of milk production and wagashi processing (Chopra and Ouaouich, 2009; FAO, 2012). Parakou, Cotonou and Abomey-Calavi are cosmopolitan municipalities where wagashi is more consumed (Dossou et al., 2006).

Choosing of respondents

Due to the lack of database during the survey respondents were

randomly selected according to the criteria of accessibility and availability to provide the information. A total of 318 producers, 164 sellers and 464 consumers of wagashi in the study areas were interviewed.

Methods of data collection

An individual survey by interviewing the respondent was performed. The data collected were related to the traditional conservation methods, the shelf life of wagashi preserved by traditional methods, the source of additives used, the quality of the water used for production and / or conservation of wagashi, the transportation conditions of wagashi.

Statistical analysis of data

The data collected were analyzed using SAS (1996). The percentages of respondents for each conservation methods were calculated by actor in the chain (producer, retailer and consumer) and study area. To show differences between these percentages calculated, pair wise comparison were made using Z bilateral test. Shelf life of wagashi were calculated by category of actor (producer, retailer and consumer). The significativity of conservation methods was determined by the F test after analysis of variance (ANOVA). The means shelf lives were calculated by method of storage for each category of actor and pair wise compared using student t test.

RESULTS

Preservation methods

Several traditional methods were used for wagashi conservation by producers, sellers and consumers in Benin (Table 2). Among these methods, the main practices were sun drying, used by 62.26% of respondents, followed by when conservation by 21.07% of the producers investigated. Sun drying of wagashi was especially used by producers of Pehunco and Djougou, whereas the whey conservation was the main method used by 100% of respondents in Save. Wagashi conservation by soaking in colored water with *Sorghum vulgaris* or *Sorghum caudatum* was also used by 8.49% of producers. This practice of preservation of wagashi

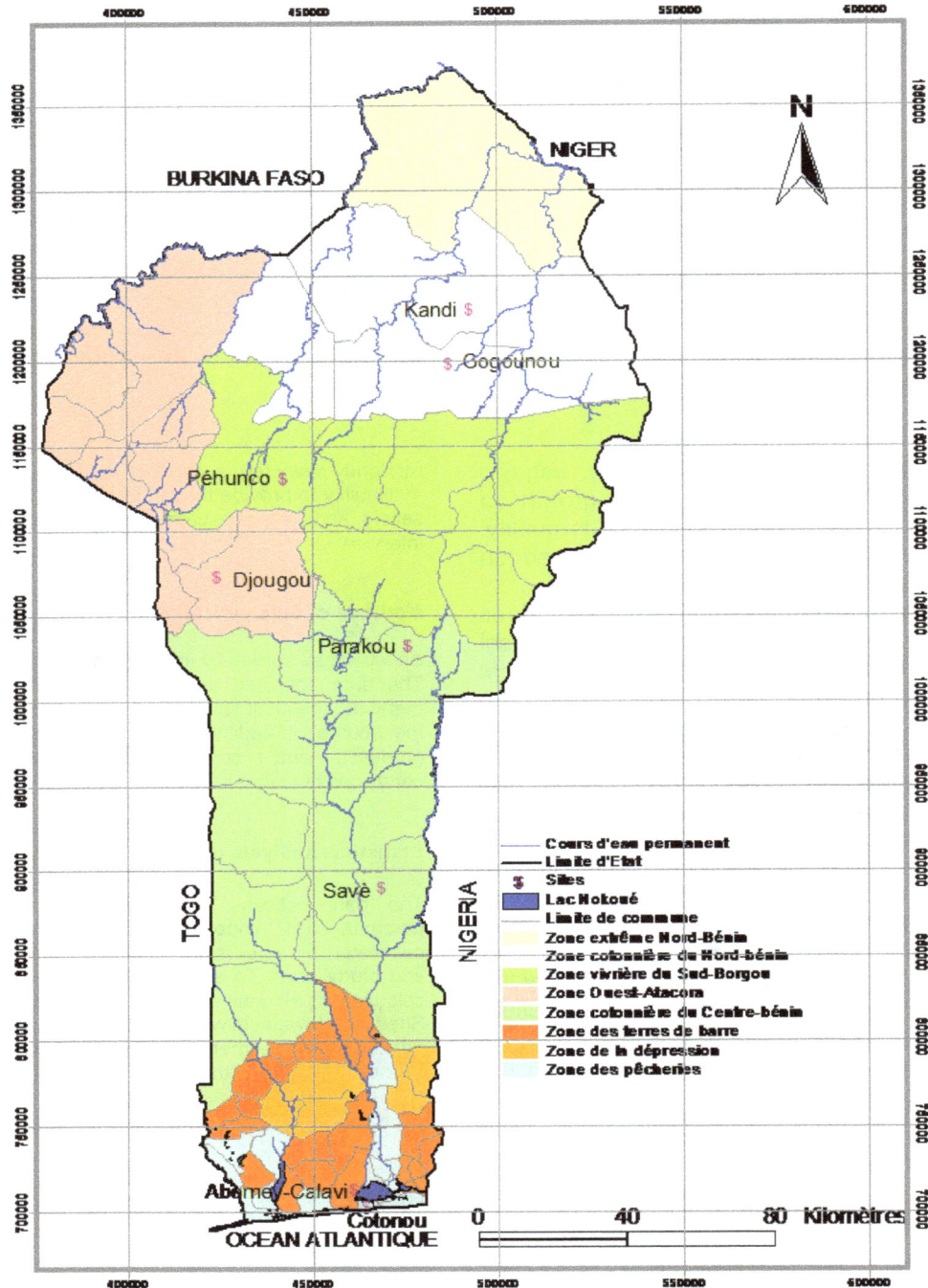

Figure 1. Benin's map showing the survey areas.

dominated in Gogounou. The method of storage of the product soaking in untreated water was, used especially by Fulani etchnic group of Parakou camping's by 8.17% of the producers. Daily cooking of wagashi at 80 tob100°C was practiced by 86.58% of respondents of the sellers followed by preservation of wagashi in untreated water (8.54% of respondents) and sun drying which is applied by 3.66% of sellers. In the category of consumers, it appeared that the sun drying of wagashi practiced by 35.13% of respondents especially in the northern., remained the most widely used method, followed by traditional smoking, daily cooking and frying of wagashi by 28.8, 25.86 and 6.03% of consumers, respectively. Figure 2 illustrates sun drying and whey

Table 2. Proportion of respondents (%) using different traditional methods for wagashi preservation in Benin.

% at the level of area	Producers					Retailers				Consumers				
	N	SD	WP	CUW	CCW	N	DC	CUW	SD	N	F	DC	TS	SD
Pehunco	143	97.20[a]	2.80[b]	0.00	0.00	22	100[a]	0.00	0.00	50	2.00[b]	36.00[b]	28.00[a]	30.00[b]
Djougou	49	95.92[a]	0.00	0.00	4.08[b]	21	95.24[a,b]	4.76[a]	0.00	36	5.56[a,b]	5.56[c]	27.78[a]	52.78[a]
Parakou	33	27.70[b]	3.03[b]	69.70[a]	0.00	60	86.66[a,b]	8.32[a]	3.33[a,b]	104	4.80[b]	32.69[b]	27.50[a]	25.96[b]
Gogounou	33	9.09[b]	6.06[b]	9.09 b	75.76[a]	03	100[a]	0.00	0.00	52	0.00	1.92[c]	28.8[a]	67.31[a]
Kandi	-	-	100[a]	-	-	09	77.78[b]	0.00	22.22[a]	59	0.00	0.00	32.20[a]	67.80[a]
Save	60	0.00	100[a]	0.00	0.00	16	81.25[b]	06.25[b]	12.50[a,b]	38	0.00	57.89[a]	26.31[a]	21.05[b]
Abomey-Calavi	-	-	-	-	-	04	75.00[b]	25.00[a]	0.00	49	14.28[a]	44.90[a,b]	30.60[a]	8.16[b,c]
Cotonou	-	-	-	-	-	29	79.31[b]	20.69[a]	0.00	76	17.10[a]	27.63[b]	28.94[a]	19.74[b,c]
Mean	318	62.26	21.07	8.17	8.49	164	86.58	8.54	03.66	464	6.03	25.86	28.88	35.13

Percentages in the same column followed by different letters are significantly different (p< 0.05), SD : Sun Drying ; WP : Whey Preservation ; CUW : Conservation in Untreated Water ; CCW : Conservation in Colored Water with *Sorghum* ; DC : Daily Cooking ; F : Frying ; TS: Traditional Smoking. N= Sample's size. -: no respondent surveyed or null.

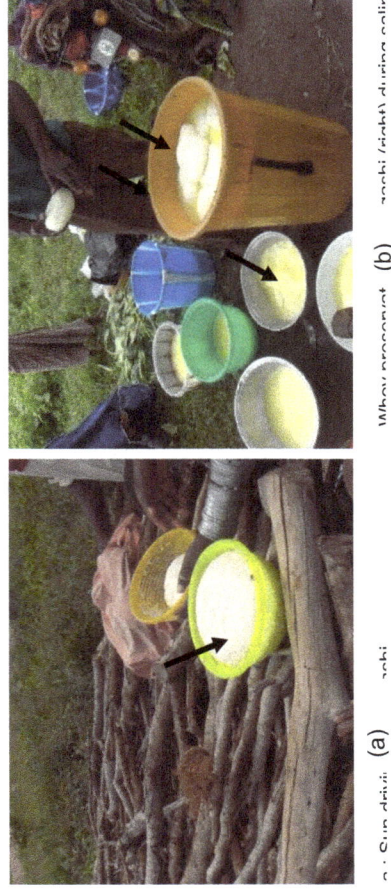

a : Sun drying (a)   Whey preservat (b)

Figure 2. (a) Sun drying and (b) whey preservation of wagashi during selling.

preservation of wagashi methods.

Shelf life of wagashi

The shelf life of wagashi preserved by the different methods collected did not exceeded one month. However, some methods, based on information from stakeholders, seem to be more effective than others, depending on the stakeholder group. In addition, according to the producers, shelf life time of wagashi preserved by sun drying (8.48 ± 0.49 days) was the most efficient method of wagashi preservation. Daily cooking with a shelf life about 8.26 ± 0.42 days was the most practiced and effective method used by the retailers, whereas smoking of wagashi whose shelf life is one month, was recognized

Table 3. Shelf life of wagashi preserved with different methods.

Actor	Preservation methods	Sample's size	Shelf-life	Standard deviation	Significativity
Producers	Solar drying	198	8.48[a]	0.49	***
	In untreated water	26	2.57[c]	1.37	
	In whey	67	3.04[b,c]	4.74	
	In coloured water	27	4.74[b]	1.34	
Sellers	Daily cooking	142	8.26[a]	0.42	NS
	In untreated water	14	5.50[b]	1.34	
	Solar drying	6	6.16[b]	2.06	
Consumers	Frying	28	4.62[c]	2.01	****
	Traditional smoking	134	30.00[a]	0.00	
	Solar drying	163	11.73[b]	9.77	
	Daily cooking	120	7.14[c]	2.82	

Means in the same column by actor followed by different letters are significantly different ($p < 0.05$).

as the most effective method for wagashi preservation by consumers. A significant difference ($p < 0.001$) between the shelf life of the different methods of wagashi preservation by the producers and consumers was observed.

Moreover, none of the methods used by producers and sellers did not contribute to enhance the quality of wagashi for more than twelve days (Table 3). The factors involved in relation to the short shelf life of wagashi by these different methods were related to the rapid decay of wagashi, the quality of the milk used , the lack of appropriate method for conservation of wagashi, the contamination by moulds, the quality of feed of animals from which the milk is obtained.

Quality of water used to process or preserve wagashi

The water used by producers, consumers and sellers for the production or storage of wagashi is often of poor hygienic quality. In fact, about 70% (69.81%) of the producers used poor quality of water for wagashi processing. Moreover, water from wells was used by producers of Pehunco for wagashi manufacturing, whereas water from backwater at Sambo-gah in Djougou was used and water from river was used by producers of Savê. About 76.22% of retailers, declared using good quality of water against 91.59% of consumers for preservation of wagashi.

Conditions of transporting and saling of wagashi

Wagashi is packaged and transported under poor hygienic conditions. Packed in stored fortune bags and in bowls or baskets and cut containers, wagashi was transported at ambient temperature (25 to 45°C) from the production area to the sellers' home, sometimes during several times (8 h).

The situation is the same with the consumer. During saling, the conservation of wagashi is often done by soaking the product in untreated water, especially at Cotonou markets (95% of sellers).

At Gogounou, wagashi remained immersed in colored water extracted from *Sorghum vulgaris* and/or *S. caudatum* during the saling period. About 100% of sellers from Pehunco, Parakou, Djougou and Save stored wagashi in open shelves, exposed to flies, dust, weather (rain, wind), insects and other pests during (Figure 3).

Conservation of wagashi by microbial decontamination and deodorization of spoiled wagashi

Several techniques for microbial decontamination and deodorization of altered wagashi (by visible moulds especially) were used including the boiling of wagashi at 80 to 100°C, followed by sun drying used by 62.89, 100 and 61.89% of producers, sellers and consumers, respectively.

The boiling coupled with the use of plants' leaves of *Vitellaria paradoxa* or *Pennisetum polystachion* or *Piliostigma thonningii* and sun drying was practiced by 19.18% of the producers, and the frying and smoking were practiced by 0.22% and 9.27% of consumers, respectively. The main plant used by producers for decontamination and deodorization of wagashi was *P. thonningii* followed by *V. paradoxa*.

The toxicity for human of extracts from these plants must be verified in order to evaluate hazards associated with their consumption.

DISCUSSION

From analysis of the data related to the traditional

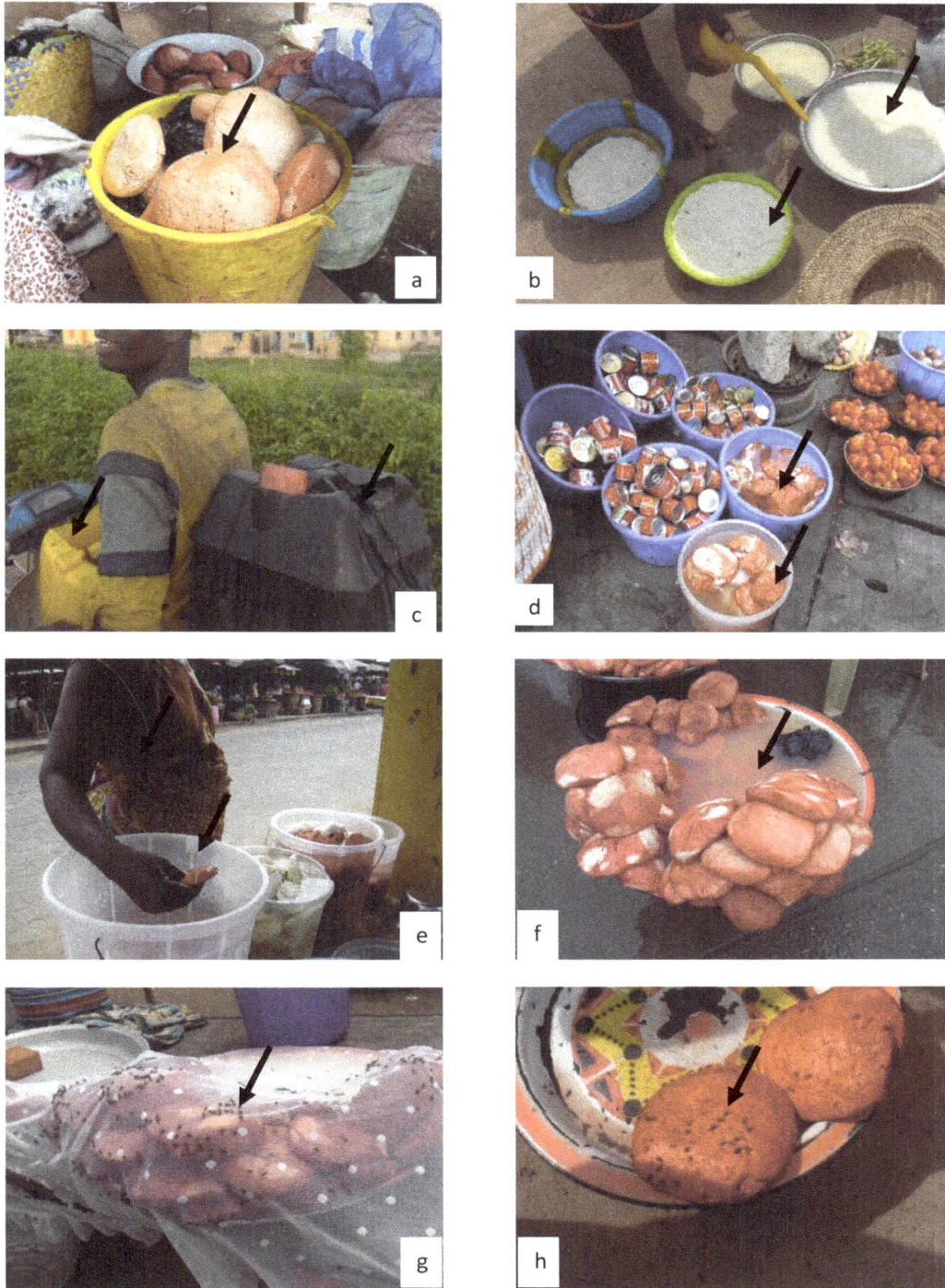

Figure 3. Illustration of wagashi preservation under unsanitary conditions of production, transportation and selling. a: wagashi pile up in a bucket in unhygienic condition; b: wagashi in a sieve, set on the land; c: wagashi transported in a can in unhygienic condition; d: wagashi pile up in untreated water at selling in unhygienic condition, e: Palpation of wagashi by the buyer prior choice f: wagashi without packaging and subjected to rainwater, g: wagashi at early of spoilage step and covered by a transparent cloth invaded by flies in the market, h: wagashi unprotected invaded by flies.

methods used for wagashi conservation, we can put out the potential limits concerning the sun drying method of wagashi which it used only in the dry season. In addition, sun drying method of wagashi preservation may provide microbiological contamination of the product mainly by *Bacillus* spores, moulds and others microorganisms. During processing, undesirable foreign materials may accidentally be mixed with wagashi (Shibamoto and Bjeldanes, 2009). This method, however, has the advantage of reducing the water activity (Aw) of wagashi when the drying has been conducted at term. The microbial load after an efficient sun drying time of wagashi would be drastically reduced(Medveďová et al., 2009). It would be better to improve this traditional method acting primarily on the integrity of wagashi throughout the drying time by limiting the direct contact of the product with the aforementioned contaminants and ensuring its position in relation to the direction of wind and maximum sunshine.

When conservation and untreated water have limits due to the fact that both methods involve the use of hands, utensils and water which could be a vehicle or source of contaminants affecting quality of the product. Hands (not disinfected or unprotected), constitute surely an immediate cause of microbial contamination, the water used often under poor hygiene quality can causes microbiological, physical and chemical contamination. Indeed, the backwater and the river water can carry pathogens such as *E. coli, Klebsiella, Salmonella, Citrobacter, Enterobacter and Enterococcus, Clostridium perfringens, Pseudomonas aeruginosa, Shigella* spp, *Candida albicans, Staphylococcus aureus, Giardia lamblia, Cryptosporidium* and human enteric viruses (Degbey et al., 2010, 2011).

Moreover, the water used in Benin for this activity can also be a vehicle of toxic heavy metals such as mercury, arsenic, cadmium and lead (Degbey et al., 2010) and pesticides such as endosulfan, DDT, dieldrin, heptachlor and lindane (Degbey et al., 2010; Agbohessi et al., 2012). The wholesomeness of utensils may not guarantee safety of the product given to cross-contamination that may occur during conservation in untreated water and whey storage. The whey preservation of wagashi certainly may not have all the required microbiological quality of a product whose chemical composition of carbohydrates, lipids, minerals and trace of protein is a favourable factor for microorganism's growth.

The method of colored water conservation represents a two-pronged approach: the use of water is a gap previously reported. However, the extraction and use of dyes of *S. caudatum* and *S. vulgaris* were reported by Akande et al (2010), Khalil et al. (2010), Agbangnan (2011), Kayodé et al. (2011, 2012) as a potential source of antioxidants such as polyphenols, 3-deoxyanthocyanidine and flavonoids found in the extracts. Also, these dyes possess anticancer properties and anti-cardiovascular disease (Kayodé et al., 2011, 2012), hepatoprotective and hematopoietic properties (Akande et al., 2010). Unfortunately, the extracts of these dyes do not possess antimicrobial properties (Belewu et al., 2005; Agbangnan, 2011) and microbial contamination

from water can affect the quality of the final product. In addition, the dyes themselves can carry spores of *Bacillus* and some pathogenic fungi and other microorganisms.

Sellers represent the intermediate chain of wagashi conservation between producers and consumers. The method of packaging and transportation of wagashi by the retailers may have significant negative effects in maintaining the integrity of the product quality due to the use of none recommended for transporting packaging wagashi at temperature ranging between 25 to 45°C. Daily cooking at which wagashi is subjected at retailer's home, if it reduces the remaining microbial load, could destroy the most heat labiles nutrients in the wagashi. Indeed, the repeated heat treatment may destroy nutrients in the product such as protein, vitamins and minerals (Shibamoto and Bjeldanes, 2009).

The frying which is relatively a good conservation method as having the advantage of destroying the microbial flora of the product when keeping it in fat state which limits its recontamination by non-lipophilic microorganisms. However, the quality of the frying oil can be no useful for the maintenance of nutritional quality of wagashi, the composition of oil rich in saturated fatty acids could give to the product, the vehicle of triglycerides which are lipoprotein and cholesterol precursors inside the wagashi (Nout et al., 2003). In case the consumer will fry several times the same product, he could expose wagashi to benzopyrenes contamination, which may be carcinogenic to the consumer (Shuguang Li et al., 1994; Shibamoto and Bjeldanes, 2009)

The traditional smoking, while offering a long shelf life to the product, present risk of being contaminated by benzopyrenes (Garcia, 1999; Kazerouni et al., 2001). Also, this method may lead to an important loss of nutrients (fatty acids, tocopherol and vitamins especially) in wagashi (Espe et al., 2002).

Practices of microbial decontamination using the leaves of *V. paradoxa, P. polystachion, P. thonningii* for deodorization of spoiled wagashi shares of deodorization and repairing fitness of wagashi in the way of deterioration according to producers, may constitute poisoning factors to be investigated.

To summarize, endogenous practices of wagashi preservation in Benin are variable and present limits for the good preservation of the product. In fact, dialing cooking, traditional smoking and frying of wagashi are heat treatment methods of wagashi preservation which may reduce the nutritional value of this cheese. The sun drying preservation like untreated and colored water and whey preservation of wagashi may alter hygienic and sanitary qualities of this cheese due to the potential microbial and chemical contaminations of the product through these methods. Preservation of cheese wagashi by these techniques may negatively affects physical, chemical, microbial and nutritional qualities of wagashi. New technique using plants extracts with very low toxicity, antioxidant, antimicrobial and biopreservative capacity is a good alternative method to be used at traditional scale for the preservation of wagashi regarding their biopreservative potential largely studied in literature and their security for consumer's health (Hsouna et al., 2011; Barkat and Bouguerra, 2012; Varona et al., 2013).

Conclusion

The study reported the traditional conservation practices of wagashi used by producers, consumers and sellers in Benin. Sun drying is the most used traditional method for wagashi preservation. This method could offer more guarantees for the safety of the product if the process was improved. Daily cooking although extends the shelf life of the product could harm thermolabile nutrients of wagashi and affects its nutritional quality. Sun drying and traditional smoking conservation methods are typically used by consumers, but they are not without risk to their health. Combination of traditional methods of conservation wagashi seems to ensure its hygienic and nutritional quality of the product.

ACKNOWLEDGMENTS

The authors are grateful to University Council of Development (CUD) for its financial support. They are thankful to Mrs Avocegamou Séverin and Wabi Mabouroukath for their technical assistance.

REFERENCES

Agbangnan CP (2011). Extraction et concentration d'extraits polyphénoliques naturels bioactifs et fonctionnels par procédés membranaires : caractérisation des structures moléculaires d'extraits du sorgho (Sorghum caudatum) du Bénin. Thèse unique des Universités d'Abomey-Calavi (Bénin) et de Pau des pays de l'Adour (France). P. 354.

Agbohessi TP, Toko II, Kestemont P (2012). État des lieux de la contamination des écosystèmes aquatiques par les pesticides organochlorés dans le Bassin cotonnier béninois Cah. Agric. 21:46-56.

Aïssi VM, Soumanou MM, Bankolé H, Toukourou F, de Souza CA (2009). Evaluation of hygienic and mycological quality of local cheese marketed in Benin. Aust. J. Basic Appl. Sci. 3(3):2397-2404.

Akande IS, Oseni AA, Biobaku OA (2010). Effects of aqueous extract of Sorghum bicolor on hepatic, histological and haematological indices in rats. J. Cell Anim. Biol. 4(9):137-142.

Aworh OC, Egounlety M (1985). Preservation of West African soft cheese by chemical treatment. J. Dairy Res. 52:189-195.

Barkat M, Bouguerra A (2012). Study of the antifungal activity of essential oil extracted from seeds of Foeniculum vulgare Mill. For its use as food preservatives. Afr. J. Food Sci. 6:239-244.

Belewu MA, Belewu KY, Nkwunonwo CC (2005). Effect of biological and chemical preservatives on the shelf life of West African soft cheese. Afr. J. Biotechnol. 4(10):1076-1079.

Chopra S, Ouaouich A (2009). Rapport de synthèse sur l'étude pour l'identification des filières agroindustrielles prioritaires dans les pays membres de l'UEMOA. Programme de Restructuration et de Mise à Niveau de l'Industrie des Etats membres de l'UEMOA. P. 105.

Degbey C, Makoutode M, Agueh V, Dramaix M, De Brouwer C (2011). Facteurs associés à la qualité de l'eau de puits et prévalence des maladies hydriques dans la commune d'Abomey-Calavi (Bénin).

Santé. 21:47-55.

Degbey C, Makoutode M, Fayomi B, De Brouwer C (2010). La qualité de l'eau de boisson en milieu professionnel à Godomey en 2009 au Bénin Afrique de l'Ouest. J. Int. Santé Trav. 1:15-22.

Diao X, Hazell P, Resnick D, Thurlow J (2006). The role of agriculture in development: implications for Sub-Saharan Africa. Int. Food Policy Res. Inst. P. 112.

Dossou J, Hounzangbe Adote S, Soulé H (2006). Production et transformation du lait frais en fromage peulh au Bénin: Guides de bonnes pratiques, version validée lors de l'atelier national du 14 juillet 2006. Consulté le 21 Mars 2012 à l'adresse suivante: http://www.repol.info/IMG/pdf/Fiche_wagashi_VF.pdf.

Espe M, Nortvedt R, Lie Ø, Hafsteinsson H (2002). Atlantic salmon (Salmo salar, L) as raw material for the smoking industry. II: Effect of different smoking methods on losses of nutrients and on the oxidation of lipids. Food Chem. 77:41-46.

FAO (2012). Estimation des effectifs d'animaux vivants (bovins) au Bénin en 2011. Consulté 04/2012 à l'adresse : http://www. Countrystat.org/ben/cont/pxwebquery/ma/053spd135/fr.

Garcia FMS (1999). Determination of benzo[a]pyrene in some Spanish commercial smoked products by HPLC-FL. Food Addit. Contam. 16(1):9-14.

Hsouna BA, Trigui M, Mansour BR, Jarraya MR, Damak M, Jaoua S (2011). Chemical composition, cytotoxicity effect and antimicrobial activity of Ceratonia siliqua essential oil with preservatives effects against Listeria inoculated in minced meat. Int. J. Food. Microbiol. 148:66-72.

Kayodé APP, Bara CA, Dalodé-Vieira G, Linnemann AR, Nout MJR (2012). Extraction of antioxidant pigments from dye sorghum leaf sheaths. LWT - Food Sci. Technol. 46(1):49-55.

Kayodé APP, Nout MJR, Linnemann AR, Hounhouigan DJ, Berghofer E, Siebenhandl-Ehn (2011). Uncommonly High Levels of 3-Deoxyanthocyanidins and Antioxidant Capacity in the Leaf Sheaths of Dye Sorghum. J. Agric. Food Chem. 59(4):1178-1184.

Kazerouni N, Sinha R, Che-Han H, Greenberg A, Rothman N (2001). Analysis of 200 food items for benzo[a]pyrene and estimation of its intake in an epidemiologic study. Food Chem. Toxicol. 39:423-436.

Kees M (1996). Le fromage peulh : facile à produire et bien apprécié, une technologie à vulgariser. Rapport de recherche GTZ, Université Eschborn. RFA. pp. 8-25.

Kèkè M, Yèhouénou B, Dahouénon E, Dossou J, Sohounhloué DCK (2008). Contribution à l'amélioration de la technologie de fabrication et de conservation du fromage peulh waragashi par injection de Lactobacillus plantarum. Ann. Sci. Agron. Bénin. 10(1):73-86.

Khalil A, Baltenweck-Guyot R, Ocampo-Torres R, Albrecht P (2010). A novel symmetrical pyrano-3-deoxyanthocyanidin from a Sorghum species. Phytochem. Lett. 3(2):93-95.

Shuguang Li MS, Pan D, Wang G (1994). Analysis of Polycyclic Aromatic Hydrocarbons in Cooking Oil Fumes. Arch. Environ. Health. 49(2):119-122.

Medveďová A, Valík Ľ, Studeničová A (2009). The Effect of Temperature and Water Activity on the Growth of Staphylococcus aureus. Czech J. Food Sci. 2:S2-28-S2-35.

Nout R, Hounhouigan DJ, van Boekel T (2003). Les Aliments: Transformation, Conservation et Qualité. Backhuys Publishers, Leiden. Netherlands. P. 279.

SAS (1996). SAS/STAT. User's guide (Ressource électronique). 4 ème éd., version 6, New-York : SAS. Inst. Inc., Cary.

Shibamoto T, Bjeldanes L (2009). Introduction to Food Toxicology, second edition. Academic Press is an imprint of Elsevier, ISBN: 978-0-12-374286-5; P. 320.

Varona S, Rojo RS, Martin A, Cocero JM, Serra TA, Crespo T, Duarte MMC (2013). Antimicrobial activity of lavandin essential oil formulations against three pathogenic food-borne bacteria. Ind. Crop. Prod. 42:243-250.

14

Effect of ultra-violet rays on growth and development of rust red flour beetle, *Tribolium castaneum* (Herbst)

R. P. Naga, Ashok Sharma, K. C. Kumawat and B. L. Naga

Department of Entomology, S. K. N. College of Agriculture (S. K. Rajasthan Agricultural University),
Jobner (Rajasthan)- 303329, India.

The adults of *Tribolium castaneum* were exposed to ultraviolet (UV) light (254 nm) for different time periods (2, 4, 6, 8, 10, 12, 14, and 16 min) and each replicated thrice. The culture of unexposed beetles was maintained for comparison. The different time period levels were maintained in the laminar flow cabinet using ultra-violet radiations (254 nm). The fecundity (113.00 eggs/female), hatchability (55.33%), pupation (53.01%) and adult emergence (F_1) (36.67) were maximum in unexposed beetles, whereas, minimum in beetles exposed for 16 min (23.50 eggs/ female, 29.33, 24.98 and 4.00%, respectively) under UV radiation.

Key words: *Tribolium castaneum* (Herbst), ultraviolet (UV) radiation, beetles, fecundity.

INTRODUCTION

Wheat (*Triticum aestivum* Linn.) is one of the important crops in Indian agriculture occupying an area of 28.0 million hectares with the production to the tune of 78.6 million tons (Anonymous, 2009). In developing countries, the grains (cereals and legumes) constitute the key resource of daily energy and protein intake. Among the wheat growing countries, India ranked second both in terms of area and production (Anonymous, 2008). A number of insect pests are found attacking the wheat and its products, among them, the rust red flour beetle, *Tribolium castaneum* (Herbst) causes both quantitative and qualitative damage (Bhargava and Kumawat, 2010). Irradiation is effective against stored grain pests as it causes mortality as well as sterility in insects depending on the dose and time of exposure. Complete reduction of *Alphitobius diaperinus* populations was achieved from 6 to 9 months storage periods, when eggs were exposed to UV-rays for 8 min (Begum et al., 2007). *Cadra cautella* eggs were less sensitive to UV-rays than *Tribolium castaneum* and *Tribolium confusum*. No adult emerged when three days old eggs of *T. castaneum* were irradiated for 16 or 24 mins, or from two and 3 days old eggs of *T. confusum* irradiated for 16 or 24 min (Faruki et al., 2007). The Ultra-low-voltage (ULV) has been found effective against many stored grain pests and has obvious trend of mortality against *T. castaneum* (Calderon et al., 1985; Collins and Kitchingman, 2010). However, meagre work has been done on the effect of UV-rays on the life processes of *T. castaneum*. The present study aimed to find out the effect of UV-rays with the search of a non-chemical method against life processes of *T. castaneum* which is a secondary pest and causes more qualitative damage than the quantitative.

MATERIALS AND METHODS

To maintain the stock culture of *T. castaneum,* it was procured from the pure culture already maintained in Department of Entomology, S. K. N. College of Agriculture, Jobner. The wheat grains were

Table 1. Fecundity, hatchability, pupation percentage and adult emergence (F_1) of *T. castaneum* as influenced by different exposure periods of UV rays.

S/N	Exposure time to UV rays	Fecundity/ female*	S.D.	Hatchability (%)**	S.D.	Pupation (%)**	S.D.	Adult emergence (F_1)***	S.D.
1	2 min	36.00 (1.56)	2.97	47.33 (43.47)	1.15	45.77 (42.57)	0.10	17.33 (4.16)	0.58
2	4 min	36.00 (1.56)	2.00	45.33 (42.32)	0.58	42.64 (40.77)	0.73	15.33 (3.92)	0.58
3	6 min	32.50 (1.51)	0.87	42.67 (40.79)	1.53	39.85 (39.14)	0.55	13.33 (3.65)	1.15
4	8 min	30.00 (1.48)	1.50	40.33 (39.42)	1.53	37.20 (37.58)	0.53	11.67 (3.42)	0.58
5	10 min	30.00 (1.48)	1.50	37.67 (37.86)	0.58	31.86 (34.36)	0.55	9.33 (3.05)	1.15
6	12 min	29.00 (1.46)	2.25	35.33 (36.47)	1.53	29.23 (32.73)	1.15	7.67 (2.77)	1.15
7	14 min	27.67 (1.44)	0.76	32.67 (34.86)	1.15	26.51 (30.99)	1.31	6.00 (2.45)	0.58
8	16 min	23.50 (1.37)	1.32	29.33 (32.79)	1.53	24.98 (29.99)	1.46	4.00 (2.00)	0.58
9	Unexposed	113.00 (2.05)	2.65	55.33 (48.06)	2.52	53.01 (46.73)	1.08	36.67 (6.06)	1.73
	SEm±	0.04		0.18		0.35		0.09	
	CD at 5%	0.11		0.52		1.05		0.26	

Data based on two pairs of adults (three replications). *Figures in the parenthesis are log X values. **Figures in the parenthesis are angular transformed values. *** Figures in the parenthesis are √X values.

cleaned, rinsed in water, sun dried and were subjected to sterilisation at 60°C temperature for five hours to eliminate any insect infestation, hidden or otherwise. These grains were conditioned for 48 h at 29 ± 1.5°C temperature and 70 ± 5.0% relative humidity and crushed. For maintaining subsequent insect culture, 20 pairs of newly emerged adults were released for oviposition in the glass jar (size 18 × 10 cm) containing 200 g crushed wheat grains. For handling the infested grains and insects, a forcep and a camel hair brush was used. Subsequent experiment was conducted at 29 ± 1.5°C temperature and 70 ± 5.0% relative humidity. The newly emerged adults of *T. castaneum* (immediately after hatched out from the eggs, unmated) were exposed to UV light for different time periods (2, 4, 6, 8, 10, 12, 14, and 16 min) and each replicated thrice. The culture of unexposed beetles was maintained for comparison. The different time period levels were maintained in the laminar flow cabinet using UV radiations (254 nm).

The UV light tube was of Philips make (30 Watt) and the equipment to hold the same was manufactured by Kirloskar Electrodyne. The distance of beetles and the UV tube was 50cm emitting energy at the 254 nm at an intensity of 600 uw/cm². Two different pairs of adults immediately after emergence (one pair male and one pair female) were exposed to the ultra-violet radiations (for each time periods mentioned above) and were transferred into the Petri dishes having black paper in the bottom and 3 g sterilized and conditioned broken wheat grains on it. The beetles were allowed to mate and oviposit. Three replications of the culture were maintained. The eggs laid by the beetles were separated every 24 h by sieving (50 mesh sieve) and were counted under the microscope. The survival of the eggs was recorded after examining the egg chorion and by recording the number of larvae that hatched out. The hatched out larvae were transferred into different glass vials having sterilized and conditioned broken wheat grains and reared at 29 ± 1.5°C temperature and 70 ± 5.0 30% relative humidity. The pupation percentage was worked out on the basis of larvae pupated out of the total larvae. The number of adults emerged were counted every 24 h. After counting, the newly emerged adults were discarded from the samples so as to avoid further counting.

The percentage data on hatching and pupation were transformed into angular values (arc sine √percentage) to convert the percentage into degrees. The data on fecundity were converted into log X values and that of adult emergence into √X for ease in analysis of variance.

RESULTS AND DISCUSSION

The results on fecundity (Table 1) indicated that the treatments, on which the *T. castaneum* was allowed to feed and breed, significantly affected the fecundity. The fecundity on different treatments differed greatly (23.50 to 113.00 eggs/female). The maximum number of eggs was laid by the unexposed female beetles (113 eggs/female) and differed significantly (p 0.05 = 0.11) from rest of the treatments. The minimum fecundity (23.50 eggs/female) was recorded in the beetles exposed for 16 min which was found at par with 14 min (27.67 eggs/female), 12 min (29 eggs/female), 10 min (30 eggs/female) and 8 min (30 eggs/female). The descending order of ovipositional potential recorded on different exposed time periods was: unexposed beetles, 2, 4, 6, 8, 10, 12, 14 and 16 min.

The hatchability on different exposure periods was in the range of 29.33 to 55.33% and differed significantly to each other. The maximum hatchability of the eggs was recorded in unexposed beetles (55.33%) followed by 2 min (47.33%) and differed significantly from rest of the treatments. The minimum hatchability (29.33%) was recorded in 16 min exposed beetles, followed by 14 min (32.67%) and 12 min (35.33%). Rest of the treatments ranked in the middle order. All these treatments differed significantly (p 0.05 = 0.52) from each other. The ascending order of hatchability was found to be in the treatments of 16, 14, 12, 10, 8, 6, 4, and 2 min exposed beetles and unexposed beetles. The pupation percentage on different exposure periods was in the range of 24.98 to 53.01%. The maximum pupation was recorded in the unexposed beetles (53.01%), whereas minimum in the

Table 2. Correlation and regression coefficient between exposure time of UV rays and life processes of *T. castaneum*.

S/N	Aspects	Correlation coefficient (r value)*	Regression equation (y = α + βx)*	Coefficient of determination (R^2)
1	Exposure to UV rays vs. fecundity	-0.66	Y = 66.374 - 3.329X	0.42
2	Exposure to UV rays vs. hatchability	-0.98	Y = 52.198 - 1.441X	0.97
3	Exposure to UV rays vs. pupation	-0.98	Y = 50.12 - 1.681X	0.97
4	Exposure to UV rays vs. adult emergence	-0.86	Y = 25.747 - 1.533X	0.75

*Significant at 5% level.

16 min exposure period (24.98%). The unexposed treatments differed significantly (p 0.05 = 1.05) with the 2 min (45.77%), 4 min (42.64%) and 6 min (39.85%) exposure periods with respect to the pupation percentage. The pupation in 16 and 14 min exposure periods differed non-significantly to each other.

The descending order of pupation percentage was recorded to be in the treatments of unexposed beetles, 2, 4, 6, 8, 10, 12, 14 and 16 min exposure periods. The (F_1) adults emerged was in the range of 4.00 to 36.67. The minimum adults were emerged from 16 min exposed beetles (4.00) and differed significantly over rest of the treatments. The maximum number of adults were emerged from unexposed beetles (36.67) and stood significantly inferior to other treatments. The rest of the treatments were in the middle order to reveal the adult emergence. The ascending order of adult emergence was: 16, 14, 12, 10, 8, 6, 4, and 2 min exposed beetles and unexposed beetles. It was demonstrated that ULV was effective in reducing development in different storage insect pests (Collins and Kitchingman, 2010). This could be possible due to the fact that the UV radiation reduced hatching of eggs and adult eclosion which gets support from the findings of Faruki et al. (2007).

The fecundity, hatchability, pupation percentage and adult emergence (F_1) was maximum in unexposed beetles, whereas, minimum in beetles exposed for 16 min under UV radiation (254 nm). The correlation analysis indicated that there was a significant negative correlation (r = -0.66, -0.98, -0.98 and -0.86) between the exposure period to UV rays and fecundity, hatchability, pupation and adult emergence of *T. castaneum* (Table 2). However, the simple correlation does not disclose the facts of significance, the regression equations Y = 66.374 to 3.329X, Y = 52.198 to 1.441X, Y = 50.12 to 1.681X and Y = 25.747 to 1.533X were obtained which permitted the amount of resultant effect for each unit period of exposure to UV rays. The equations indicated that 66.374, 52.198, 50.12 and 25.742 are the intercepts and exposure to UV rays for one minute was responsible to cause reduction in fecundity, hatchability, pupation and

adult emergence by 3.329, 1.441, 1.681 and 1.533%, respectively. A decreasing trend in these parameters was evident when period of exposure was increased. Mohan and Kumar (2010) reported the UV irradiation as a promising agent for controlling the cotton stainer, *Dysdercus koenigii* and observed significant decrease in survival with increasing exposure time, delayed moulting into adult, and morphological deformities in adults and nymphs.

REFERENCES

Anonymous (2008). Partiyogita Darpan, March P. 1549.
Anonymous (2009). Economic Survey. Directorate of Economics and Statistics, Government of India. P. 183.
Begum M, Parween S, Faruki SI (2007). Combined effect of UV-radiation and triflumuron on the progeny of *Alphitobius diaperinus* (Panger) (Coleoptera: Tenebrionidae) at different storage period. Univ. J. Zool. Rajshahi Univ. 26:45-48.
Bhargava MC, Kumawat KC (2010). Insect pests of stored grains and grain products. Pests of stored grains and their management, New India publishing Agency, New Delhi. pp. 29-70.
Calderon M, Bruce WA, Leesch JG (1985). Effect of UV radiation on eggs of *Tribolium castaneum*. Phytoparasitica 13(3-4):179-183.
Collins DA, Kitchingman L (2010). The effect of ultraviolet radiation on stored product pests. 10th International Working Conference on Stored Prod. Protect. pp. 632-636.
Faruki SI, Das DR, Khan AR, Khatun M (2007). Effects of ultraviolet (254 nm) irradiation on egg hatching and adult emergence of the flour beetles, *Tribolium castaneum*, *T. confusum* and the almond moth, *Cadra cautella*. J. Insect Sci. 7:7-36.
Mohan S, Kumar D (2010). Effects of UV irradiations on the survival of the red cotton bug, *Dysdercus koenigii* (Heteroptera, Pyrrhocoridae). Aust. J. Agric. Eng. 1(4):132-135.

15

Rheological characterization and texture of commercial mayonnaise using back extrusion

Rene Maria Ignácio and Suzana Caetano Da Silva Lannes

Biochemical-Pharmaceutical Technology Department, Pharmaceutical Sciences Faculty, São Paulo University, Av. Prof. Lineu Prestes, 580/Cidade Universitária, São Paulo-SP, CEP 05508-900, Brazil.

When creating or redesigning food product researchers need to pay special attention to textural and rheological properties. Texture is a sensory perception derived from the structure of food and is related to viscosity and elasticity. Mayonnaise is semi-solid oil-in-water emulsion with starch in its formulations when fat-reduced. Under shear forces it exhibits different types of macroscopic flow behavior, such as shear thinning, yield stress behavior, thixotropy, and viscoelasticity. One problem with the rheological characterization of mayonnaise is that it is particularly difficult to isolate and quantify individual effects. Only sensory evaluation is capable to fully detect and describe them. Hence the ability to measure and to characterize them accurately and quickly enables the food industry to set standards for quality and to monitor deterioration during storage and distribution. Back extrusion is a method used to quantify the behavior of thixotropic fluids. The relatively short time and low cost required to conduct the test makes it a suitable technique for quality control in product development. This paper evaluates back extrusion as method to analyze mayonnaise. It is performed at 5, 25 and 40°C in a Stable Micro Systems TA-XT2 using twelve commercial mayonnaises. The method has provided qualitative and quantitative characteristics of samples in an accurate and quick way.

Key words: Rheology, back extrusion, texture, mayonnaise.

INTRODUCTION

Texture is a difficult term to define since it means different things to different people. Food texture is a sensory perception derived from the structure of food at molecular, microstructure, and macroscopic levels (Chen, 2009; Weenen et al., 2003; Rosenthal, 1999). Szczesniak (2002) pointed out the multidimensional nature of texture and its importance to the consumer saying that only a human being can perceive and describe it and the texture testing instruments can detect physical parameters which then must be interpreted in terms of sensory perception, the most important ones being the sense of touch and pressure.

When creating or redesigning food product researchers need to pay special attention to textural and rheological properties. The science of rheology has many applications in the field of food acceptability, food processing and handling. A number of food processing operations depend heavily upon rheological properties of the product at an intermediate stage of manufacture because this has a profound effect upon the quality of the finished product. Plasticity, pseudoplasticity, and the property of shear thinning are important quality factors in foods and the study of these properties is part of the science of rheology. Some foods, like mayonnaise, are either plastic or pseudoplastic in nature. They are required to spread and flow easily under a small force but to hold their shape when not subjected to any external force other than gravity. All of these properties fall within the field of rheology (Bourne, 2002).

Mayonnaise is a mixture of egg, vinegar, oil and spices, with 70 to 80% fat. It is a semi-solid oil-in-water emulsion formed by first mixing the eggs, vinegar and spices and then slowly blending in the oil resulting in closely packed 'foam' of oil droplets (Depree and Savage, 2001; Szczesniak, 2002). Mayonnaise rheology and texture can depend on the distribution of the oil, the interaction between oil droplets, and the egg yolk emulsifier. Alteration of the mayonnaise's microstructure leads to different rheological properties. This means that the rheology is sensitive to any change in the materials microstructure (Stokes and Telford, 2004).

In emulsions, droplet size is perhaps the most important factor in determining properties like consistency, rheology, shelf life stability, color and taste. In general, emulsions with smaller droplet size result in great stability. Current equipment used for emulsion preparation includes colloid mills, sonicators or high pressure homogenizers. The advantage of high pressure homogenizers over other technologies is that more uniform droplets size distribution are obtained since the product is subjected to strong shear and cavitation forces that efficiently decrease the diameter of the original droplets (Martin-Gonzalez et al., 2009)

Rheology of mayonnaise can provide valuable information that can be used in quality control of commercial production, storage stability, sensory assessments of consistency, knowledge and design of texture, design of unit operations, and knowledge of the effects of mechanical processing on the structure of the emulsions (Gallegos and Berjano, 1992). Mayonnaise has been shown to be shear thinning, viscoelastic and thixotropic and it has a yield stress. The yield stress is the minimum stress required to enable flow or fracture due to the breakdown or alteration of a materials microstructure. One of the many difficulties encountered during the characterization of such a complex material is to isolate and quantify individual effects (Goshawk and Binding, 1998). Ingredients used in mayonnaises interact with each other either physically or chemically and determine the quality of the final products. The necessity in reduced-fat/cholesterol also in mayonnaises has made it necessary to identify appropriate ingredients for formulating these products as the removal of certain ingredients like fat may have a significant influence on the quality and taste of food emulsions (Ma and Boye, 2013).

Back extrusion is a method used to quantify the behavior of thixotropic fluids. The relatively short time and low cost required to conduct the test makes it a suitable technique for quality control in product development. Yield stresses can be easily determined (Brusewitz and Yu, 1996). It involves two physical movements: (a) a cylindrical plunger is forced down into the fluid and (b) the fluid flows upward through a concentric annular space. Usually, the cylindrical plunger is a solid rod and the fluid is contained in a cylindrical cup. The force required to extrude the

fluid in the direction opposite to that of the rod is measured (Gujral and Sodhi, 2002; Paoletti et al., 1995; Kaneda and Takahashi, 2011).

Since mayonnaise rheological properties are related to the formulation, process parameters, the storage temperature and commercial mayonnaises present differentiated formulations analysis performed in distinct temperatures demonstrates such differences in its rheology. Moreover, as the complexity of texture can only be fully detected and described by sensory evaluation, the ability to measure it accurately and quickly enables the food industry to set standards for quality and to monitor deterioration that occurs during storage and distribution.

This paper evaluates back extrusion as a method to analyze mayonnaise rheologically and texturally, discussing the results in different formulations.

MATERIALS AND METHODS

Twelve commercial mayonnaises were analyzed. Eight of them were traditional type (codes I, II, III, IV, V, VI, VII and VIII) and four of them were light type (codes III Light, IV Light, V Light and VII Light) as shown in Table 1. The exact composition of these commercial mayonnaises is not known. All of them were bought in local supermarket.

Tests were carried out in a TA-XT2 texturometer (Stable Micro Systems, UK) using Back Extrusion Cell (A/BE) using 5 kg load cell. Dimensions of the back extrusion probe are: Inner cup diameter 55 mm; cup height 70 mm and compression plate diameter 45 mm (Figure 1). The data were captured by the program "Texture Expert Exceed" (Stable Micro Systems, UK). The parameters used were tested in advance and are shown in Table 2. Back extrusion was performed at 5, 25 and 40°C. Mayonnaises were stored at each one of these temperatures in a controlled temperature refrigerator for 24 h prior to analysis. By the time of the analysis the samples were removed from the refrigerator, filled in the cup and tested immediately. At least two independent tests were made in each mayonnaise.

RESULTS AND DISCUSSION

Force as a function of time curves (also Force-Time curve) obtained using probe back extrusion for commercial mayonnaises at 25°C are in Figure 2. Since similar overall behavior was seen in all samples only a few curves are shown in order to make the understanding easier.

The form of these curves is consistent with a typical force-distance curve obtained from back extrusion test described by Bourne (2002), who also divided the curve in sections explaining what is happening in each one of them. The sections displayed in Figure 2 are described in Table 3 using Bourne's as a base.

The reader should be warned that the use of Force-Time graphic instead of Force-Distance is due to the fact that time and distance parameters are numerically equal (25 s and 25 mm, respectively) and the velocity parameter is 1 mm/s (Table 2), therefore, the two

Table 1. Sample codes for commercial mayonnaises used.

Commercial mayonnaise	Sample code
Traditional	I, II, III, IV, V, VI, VII and VIII
Light (same brand as III)	III Light
Light (same brand as IV)	IV Light
Light (same brand as V)	V Light
Light (same brand as VII)	VII Light

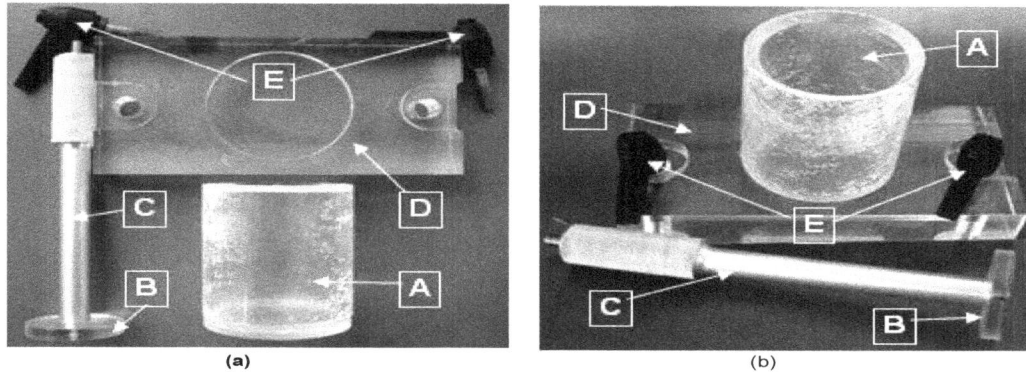

(a) (b)

Figure 1. Back extrusion probe set (a) cup and base disassembled and (b) cup and base assembled. Letters A to E mean: (A) inner cup, (B) compression plate, (C) extension bar, (D) base and (E) screw-nut.

Table 2. Parameters settings used in the stable micro systems TA-XT2.

Parameter	Setting
Test mode and option	Measure force in compression/return to start
Pre-test speed	1.0 mm/s
Test speed	1.0 mm/s
Post-test speed	1.0 mm/s
Distance	25 mm
Time	25 s
Trigger type	Auto
Trigger force	10 g
Data acquisition rate	200 pps

graphics produce curves with exactly same shape.

From section A to B the mayonnaise is immediate compressed with an instantaneous elastic deformation, which means its original form can be recovered instantaneously and completely if the force is removed, theoretically speaking, because few foods are perfectly elastic possessing flow properties in addiction to elasticity, most frequently 'plastic' and 'viscoelastic' (Bourne, 2002).

In section B the yield value is reached. When expressed in terms of shear stress this represents the 'yield stress' the minimum force that must be exceeded before flow begins. This type of flow is often found in foods and mayonnaise is a typical example of them.

Permanent deformation starts in section B going through section C because now original form cannot be recovered neither instantaneously nor completely if the force is removed because the flow has started through the annular concentric space and the behavior starts to changes from solid to liquid.

In section C maximum force is reached and a viscous deformation with a constant force from section C to D starts showing the viscoelatic property of mayonnaise. This section is a Newtonian region since force is independent of distance or time (or shear stress), leading to a constant viscosity.

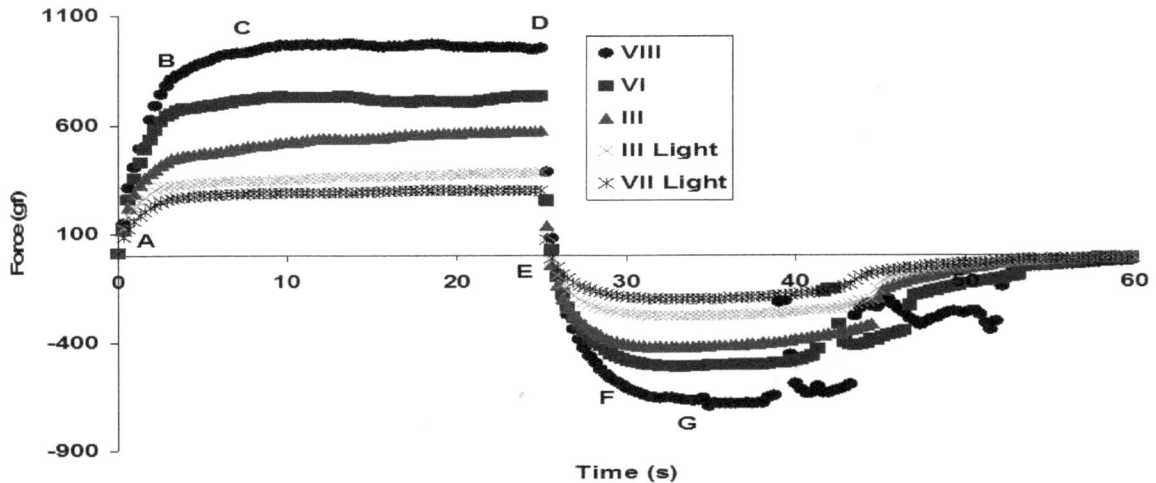

Figure 2. Force-Time curve obtained using back extrusion method for commercial mayonnaises at 25°C. Symbols represent samples (●)VIII, (■)VI, (▲)III, (×)III Light and (∗)VII Light (Table 1). Sections A to G are explained in Table 3.

Table 3. Steps that can be indentified in a typical force-distance/time curve obtained from back extrusion test.

S/N	Step	Section in Figure 2
1	Compression plate begins to plunge into the sample;	A
2	Force rises steeply over a short time of movement compressing the sample;	A to B
3	Yield value is reached and elastic deformation begins;	B
4	Extrusion starts through the annular concentric space - behavior change region "from solid to liquid"	B to C
5	Maximum force is reached and constant stress region begins with viscous deformation;	C
6	Newtonian region, the shear stress is independent of shear rate, leading to constant viscosity (in point D, the compression plate starts to go up);	C to D
7	Compression plate reverses direction and starts to move upward and force falls to zero	D
8	Force falls steeply over a short time of movement decompressing the sample	E
9	Yield value is reached and elastic deformation begins	F
10	Extrusion starts through the annular concentric space – behavior change region "from solid to liquid";	G

Szczesniak (2002) defined 'Hardness' as the force necessary to attain a given deformation, a resistance of the food to the applied compressive forces and Bourne (2002) defined it as the peak force in a force-time curve resulting from a test that compresses a bite-size of food that imitates the action of the jaw, therefore, section C can be termed as 'hardness'.

After section D same sequence of phenomena is observed in a mirror form between sections E and G with negative values of force. These forces in module are smaller than the ones observed in respective A-D sections, which means mayonnaise exhibits shear-thinning and thixotropy characteristics.

By backing extrusion twice the same sample, one following immediately the other, mayonnaise thixotropy and dependency of shear stress and of shear historical became clear as can be seeing in Figure 3. Here two different samples from brand V, V1 and V2, are back extruded twice: V11 and V21 by the first time and V12 and V22 by the second time. Force and viscosity

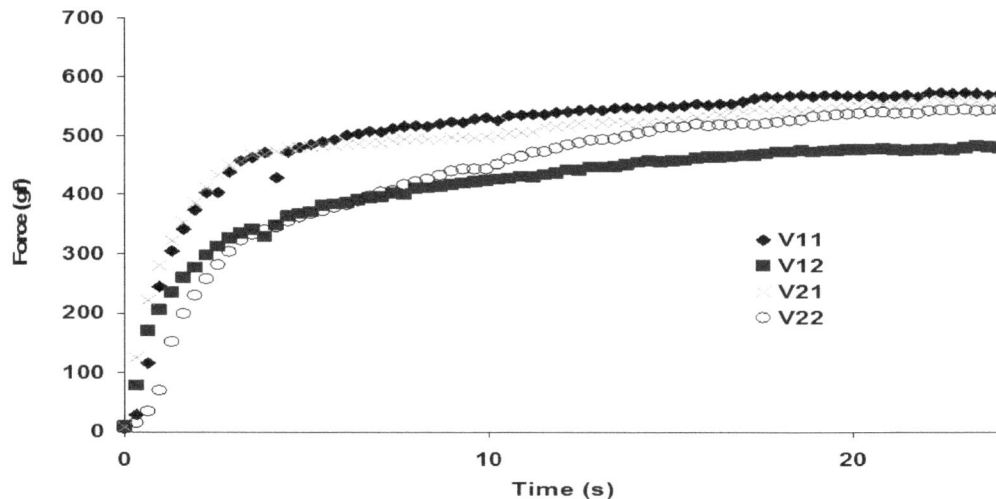

Figure 3. Force-time curve obtained using back extrusion method for brand V at 25°C in a repeated extrusion. (◆)V11: First sample, first extrusion; (■) V12: First sample, second extrusion; (×)V21: Second sample, first extrusion; (○)V22: Second sample, second extrusion.

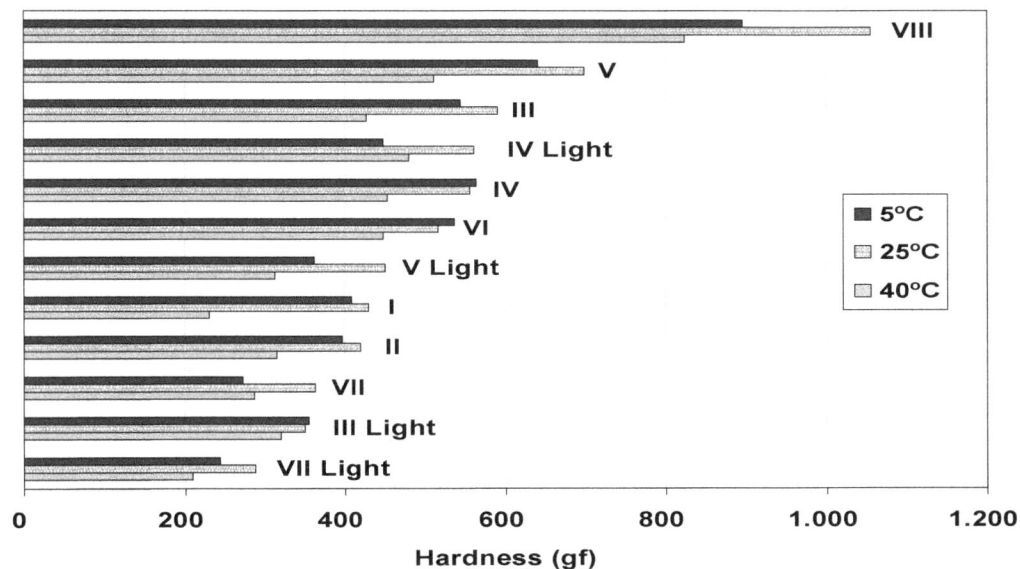

Figure 4. Hardness obtained from back extrusion test of mayonnaise samples at 5, 25 and 40°C.

reduction presented in second extrusion also indicates that compression force affect mayonnaise structure irreversibly by breaking it down.

The non-linear rheological properties of commercial mayonnaise were studied by Kaneda and Takahashi (2011). The stress growth behavior under constant shear flow was observed using a strain control-type rheometer. As the stress-strain curves of commercial mayonnaise showed non-linear behavior, the curves were analyzed with a polynomial model that is a power series of strain. Two types of commercial mayonnaise were investigated - traditional and low-fat mayonnaise. The second order

coefficient of the polynomial model revealed that the structure fracture behavior of these types of mayonnaise was very different. Because the fracture behavior of semisolid foodstuffs is considered to be an important mechanical property, the approach may be useful for quantitative estimation of the texture of such foods.

Mayonnaise rheological properties are related with formulation, process parameters and storage temperature. By performing rheological analysis in distinct temperatures it is possible to demonstrate such differences. Figure 4 and Table 4 present hardness values (medium) obtained at 5, 25 e 40°C using back

Table 4. Hardness obtained from back extrusion test of mayonnaise samples at 5, 25 and 40°C.

Sample/brand	5°C	25°C	40°C
I	407.22[a,b]	430.03[a,b]	229.99[a]
II	396.48[a,b]	419.20[a,b]	314.88[a,b]
III	544.04[b,c]	590.25[b,c]	427.06[b,c]
III Light	355.25[a,b]	349.65[a,b]	320.28[a,b]
IV	563.70[b,c]	556.25[b,c]	452.32[c]
IV Light	447.59[a,b,c]	561.35[b,c]	479.80[c]
V	640.37[c]	698.26[c]	510.02[c]
V Light	361.95[a,b]	450.27[a,b]	312.22[a,b]
VI	536.07[b,c]	515.78[a,b,c]	447.51[c]
VII	272.89[a]	362.96[a,b]	287.20[a]
VII Light	243.91[a]	287.67[a]	209.32[a]
VIII	894.75[d]	1054.83[d]	823.20[d]

[a,b,c] Same letter shows no significative difference at 5% level at Tukey-HSD Test.

extrusion method for all samples. It is possible to see the variation due to temperature and brand.

Laverse et al. (2012) studied the microstructure and quantification of fat in four types of mayonnaise. The dynamic-mechanical properties of the mayonnaise samples were also studied using a controlled-strain rotational rheometer. Four types of commercially produced mayonnaises, chosen to exhibit variability in terms of visible structure of fat, were used for the experiments. Appropriate quantitative three-dimensional parameters describing the fat structure were calculated. With regards to the microstructural and rheological relationship, results from the correlation carried out show that a correlation exists among some microstructural and rheological parameters of the mayonnaise samples. The results showed that X-ray microtomography technique was used by is a suitable technique for the microstructural analysis of fat as it does not only provide an accurate percentage volume of the fat present but can also determines its spatial distribution.

Marinescu et al. (2011) studied the application of beta-glucan prepared from spent brewer's yeast as a fat replacer in mayonnaise. Fat was partially substituted by P-glucan at levels of 25, 50 and 75%. The results indicated that all mayonnaises exhibited thixotropic shear thinning behavior under steady shear tests and were rheologically classified as more solid like gels. The mayonnaise with 50% beta-glucan showed higher storage stability than the other samples. It has been demonstrated that spent brewer's yeast beta-glucan can be used as a fat replacer in mayonnaise as well as an emulsion stabilizer.

Brand VIII presents the highest hardness values in the three temperatures. Brand VII Light presents the smallest hardness in the three temperatures. Brand VIII is the harder and brand VII Light is the softer in terms of texture.

A large number of ingredients called "texturizing agents" (also described as texturizers, thickeners, viscosity modifiers, bodying agents, gelling agents and stiffening agents) are available to help bring the texture of foods in to the range preferred by consumers (Bourne, 2002). A thickener is a chemical component or mixture of components that can impart long-term emulsion stability by thickening a food system, reducing the movement of the system and by forming viscous, ordered networks in the continuous phase to prevent oil separation (Ma and Boye, 2013). Brand VIII has higher hardness due to the starch (texturizing agents) in its formulation, possibly starch or modified starch that can be used combined with non-starch hydrocolloids to obtain products such as reduced-fat mayonnaise with desirable textural properties. Textural change can be obtained due to specific interactions between starch and emulsifiers and non-starch hydrocolloids in these products, and the starch and hydrocolloids combination can reduce the long and slimy texture of reduced-fat products compared to when hydrocolloids are used alone (Ma and Boye, 2013).

Nikzade et al. (2012) optimized mixture proportions of low cholesterol-low fat mayonnaise contained soy milk as an egg yolk substitute (10%) with different composition of xanthan gum, guar gum and mono- and diglycerides emulsifier (0 to 0.36% of each component) were determined by applying the simplex-centroid mixture design textural and rheological properties and sensory characteristics for effective formulation process. Results revealed that the best mixture was the formulation contained 6.7% mono- diglycerides, 36.7% guar gum and 56.7% xanthan gum. The xanthan gum was the component showing the highest effect on all the properties of mayonnaise samples. In addition, an increase of xanthan gum followed by guar gum caused greater values for the stability, heat stability, consistency coefficient, viscosity, firmness, adhesiveness, adhesive force and overall acceptance and lower value for flow behavior index.

Except for brands III Light, IV and VI all other brands exhibit a reduction in their hardness from 25 to 5°C. This can be justified by a small break down in emulsion with low temperatures. The bigger stability was showed at 25°C.

Brand IV do not show difference between its traditional and light form (VI and VI light, respectively) at 1% of significance level (Tukey test, LSD). It is the bestselling brand which means it can be used as a standard for hardness comparison since its texture is already well accept by consumers.

Most brands show biggest hardness values at 25°C which demonstrate more stability since this is a normal storage temperature before selling. Market offers mayonnaises with a big variety of textural characteristics.

Conclusion

The back extrusion method to determine rheological properties of mayonnaises showed sensible and very efficient for providing quantitative and qualitative characteristics. However, it is necessary a standardization of samples to get repeatability of measurements. Most results showed no differences between traditional and light mayonnaises texture, meaning the standardization and quality of the formulations.

REFERENCES

Bourne MC (2002) Food texture and viscosity: Concept and Measurement. London: Academic Press. P. 427.

Brusewitz GH, YU H (1996) Back extrusion for determining properties of mustard slurry. J. Food Eng. 27:259-265.

Chen J (2009). Food oral processing-A review. Food Hydrocolloids 23(1):1-25.

Depree JA, Savage GP (2001) Physical and flavour stability of mayonnaise. Trends. Food. Sci. Technol. 12(5/6):157-163.

Gallegos C, Berjano M (1992) Linear viscoelastic behavior of commercial and model mayonnaise. J. Rheol. 36(3):465-478.

Goshawk JA, Binding DM. (1998). Rheological phenomena occurring during the shearing flow of mayonnaise. J. Rheol. 42(6):1537-1553.

Gujral HS, Sodhi NS (2002) Back extrusion properties of wheat porridge (Dalia). J. Food Eng. 52:53–56.

Kaneda I, Takahashi S (2011) Stress Growth Behavior of Commercial Mayonnaise under Constant Shear Flow. Food Sci. Technol. Res. 17(4):381-384.

Laverse J, Mastromatteo M, Frisullo PDELI, Nobile MA (2012). X-ray microtomography to study the microstructure of mayonnaise. J. Food Eng. 108:225-231.

Ma Z. Boye JJ (2013). Advances in the design and production of reduced-fat and reduced-cholesterol salad dressing and mayonnaise: a review. Food Bioprocess Technol. 6:648-670.

Marinescu G, Stoicescu A, Patrascu L (2011) The preparation of mayonnaise containing spent brewer's yeast beta-glucan as a fat replacer. Roman. Biotechnol. Letters, 16:6017-6025.

Martin-Gonzalez MFS; Roach A, Harte F (2009) Rheological properties of corn oil emulsions stabilized by commercial micellar casein and high pressure homogenization. Lebensm.-Wiss. Technol. 42:307-311.

Nikzade V, Tehrani MM, Saadatmand-Tarzjan, M. (2012) Optimization of low-cholesterol-low-fat mayonnaise formulation: Effect of using soy milk and some stabilizer by a mixture design approach. Food Hydrocol. 28:344-352.

Paoletti F, Nardo N, Saleh A, Quaglia GB (1995) Back extrusion test on emulsions stabilized whey protein concentrates. Lebensm.-Wiss. Technol., 28:616-619.

Rosenthal AJ (1999). Food texture measurement and perception. Gaitherburg: Aspen Publishers,. P. 311.

Stokes JR, Telford JH. (2004) Measuring the yield behaviour of structured fluids. J. Non-Newtonian Fluid Mech. 124:137-146.

Szczesniak AS (2002). Texture is a sensory property. Food Qual. Prefer. 13:215–225.

Weenen H, Van Gemert LJ, Van Doorn JM, Dijksterhuis GB, DE Wijk RA (2003) Texture and mouthfeel of semisolids foods: commercial mayonnaises, dressings, custard dressings and warm sauces. J. Texture. Stud. 34:159-179.

16

Assessment on challenges and opportunities of goat farming system in Adami Tulu, Arsi Negelle and Fantale districts of Oromia Regional State, Ethiopia

Arse Gebeyehu*, Feyisa Hundessa, Gurmessa Umeta, Merga Muleta and Girma Debele

Adami Tulu Agricultural Research Center, P. O. Box 35, Batu, Ethiopia.

This study was conducted in Adami Tulu, Arsi Negelle and Fantale districts of Oromia Regional State, Ethiopia. The objective of this study was to assess and identify the challenges and opportunities of goat production under farmers' management system. For this study, 6 Peasant Associations (PA: is the smallest administrative unit in Ethiopian government structure), 2 from each district, were selected. From each PA, a group of 15 to 20 farmers were organized at each study site and different types of Participatory Rural Appraisal were conducted. The data collected were analyzed using PRA tool of pair-wise rankings and results expressed in simple descriptive statistics. Goat production system in these study areas were identified as mixed crop-livestock production systems in Arsi Negelle and Adami Tulu districts. Farmers in the Fantale administrative district practiced pastoral and agro-pastoral production systems. Goat production purposes were also identified as mainly to provide milk and meat for home consumption and cash income generation. The rankings of these purposes varied across study areas. Goat fattening is a recent appearing practice in these rural communities. Farmers in the Arsi Negelle and Adami Tulu districts practiced traditional fattening system mostly from July to September. A common operation in traditional fattening system was castration, which improves body condition. Farmers identified feed shortage from November to December, disease incidences, predatory attacks and water shortage as challenges to goat production. Farmers' major sources of income were mainly crop and animal production (goats, sheep and cattle). Animal feeds were mostly available from late May to October. Great oopportunities exists for goat production and productivity improvement in surveyed administrative districts, in which the environment was still conducive and animal productivity can be improved by improved management.

Key words: Goat fattening, Income sources, feed shortage.

INTRODUCTION

Ethiopia has diverse agro-ecological zones suitable for livestock production. Agricultural scenario in Ethiopia is characterized by the pastoralism in low land area, and mixed farming systems in mid and highland areas. Livestock have traditionally been an important component of the agricultural industry in Ethiopia. A 2011/2012 livestock census puts the goat population in Ethiopia at 22.6million (CSA, 2012) of which 32% goats were found in Oromia Regional State. Mid-rift-valley areas of Ethiopia are known to have a high population of sheep and goats. Majority of Ethiopian goat farmers are subsistence farmer. In East Shoa administrative zone, there are over 488.5 thousand goats and in West Arsi administrative zone, there are over 370 thousand goats. The recent data from CSA (2012) indicated that country's goat population growth rate is 1.1% with off-take rate of 35%.

*Corresponding author. E-mail: arse.gebeyehu@yahoo.com.

Abbreviation: PA, Peasant association.

In Ethiopia goat production accounts for 16.8% of total meat supply (Ameha, 2008) and 16.7% of milk consumed in the country (Tsedeke, 2007).

Ethiopian indigenous goats are genetically less productive as compared to temperate breeds (Mohammed et al., 2012). Indigenous goat breeds constitute over 95% of the small ruminant population of Africa and that of Ethiopia is 99.77% that are indigenous breeds (CSA, 2012). Although Ethiopia has tremendous amount of large and small ruminants their productivity has been constrained by complex challenges. Goats are owned by the majority of smallholder rural farmers for whom this resource is critical for nutrition and income. Goats are also an important and secure form of investment, which happens to be major farming activity on vast areas of natural grasslands in regions where crop production is impracticable (Tadelle and Workneh, 2007).

Chronic feed shortage during short rainfall and primitive type of feeding are considered major limiting factors to animal production in Ethiopia. Because of the above problems, Ethiopian livestock production systems are mostly characterized by very low reproduction and production performance managed by resource limited farmers for subsistence. To this augment, this low production and productivity inventory and appraisal of basic resources are the first step towards the over-all process of resources management. This is why emphasis is being increasingly paid on the detailed analysis of the existing and potential resources and formulation of scientific plans for their development, sustained production and utilization. The primary purpose of conducting this Participatory Rural Appraisal (PRA) was to assess and identify the production systems, with their challenges and opportunities for goat production in and around the study areas.

MATERIALS AND METHODS

Description of the study area, site and farmers selection

This study was conducted in six selected PAs of three administrative districts in 2010. The three administrative districts were Arsi Negelle district of west Arsi administrative zone and Adami Tulu and Fantale districts of East Shoa administrative zone of Oromia Regional State. Adami Tulu and Arsi Negelle districts are located at 160 and 210 km south of Addis Ababa, respectively and Fantale district is located on 200 km east of Addis Ababa. Arsi Negelle district is located at 7° 17' N to 7° 66' N and 38° 43' E to 38° 81' E, Adami Tulu district is located at 7° 60' N to 8° 03' N and 38° 51' E to 38° 69' E and Fantale district is located at 8° 70' N to 9° 13' N and 39° 75' E to 40° 04' E (Google Earth, 2012).

A two stage sampling technique was implemented. In the first stage, Arsi Negelle, Adami Tulu and Fentale districts were selected using purposive sampling and followed by identification of potential PAs (Hailu, 2008; Tesfaye, 2010; Assen and Aklilu, 2012). Goat population and accessibility of the areas were used as criteria in selecting districts and PAs. Accordingly, two PAs from each district were selected for this study. The selected PAs include O'itu and Abijata from Adami Tulu administrative district, Aliwoyo and Daka from Arsi Negelle administrative district and Gidara and Kobo from

Fantale administrative district. In the second stage, at each selected PA 15 to 20 farmers (Arse et al., 2010) were randomly selected for group discussion using systematic sampling procedure. Finally, a multi-disciplinary team composed of Nutritionist, Breeders, Productionist, Economist, Extensionist as well as Development Agents of the respective PA was established to conduct the survey using different PRA tools.

Data collection and analysis

Different PRA tools were employed to collect information on different aspects of goat production practices. The PRA tools used to generate information from participants were: Focused group discussion, Pair-wise ranking of preferences of income sources, goat production challenges, purposes of goat production, gender based activities to identify gender related activities, scheming seasonal calendar of feed shortage and availability. The data collected were analyzed using PRA tool of pair-wise rankings and results expressed in simple descriptive statistics.

RESULTS AND DISCUSSION

Goat production system

Farmers in this study area were practicing different type of production systems. Farmers in these study areas do not know about the breed type of their goat. The goat breed type in these study areas belongs to rift valley family and are called Arsi breed (Kassahun and Solomon, 2008). The Arsi breed goat is distributed throughout the Arsi, West-Arsi, Bale and parts of East-Shoa, South-Shoa and West Hararghe administrative zones, in altitude that range from 300 m lowland up to 4000 m above sea level, highland (Kassahun and Solomon, 2008). Solomon et al (2008) reported that the Arsi goats were kept in small flocks in mixed farming systems in the highlands as well as in agro-pastoral systems at lower altitudes. Majority of farmers rear their goats under predominantly free grazing systems. Tethered feeding was practiced around perennial crop growing areas. Some arable farmers provided crop residues, thinning of maize and sorghum, kitchen waste and chopped browse to their goats. The use of crop residues as feed source was also indicated by CSA (2012) and Assen and Aklilu (2012).

Arsi Negelle and Adami Tulu districts were more dependent on crop production. Farmers in Arsi Negelle and Adami Tulu districts were practicing mixed crop-livestock production system and in which goat production was the major share in livestock production. The results are in agreement with findings of Solomon et al. (2008) indicated that the goat production played important role in mixed crop-livestock production system. Pastoral and agro-pastoral farmers of Fantale administrative district rear goat in mixed livestock system. In this society, goat production was the dominant livestock, and income from it play important role in covering home consumption expenses. The results obtained in this present study are in agreement with the results reported by Markos (2006) and Solomon et al (2008).

Table 1. Purpose of goat rearing.

Production purposes	Arsi Negelle district		Adami Tulu district		Fantale district	
	Daka	Aliwoyo	Abijata	O'itu	Gidara	Kobo
Milk	1st	1st	1st	1st	3rd	2nd
Meat	3rd	5th	5th	5th	4th	5th
Cash income	2nd	2nd	2nd	2nd	1st	1st
Saving value	4th	4th	4th	4th	5th	4th
Wealth indicative	5th	3rd	3rd	3rd	2nd	3rd

Table 2. Major sources of income.

Income sources	Arsi Negelle district		Adami Tulu district		Fantale district	
	Daka	Aliwoyo	O'itu	Abijata	Gidara	Kobo
Crop production	2nd	1st	1st	1st	3rd	5th
Goat production	3rd	3rd	2nd	2nd	1st	1st
Sheep production	5th	5th	6th	5th	4th	2nd
Cattle production	1st	2nd	3rd	3rd	2nd	3rd
Donkey cart service	4th	4th	5th	4th	5th	6th
Horse cart service	9th	8th	8th	7th	9th	9th
Poultry production	6th	6th	4th	6th	8th	8th
Beekeeping	8th	7th	9th	8th	*	*
Camel production	*	*	*	*	6th	4th
Petty Trade	7th	*	7th	*	7th	7th

* = Unknown or does not exist in the area.

Purpose of goat farming and sources of income

The primary purpose of goat rearing was to sustain their livings. The pair-wise ranking of purpose of goat farming in both pastoral and agro-pastoral farming systems is presented in Table 1. In these study areas of pastoral and agro-pastoral farming system the primary purpose of keeping goat was to generate income, followed by milk and meat for home consumption (Table 1). In Arsi Negelle and Adami Tulu where mixed crop-livestock production farming system is dominant farming system, farmers reared goats to generate cash income followed by meat for consumption, savings and holiday festive and ceremonies. The purpose of goat production in all the three districts was almost similar. The present reports are in agreement with Muluken (2007) who reported about goat production system in Metama and Sekota administrative districts of Ethiopia. The result of this study also agreed with Addis (n.d) who reported that farmers in developing country like Ethiopia raise goats for meat and milk production. Endashaw (2007) and Tsegahun et al (2000) also reported that farmers of highlands in reared goat mainly for milk and meat for home consumption and for cash income generation. Tesfaye (2009) and Mekuriaw et al (2012) also reported that the farmers keep goat for cash income generation.

The major income sources ranked in Table 2 was adopted from original pair-wise comparison matrix of income sources in these study areas. Farmers in the mixed crop-livestock farming system ranked crop production as top income generator and followed by income generated from livestock productions. This report is in agreement with Endashaw (2007) in which farmers of Wonsho PA in Sidama administrative zone ranked income from crop production on top. But in the pastoral and agro-pastoral areas livestock production was given the highest priority than crop due to frequent crop failure by insufficient rain fall. According to extrapolated data from pair-wise ranking results, in Arsi Negelle and Adami Tulu districts crop production was the major source of income because these two districts practice mixed crop-livestock production system. Therefore, in mixed crop-livestock farming system crop production was the most reliable and it generates significantly high income than livestock production does. This study results also indicated that the pastoral and agro-pastoral farmers in Fantale administrative district ranked livestock production as the major income generating sector. Farmers in the two administrative districts namely, Arsi Negelle and Adami Tulu their income source preference varies between and within the administrative districts though they practice the same type of farming system. The farmers in

Table 3. Seasonal calendar of fattening and feed shortage.

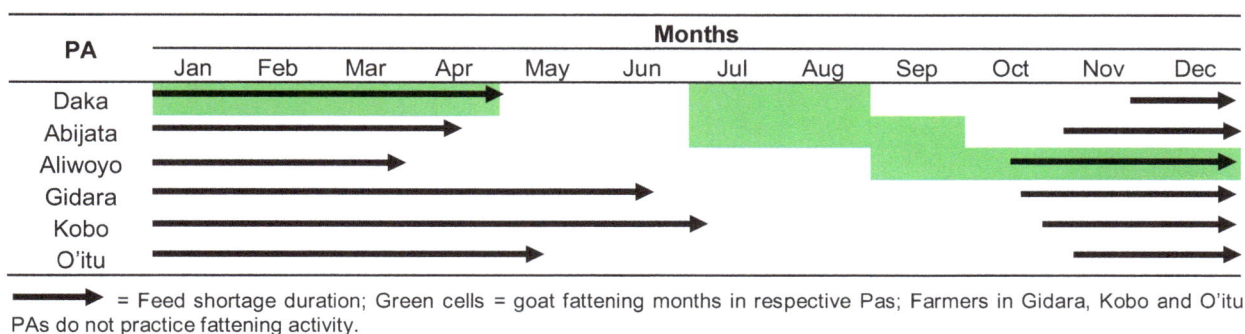

PA	Months											
	Jan	Feb	Mar	Apr	May	Jun	Jul	Aug	Sep	Oct	Nov	Dec
Daka												
Abijata												
Aliwoyo												
Gidara												
Kobo												
O'itu												

➤ = Feed shortage duration; Green cells = goat fattening months in respective Pas; Farmers in Gidara, Kobo and O'itu PAs do not practice fattening activity.

the two PAs of Fantale administrative district unanimously ranked goat production as the major source of their cash income (Table 2). This present findings are in agreement with Solomon et al. (2010) who reported that the primary objective of goat rearing are cash income, savings and meat for household consumption in pastoralist communities.

Seasonal calendar of feed availability and goat fattening activity

Farmers reported the severe feed shortage appearing in their area was recent years' phenomena. They reported that goat farming is being challenged by the currently appearing feed shortage. Farmers reported this recently that worsening feed shortage was due to untimely and erratic rainfall as a result of climate change. Feed shortage was the major goat production constraints in all study areas. Seasonal feed shortage was also reported by Solomon et al. (2010) in Gomma, Metema, Mieso and Fogera districts of Ethiopia. Farmers indicated that the worst feed shortage happens during dry season of the year. The major feed shortage months in the six PAs is given in Table 3. These months were from November to May. The degrees of severity of feed shortage vary across the administrative districts and even within a district. Due to problems mentioned in Table 3, the respondents reported that the goat production trends was declining and risking their livelihood. Farmers also indicated that the untimely and erratic type of rain fall caused by climate change is affecting shrubs and bushes growth adversely. They also reported that diseases were spreading more rapidly by this climate change. The length of array shows feed available season. As indicated in Table 3, farmers of Aliwoyo PA have longer feed shortage season than others.

Moderately improved fattening practices in Adami Tulu and Arsi Negelle were the recent development. This is in agreement with the report of Tsegahun et al. (2000) that indicated goats usually receive little to no supplementary feeding in the highland and semi-arid mixed farming

systems. The farmers in Fantale administrative district do not practice fattening of goat; they simply rear animal mainly for milk. They keep their wethers with other flock and sell when they need cash. But farmers around peri-urban and in nearby urban areas and farmers in the mixed crop-livestock production system practices seasonal fattening targeting big markets of the year and this is in agreement with CSA (2012) report. Farmers in Daka, Abijata and Aliwoyo PAs were from mixed crop-livestock production system area and they practice seasonal fattening. Gidara and Kobo PAs were from nomadic-pastoral area and they do not practice seasonal fattening. O'itu PA farmers were from mix crop-livestock production system but they were far away from main highway located in the raged swatch of rift valley.

But Kobo and O'itu PA farmers practice traditional type of fattening system in which they fatten their candidate wether targeting big holidays festive. Here the most important point was that farmers who fatten wether do not put especial attention on the wether aimed for selling except castrating; which they believe castration would enhance rate of weight gain. This traditional fattening system takes long time. Gidara's farmers do not practice any kind of fattening. The farmers were entirely pastoral and they practice pastoral production system. Whenever they need cash they sell any of wethers in nearby markets.

Major challenges to goat production

Climate change was affecting and challenging the life of farmers. The pasture production potential was declining because of climate change. The prevailing goat pro-duction challenges in these study areas were complex. Farmers in these study areas were practicing climate sensitive type of agriculture. Goats and other livestock were largely reared in extensive system and fully dependent on rain-fed pasture and this is in agreement with Solomon et al. (2010). The major challenges to goat production in the six selected study areas were severe feed shortage, high disease prevalence, high predatory,

Table 4. Major challenges to goat production.

Challenges	Arsi Negelle district		Adami Tulu district		Fantale district	
	Daka	Aliwoyo	Abijata	O'itu	Gidara	Kobo
Feed shortage	5th	2nd	1st	2nd	1st	3rd
Disease	1st	1st	3rd	1st	4th	1st
Predatory	4th	6th	6th	6th	6th	7th
Market	6th	5th	4th	3rd	2nd	4th
Breed	2nd	3rd	2nd	5th	5th	5th
Labor shortage	3rd	4th	5th	4th	3rd	6th
Water Shortage	7th	7th	*	*	*	2nd

Breed = Less productivity of breed (of goat); Market = Market problem, * = Does not exist.

poor market, genetically less productive breed, severe water shortage and high shortage of laborer. These findings are in agreement with reports by Beruk and Tafesse (2000), Fisseha et al. (2010) and Assen and Aklilu (2012). The participant farmers in group discussion ranked the major challenges in their area in relation to goat production (Table 4). The major goat diseases reported were anthrax, liver fluke, orf (disease like Foot-and-mouth disease), pneumonia and internal parasites.

Early kid mortality was also a serious problem of the farmers. The ranking of the challenges is made from pair-wise comparison of all. For Arsi Negelle district farmers the first ranked challenge is goat disease prevalence. On subsequent ranking of the challenges farmers of the same district of Arsi Negelle but at two different site (Daka and Aliwoyo) fall apart are having site specific challenges. In the same way farmers of Adami Tulu district and in the two study site ranked their production challenges (Table 4). In the same district of Adami Tulu, in Abijata PA feed shortage was the first ranked challenge but in o'itu disease incidence was ranked first. Farmers of Gidara and Kobo peasant associations ranked the goat production challenges differently though they were in the same district of Fantale.

Challenges and opportunities of goat keeping

Goat production can significantly benefit the owners or producers since they can be used for milk and meat for home consumption. In addition to production for milk and meat for home consumption goat was also used as major source of cash income. It was also used as insurance for crop production. In pastoral and agro-pastoral communities, goat were given higher value than other type of livestock because goat as are highly adaptive to harsh environment in pastoral areas. Farmers in all the selected study areas were willing and eager to used modern technologies because they were quite sure that improved management will improve production and productivity of their animals. Improved environment improves the genetic potential with the result of improved productivity. Goat have great role in reducing unnecessary expansion

of bushes and shrub by browsing on it and in turn maintaining the equilibrium of the nature in pasture lands. At federal government level, there is pastoral standing committee which works on improving life pastoral communities. There are also many Non-Governmental Organizations (NGO) that are working on improving living and livelihood through improving the production and productivity of their animals. Government is also working to solve the challenges.

Conclusion

Farmers were working on primitive type of production system. As a result, the production and productivity as well as income from sale of the animals were very low. Farmers in Arsi Negelle and Adami Tulu districts were practicing mixed crop-livestock production system. Arsi Negelle and Adami Tulu districts were more dependent on crop production. Farmers in Fantale administrative district practice pastoral and agro-pastoral production system. The main purposes of goat production were for milk, meat and cash generation. All the farmers castrate their billy assuming castration per se improves body condition. Only few farmer closer to urban areas practice moderately modern goat fattening system with special supplementation of conventional feeds. The farmers in the Fantale administrative districts do not have good experience of fattening before marketing their wethers. Moderately improved fattening practices in the two study areas were the recent development. Major challenges in goat production were feed shortage, disease prevalence and etc. The major income sources in the area were income from cattle production, crop production, goat production and etc. Severe feed shortage happens during dry season in the months from Novemeber to May.

REFERENCES

Addis A (n.d). Production objectives and market forces. Retrieved November 9, 2012, from http://www.ilri.cgiar.org/InfoServ/Webpub/fulldocs/AnGenResCD/docs/X5541E/X5541E04.HTM

Ameha S (2008). Sheep and Goat Meat Characteristics and Quality. In: Sheep and Goat Production Handbook for Ethiopia. Ed Alemu Yami and R.C. Merkel. http://www.esgpip.org/handbook/Handbook_PDF/Chapter%2012_%2 0Sheep%20and%20Goat%20meat%20characteristics%20and%20qu ality.pdf

Arse G, Tesfaye K, Sebsibe Z, Tekalign G, Gurmessa U, Tesfaye L, Feyisa H (2010). Participatory rural appraisal investigation on beekeeping in Arsi Negelle and Shashemene districts of West Arsi zone of Oromia, *Ethiopia*. Livestock. Res. Rural. Dev. 22, Article #120. http://www.lrrd.org/lrrd22/7/gebe22120.htm

Assen E, Aklilu H (2012). Sheep and goat production and utilization in different agro-ecological zones in Tigray, *Ethiopia*. Livestock. Res. Rural Develop. *p.24, Article #16*. Retrieved November 9, 2012, from http://www.lrrd.org/lrrd24/1/asse24016.htm

Beruk Y, Tafesse M (2000). Pastoralism and Agro-pastoralism: past and present. In Pastoralism and Agro-pastoralism which way forward? Proceeding 8th annual conference of the Ethiopian Society of Animal Production (ESAP) held in Addis Ababa, Ethiopia. August 24-26, 2000

CSA (2012) Central Statistical Agency of the Federal Democratic Republic Of Ethiopia. Agricultural Sample Survey of 2011/12 (2004 E.C). Volume II. Report on Livestock and Livestock Characteristics (Private Peasant Holdings), Central Statistical Agency, Addis Ababa, Ethiopia

Endashaw A (2007). Assessment of Production and Marketing System of Goats in Dale District, Sidama Zone. M.Sc. thesis, University of Hawassa, Department of Animal Production and Range Sciences, Hawassa, Ethiopia

Fisseha M, Azage T, Tadelle D (2010). Indigenous chicken production and marketing systems in Ethiopia: Characteristics and opportunities for market-oriented development. IPMS (Improving Productivity and Market Success) of Ethiopian Farmers Project Working ILRI Nairobi, Kenya. p. 24.

Google Earth (2012). Google Earth version 6.1.2 Geographical map information tracking system. US Department of state geographer. Downloaded from: http://www.google.com/earth/download/ge/agree.html

Hailu BA (2008). Adoption of improved *Tef* and Wheat production technologies in crop-livestock mixed systems in northern and western shewa zone of Ethiopia. PhD Dissertation, University of Pretoria, Pretoria, SA

Kassahun A, Solomon A (2008). Breeds of sheep and goats. In: Sheep and Goat Production Handbook for Ethiopia. Ed Alemu Yami and R.C. Merkel. Addis Ababa, Ethiopia.

Markos T (2006). Productivity and health of indigenous sheep breeds and crossbreds in the central Ethiopian Highlands. Doctoral Thesis. Swedish university of agricultural science, Department of Animal breeding and Genetics. Uppsala, Sweden

Mekuriaw S, Mekuriaw Z, Taye M, Yitayew A, Assefa H, Haile A (2012). Traditional management system and farmers' perception on local sheep breeds (Washera and Farta) and their crosses in Amhara Region, Ethiopia. *Livestock* Res. Rural Develop. p.24, Article #4. *Retrieved November 9, 2012,* from http://www.lrrd.org/lrrd24/1/meku24004.htm

Mohammed B, Aynalem H, Hailu D, Tesfaye AT (2012). Estimates of genetic phenotypic parameters of milk traits in Arsi-Bale goat in Ethiopia. Livestock.Res.Rural Develop. p.24, Article # 98. Retrieved November 7, 2012, from http://www.lrrd.org/lrrd24/6/bedh2408.htm

Muluken A (2007). Goat Husbandry Practices and Productive Performance in Sekota District of Amhara Region. M.Sc. thesis, Haramaya University, Haramaya, Ethiopia

Solomon A, Girma A, Kassahun A (2008). Sheep and Goat production systems in Ethiopia. In: Sheep and Goat Production Handbook for Ethiopia. Ed Alemu Yami and R.C. Merkel. Addis Ababa, Ethiopia

Solomon G, Azage T, Berhanu G, Hoekstra D (2010). Sheep and goat production and marketing systems in Ethiopia: Characteristics and strategies for improvement. Working Paper No. 23, International Livestock Research Institute http://cgspace.cgiar.org/bitstream/handle/10568/2238/IPMS_Working _Paper_23.pdf?sequence=1

Tadelle D, Workneh T (2007). Ethiopia goat production: A case study of CD-ROM encyclopedia use. ILRI

Tesfaye KB (2010). Assessment of On-Farm Breeding Practices and Estimation of Genetic and Phenotypic Parameters for Reproductive and Survival Traits in Indigenous Arsibale Goats. MSc Thesis, Haramaya University, Haramaya, Ethiopia

Tesfaye T (2009). Characterization of goat production systems and on-farm evaluation of the growth performance of grazing goats supplemented with different protein sources in Metema Woreda. M.Sc. thesis, Haramaya University, Haramaya, Ethiopia

Tsedeke KK (2007). Production and Marketing Systems of Sheep and Goats in Alaba, Southern Ethiopia. MSc Thesis submitted to University of Hawassa, Hawassa on April 2007.

Tsegahun A, Lemma S, Sebsbie A, Mekoya A, Sileshi Z (2000). National goat research strategy in Ethiopia. In: Merkel RC, Abebe G, Goetsch AL (eds.). The Opportunities and Challenges of Enhancing Goat Production in East Africa. Proceedings of a conference held at Debub University, Awassa, Ethiopia from November 10 to 12, 2000.

17

Annual and seasonal variation in nutritive quality and ruminal fermentation patterns of diets in steers grazing native rangelands

Manuel Murillo Ortiz, Osvaldo Reyes Estrada, Esperanza Herrera Torres, José H. Martínez Guerrero and Guadalupe M. Villareal Rodriguez

Faculty of Veterinary Medicine, Juárez University of the State of Durango, Castaña 106, Fraccionamiento: Nogales, CP: 34162, Durango, Dgo, México.

The objective of the study was to evaluate the annual and seasonal changes in the nutritive quality and ruminal fermentation patterns of diet consumed by grazing cattle. Diet samples were collected with four esophageal cannulated steers and four ruminally cannulated steers were used to evaluate the ruminal fermentation patterns. Data were analyzed with a repeated measurements design. No year x season interactions were observed for nutritive quality and ruminal fermentation ($p > 0.05$). However, the crude protein (CP), *in vitro* digestibility dry matter (IVDMD), Calcium (Ca), metabolizable energy (ME), ammonia nitrogen (NH_3N), volatile fatty acids (VFA) and propionate were higher in 2008 and wet season and lowest in 2006 and dry season ($p < 0.05$). Results confirm that the nutritive quality and ruminal fermentation patterns of diet selected for grazing cattle were affected by year and season and that protein an energy supplementation is necessary to improve productive performance of cattle under these management conditions.

Key words: Grazing cattle, diets, nutritive quality, ruminal fermentation.

INTRODUCTION

Native rangelands are the main forage resources for beef cattle production in north Mexico. However, rangelands in semiarid environments tend to vary greatly in quality and quantity, which subsequently affects diet composition and selectivity of grazing cattle (Obeidat et al., 2002). Some studies report that, as a result of drastic climate change, animals in the northern region of México have periods of 90 to 100 days of favorable grazing conditions and if the number of days is reduced, the survival of these animals may be in jeopardy (Navarro et al., 2002). Cattle grazing native rangelands may require nutrient supplementation to optimize their performance. Evaluation of nutritive

quality of the diet selected by grazing cattle across seasons is essential to make strategic supplementation management practices. However, these evaluations may be complemented with additional studies to more precisely establish dietary supplementation needs. Although the nutritive quality of the diet in grazing cattle is widely known, there is very little information about ruminal fermentation of the diet consumed by grazing cattle in native rangelands. pH of ruminal contents can influence fiber digestion (Mehrez et al., 1977), microbial protein synthesis (Russell et al., 1992) and finally, post-ruminal supply of energy and protein for the grazing cattle. Consequently, understanding the impacts of advancing season on nutrient supply, will aid in developing improved supplementation strategies. Therefore, information concerning seasonal changes in nutritive quality diet when coupled with estimates of ruminal fermentation provides a foundation for supplementation practices and sound nutritional programs. We hypothesized that quality nutritive and ruminal fermentation of diet consumed by grazing cattle are affected by seasonal changes. Objectives for this study were to evaluate diets consumed by grazing cattle across seasons for four years on nutritive quality and ruminal fermentation.

MATERIALS AND METHODS

Study area

The study was carried out over four years (2005, 2006, 2008, 2009) in a medium grassland located at the east of the city of Durango, Mexico (24° 22' N, 104° 32' W), at an altitude of about 1938 m above sea, which has a dry temperate (BS$_1$k) climate with average annual temperature and rainfall of 17.5°C and 450 mm, respectively. The study area covers 2,000 ha (6 ha/AU) with an average of forage biomass of 1,796 kg DM ha^{-1}. During the four years of the study, the vegetation cover was estimated using minimum area sampling with nested points (Bonham, 1989). Dominant grass species included: *Melinis repens* Willd (rose natal grass), *Chloris virgata* (feather fingergrass), *Bouteloua gracilis* (blue grama), *Aristida adscensionis* (six weeks threeawn) and *Andropogon barbinodis* (cane bluestem); bushes: *Acacia tortuosa* (poponax), *Prosopis juliflora* (mezquite), *Opuntia spp* (prickly pears and chollas), *Mimosa biuncifera* (cat claw); plus a wide variety of annual herbs. The pasture had not been grazed by previous four years and forage availability was never limiting during any of the sampling periods.

Animals and experimental periods

During the four years of study, twelve sampling periods, each 6 d long were conducted: January, February, March, April, May and June were considered to be dry season and July, August, September, October, November and December as wet season. During each sampling period, four steers Angus with esophageal cannulae of 350 ± 3 kg BW initial and 18 months age and four steer Angus cannulated of the rumen of 406 ± 5 kg BW initial and 20 months age were allowed to graze freely the study pasture. Surgery was performed on the steers according to procedures approved by the University of Durango Laboratory Care Advisory Committee. At end of the first two years study (2005, 2006), the steers with esophageal cannulae were replaced by others of same breed and body weight.

Nutritive quality

Diet samples were collected with the esophageal cannulated steers, during the first 4 days of each sampling period. Collections were made at 07. 00 h when steers were grazing most intensely. Steers were fitted with screen wire bottom collection bags and allowed to graze for 30 to 45 min periods. The samples from each steer were drained through 40 mm screen to remove saliva. Subsequently, samples (300 g) were thawed and pooled across the

4 d collection period for each animal (Holechek et al., 1982). The samples dried at 60°C for 48 h and later ground through a 1 mm screen in a Wiley mill. The esophageal samples were used for the determination of nutritive quality and mineral contents. Dry matter (DM), CP, (AOAC, 2005), Neutral Detergent Fiber (NDF) (Van Soest et al., 1991) and *in vitro* dry matter digestibility (ANKOM, 2008) were determined. We estimated ME content with the formulas used by Waterman et al. (2007): digestible energy (DE; Mcal/kg) = [0.039 x (OMD %)] – 0.10; ME (Mcal(kg) = DE (Mcal/kg) x 0.82. Where: OMD is the organic matter degradability obtained after 48 h incubation in the rumen. A subsample was incinerated in a muffle oven at 600°C during 5 h. The ashes obtained were digested in a solution HCl-HNO$_3$ and concentrations of Ca was determined by atomic absorption spectrophotometry; whereas, P content was quantified by ultraviolet-visible spectrophotometry (Cherney, 2000).

Ruminal fermentation

At 1200 h on days 5 and 6 of each sampling, ruminal liquid samples were extracted from the rumen of cannulated steers. Approximately 100 ml of whole ruminal liquid was extracted from the middle and ventral rumen each steers and the pH measured immediately with a combination electrode. The collected ruminal content was strained through four layers of cheesecloth and divided into two subsamples. The first subsample (10 ml) was acidified with 0.3 ml of 50% H$_2$SO$_4$ and frozen immediately at -40°C. After sample ruminal were thawed at room temperature and centrifuged at 10,000 x g for 10 min. Supernatant fluid was analyzed for NH$_3$N by phenol-hypoclorite procedure (Broderick and Kang, 1980) using ultraviolet spectrophotometer (Spectronic Genesys 2GP); the second subsample, (10 ml) was acidified with 2.5 ml of 25% metaphosporic acid and frozen at -40°C for later VFA analysis by gas chromatography (Autosystem XL, Perkin Elmer) (Galyean, 1997).

Statistical analyses

Data were analyzed as a repeated measurements design using MIXED procedure of SAS (2003). The model included effects for year, season and their interactions. Month was included as repeated effect, and animal within year x season was the subject for analysis. The variation between animals was specified by the RANDOM statement as animal within year x season. Individual animal was the experimental unit in all analyses. Least squares means and standard error of mean (SEM) were calculated and statistically separated by the PDIFF options of SAS (2003). Autoregressive Order 1 was used as the covariance structure, because it was better fitting structure, based on comparison of covariance structures with Akaike and Bayesian information criterions (Littell et al., 2006).

RESULTS

Climatic conditions

In this study, seasonal rainfall and mean air temperature that were recorded in the land meteorological stations located in study area are shown for reference in Table 1. Rainfall in 2008 was 793.2 mm and 2006 was drier than normal with 238.0 mm.

Table 1. Monthly mean temperature (oC) and accumulated rainfall (RF; mm) registered in the study area.

Month	2005		2006		2008		2009	
	T (°C)	RF (mm)	T (°C)	RF (mm)	T (°C)	RF (mm)	T (°C)	RF (mm)
January	13.0	22.9	13.3	1.6	9.3	0.8	13.1	0.0
February	16.2	3.1	13.7	1.3	12.0	0.6	14.7	0.0
March	17.9	2.2	15.1	0.7	12.8	0.0	16.8	0.4
April	22.5	0.2	20.9	0.0	18.3	1.4	18.8	0.0
May	23.0	18.8	22.3	0.5	20.4	3.2	21.3	12.8
June	22.2	67.6	25.2	5.4	24.1	56.4	22.0	51.8
July	21.3	68.1	22.2	96.8	21.3	273.0	21.6	64.8
August	20.2	161.1	20.8	66.4	20.5	291.0	20.8	116.4
September	19.9	47.5	21.0	61.0	18.3	150.0	18.6	254.3
October	18.2	48.1	20.8	2.1	15.6	16.8	17.1	33.4
November	14.9	0.1	15.5	2.2	11.7	0.0	12.5	5.2
December	11.6	15.4	13.1	0.0	9.9	0.0	10.7	6.2
Total	18.4	455.1	18.6	238.0	16.1	793.2	17.3	545.3

Table 2. Nutritive quality of the diet consumed by grazing steers.

Year (Y)	CP% DM	NDF	DIVMS	ME (Mcal kg^{-1})	Ca (g Kg DM^{-1})	P
2005	8.9^c	75.6^b	64.2^c	1.5^c	5.2^c	1.2^c
2006	6.3^d	78.5^a	60.1^d	1.2^d	4.8^d	0.91^d
2008	12.7^a	69.9^d	70.1^a	2.3^a	6.8^a	1.6^a
2009	9.6^b	70.9^c	67.2^b	1.9^b	5.5^b	1.4^b
SEM	2.1	2.2	1.1	1.4	1.1	0.42
Season (S)						
Wet	12.1^a	68.2^b	66.1^a	2.2^a	6.2^a	1.3^a
Dry	5.8^b	74.3^a	57.3^b	1.9^b	4.7^b	0.97^b
SEM	1.3	1.8	2.4	1.2	1.6	0.58
Effects	P<	P<	P<	P<	P<	P<
Y	**	**	**	**	**	**
S	**	**	**	**	**	**
Y*S	NS	NS	NS	NS	NS	NS

[abcd]Means within a column with different superscripts differ **($p < 0.01$). SEM: Standard Error Mean.

Nutritive quality

No year x season interactions ($p > 0.05$) were observed for CP, NDF, IVDMD and ME values (Table 2). However, the CP, NDF, IVDMD and ME, values were different among years ($p < 0.01$) and seasons ($p < 0.01$). The CP, IVDMD and ME, content, were higher in 2008 and lowest in 2005, 2006 and 2009 ($p < 0.01$). The CP, IVDMD and ME, values were greater in wet season as compared to dry season ($p < 0.01$); while NDF content was higher in dry season as compared to wet season ($p < 0.01$). Both year and season had a significant ($p < 0.01$) effect on the

contents of Ca and P. There was no interaction of years and seasons ($p > 0.5$). The Ca and P contents were greater in 2008 versus 2005, 2006 and 2009 ($p < 0.01$); while Ca and P contents was higher in wet season as compared to dry season ($p < 0.01$).

Ruminal fermentation

The ruminal fermentation patterns are shown in Table 3. The pH value was lower in wet season as compared to dry season ($p < 0.05$). The ruminal NH_3N concentration

Table 3. Ruminal fermentation patterns of the diet consumed by grazing steers pH NH3N VFA Acetate Propionate Butyrate mg dL-1 mM L-1 mol 100 moles-1.

Year (Y)						
2005	6.8[a]	5.2[c]	91.3[c]	66.6[b]	12.1[c]	7.8[a]
2006	6.7[a]	4.0[d]	87.4[d]	68.0[a]	11.5[d]	8.2[a]
2008	6.5[a]	12.7[a]	103.1[a]	61.8[d]	18.3[a]	5.3[c]
2009	6.7[a]	10.9[b]	98.7[b]	63.6[c]	15.3[b]	6.0[b]
SEM	0.1	0.5[8]	2.6	1.5	1.1	2.7
Season (S)						
Wet	6.4[b]	11.9[a]	101.0[a]	62.5[b]	15.8[a]	4.6[b]
Dry	6.7[a]	4.7[b]	89.3[b]	65.7[a]	12.8[b]	6.8[a]
SEM	0.72	0.83	2.6	1.3	2.6	2.2
Effects	P<	P<	P<	P<	P<	P<
Y	NS	**	**	**	**	**
S	*	**	**	**	**	**
Y*S	NS	NS	NS	NS	NS	NS

[abcd]Means within a column with different superscripts differ **($p < 0.01$). SEM: Standard Error Mean.

was greater in 2008 versus 2005, 2006 and 2009 ($p < 0.01$); whereas NH_3N concentration was higher in wet season ($p < 0.01$). Volatile fatty acids concentrations were affected by years and seasons ($p < 0.05$). Both year and season had effect ($p < 0.01$) on the acetate, propionate and butyrate ruminal concentrations. Acetate concentration was greater in 2006 than 2005, 2008 and 2009 ($p < 0.01$) and lowest during wet season. The propionate concentration was greater in 2008 than 2005, 2006 and 2009 ($p < 0.01$). The highest propionate concentration was observed in wet season and the lowest in dry season ($p < 0.01$). Butyrate concentration was greater in 2006 than 2005, 2008 and 2009 ($p < 0.01$). The highest butyrate concentration was obtained in dry season and the lowest in wet season ($p < 0.01$).

DISCUSSION

Nutritive quality

The differences observed in chemical composition of diets may be induced by registered rainfall in the four years of study (Cline et al., 2009). Seasonal variations in diet quality are attributed to corresponding variations in the chemical composition of forages utilized by grazing ruminants (Van Soest, 1994). During the wet season, diet consumed by grazing cattle normally meeting nutritional requirements because plants are actively growing and have access to succulent leafy green forage (Minson, 1990). An 8% CP level is considered as an adequate forage quality for grazing ruminant (NRC, 2000). The low CP value found in the present study in 2006, and dry season is insufficient for meeting the minimum

requirements of degradable protein for grazing ruminants (Kearl, 1982). Therefore, except for 2006 and dry season, the CP content can be considered to be adequate nutritive quality for grazing cattle. Nevertheless, a minimum of 12% CP is required for growing and finishing cattle (NRC, 2000). In the present study, all the values of CP among years and seasons, resulted in CP below this recommended value. According to Paterson et al. (1996) feedstuffs with CP content lower than 7% require a supplementation of nitrogen to improve their intake and digestion by the ruminant. The NDF content was 10.9% higher in 2006 than in 2008, while IVDMD was 16.6% lower in 2006 as compared to 2008. Increasing fiber content across the growing season has been observed by Johnson et al. (1998). In addition, similar responses have been attributed to increased plant maturity by other researchers (Brokaw et al., 2001).

Normally, advancing maturity is associated with increased cell wall constituents (Van Soest, 1982). The value obtained of ME in 2005, 2006, 2009 and dry season indicated that the energy requirements for maintenance of growing beef cattle (2.0 Mcal/Kg DM; NRC, 2000) would not be satisfied. Gutiérrez (1982), using similar methods of sample collection as employed in the present study, found that during the wet season increased CP, IVDMD and ME compared with dry season and attributing the differences to the phenology of rangelands plants. With the exception of P, the diet consumed for cattle throughout the years and seasons had appropriate amounts of Ca to meet requirements of beef cattle grazing native rangelands (2.3 g of P/ Kg DM and 4.6 g of Ca/Kg DM for beef cattle; McDowell (2003). Under conditions similar to those of our study, Arthington and Swenson (2004) found differences between seasons in contents of Ca and P. According to McDowell (1997) and Haenlein and Ramirez (2007) these differences in the mineral content of diets may be attributed to the interaction of a number of factors including soil, plant species, yield, pasture management, climate (temperature and rainfall) and stages of maturity. A sizable amount on the total Ca is associated with the NDF fraction (Van Soest, 1994). The association of certain minerals with fiber or other insoluble plant components could also decrease the rate and extent of mineral release from forages in the ruminant gastrointestinal tract (Van Soest, 1992).

Ruminal fermentation

Differences in ruminal pH were observed among season and similar pH values have been reported in grazing cattle (Gunter et al., 1993). The pH values below 6.2 can reduce ruminal digestion fiber (Orskov, 1982). Therefore, the ruminal pH observed during all years and seasons in this study, can be regarded as appropriate for fiber digestion and cellulolytic bacterial growth. Similar results to this study with respect to ruminal NH_3N concentrations in grazing cattle were reported by Choat et al. (2003). Other researchers have observed decreases in ruminal NH_3N concentrations when diet CP content decreases with advancing rangelands maturity (McCracken, 1993; Park et al., 1994). The suggested concentration of ruminal NH_3N for microbial growth is 5 mg/dl (Satter and Slyter, 1974) and 1 to 2 mg/dl for degradation fiber (Petersen, 1987). In this study, with the exception of 2006 and dry season, the diet consumed for cattle allowed appropriate concentrations of NH_3N to meet requirements of microorganisms that inhabiting in the rumen. These findings for acetate, propionate and VFA are in general agreement with other reports for grazing cattle (Funk et al., 1987; Krysl et al., 1987). Decreases in total ruminal VFA concentrations with advancing of the season have been reported by Adams et al. (1987).

Acetate is considered to be reflective of cell wall fermentation and increased acetate levels are normally associated with declining forage quality (Branine and Galyean, 1994); while the propionate concentration is associated with soluble carbohydrate ruminal fermentation during periods of active rangelands growth (Van Soest, 1994). Acetate is a necessary component for the formation of milk fat, whereas propionate is used for glucose production through hepatic gluconeogenesis and is needed for synthesis of milk lactose (Bergman, 1990). In this study, a low molar concentration of acetate and high molar concentration of propionate observed in wet season suggested the supply of glucogenic precursor in the rumen for glucose production in the grazing cattle. McCollum et al. (1985) reported somewhat low the butyrate concentration in steers grazing during the wet season than in this study. However, higher butyrate levels generally are associated with ruminant consuming

actively growing rangelands (Langlands and Sanson 1976).

Conclusion

Our results allow us to conclude that nutritive quality and ruminal fermentation of diet selected for grazing cattle were affected by annual and seasonal weather changes. Rainfall and temperature influenced nutritive quality and ruminal fermentation patterns. The drought induced by low rainfall, results in decreased diet nutritive quality because of the decrease of CP and increase of NDF. These changes were accompanied by decrease in the ammonia nitrogen and volatile fatty acids concentrations. A significant decreased in nutritive quality and ruminal fermentation patterns of diet consumed cattle were observed during the dry season. Therefore, the protein and energy supplementation as well as P might be beneficial for grazing cattle during this season to improve their productive and reproductive performance since these nutrients are related to the secretion of reproductive hormones (LTH, Leptin) and metabolites (glucose, urea-N, non-esterified fatty acids).

ACKNOWLEDGMENTS

This investigation couldn't have been possible without funding from FOMIX-CONACYT (Projects: DGO-2002-CO1-2522; DGO-2007-CO1-66559) and the collaboration of the academic group: Sustainable Production of Beef and Dairy Cattle (FMVZ-UJED).

REFERENCES

Adams DC, Cohran RC, Currie PO (1987). Forage maturity effects on rumen fermentation, fluid flow and intake in grazing steers. J. Range. Manage. 40(6):404-408.
ANKOM (2008). Procedures for fiber and *in vitro* analysis.
AOAC (2005). Official Methods of Analysis, AOAC International. 18th Edn., Association of Official Analytical Chemist, Washington DC., USA., pp. 212-220.
Arthington JD, Swenson CK (2004). Effects of trace mineral source and feeding method on the productivity of grazing Braford cows. Prof. Anim. Sci. 20(2):155-161.
Bergman EN (1990). Energy contribution of volatile fatty acids from the gastrointestinal tract in various species. Physiol. Rev. 70(2):567-590.
Bonham CHD (1989). Measurements for terrestrial vegetation, John Wiley and Son, P. 91.
Branine ME, Galyean ML (1990). Influence of grain and monensin supplementation on ruminal fermentation, intake digesta kinetics and incidence and severity of frothy bloat in steers grazing winter wheat pastures. J. Anim. Sci. 68(4):1139-1150.
Broderick GA, Kang JH (1980). Automated simultaneous determinations of ammonia and total amino acids in ruminal fluid and *in vitro* media. J. Dairy Sci. 63(1):64-45.
Brokaw L, Hess BW, Rule DC (2001). Supplemental soybean oil or corn for beef heifers grazing summers pasture: Effects on forage intake, ruminal fermentation and site and extent of digestion. J. Anim. Sci. 79(8):2704-2712.
Cherney DJR (2000). Characterization of Forages by Chemical

Analysis. In: Axford RFE, Omed HM (eds) Forage Evaluation in Ruminant Nutrition, Wallingford: CAB Publishing. pp. 275-300.

Choat WT, Krehbiel CR, Duff GC, Kirksey RE, Lauriault LM, Rivera JD (2003). Influence of grazing dormant native range or winter pasture on subsequent finishing cattle performance, carcass characteristics and ruminal metabolism. J. Anim. Sci. 81(12):3191-3201.

Cline HJ, Neville BW, Lardy GP, Caton JS (2009). Influence of advancing season on dietary composition, intake, site of digestion and microbial efficiency in beef steers grazing a native range in western North Dakota. J. Anim. Sci. 87(1):375-383.

Funk MA, Galyean ML, Branine ME, Krysl LJ (1987). Steers grazing blue grama rangeland and throughout the growing season. I. Dietary composition, intake, digesta kinetic and ruminal fermentation. J. Anim. Sci. 65(5):1342-1348.

Galyean ML (1997). Laboratory Procedures in Animal Nutrition Research, Texas Tech University, 77 pp. Available from: http://www.app.depts.ttu.edu (Accessed 17 June, 2012).

Gunter SA, McCollum FT, Gillen RL, Krysl LJ (1993). Forage intake and digestion by cattle grazing midgrass praire rangelands or sideoats grama/sweetclover pasture. J. Anim. Sci. 71(12):3432-3441.

Gutiérrez JL (1982). Nutritive value of diets selected by grazing cattle in Northwest Chihuahua. PhD. Thesis, New Mexico State University. USA.

Haenlein GJW, Ramirez RG (2007). Potential mineral deficiencies on arid rangelands for small ruminants with special reference to Mexico. Small Rumin. Res. 68(1):35-41.

Holechek JL, Vavra M, Pieper RD (1982). Methods for determining the nutritive quality of range management diets: a review. J. Anim. Sci. 54(2):363-376.

Johnson JA, Caton JS, Poland W, Kirby DR, Dhuyvetter DV (1998). Influence of season in dietary composition. Intake and digestion by beef steers grazing mixed-grass prairie in the northern Great Plains. J. Anim. Sci. 76(6):1682-1690.

Kearl LC (1982). Nutrient requirements of ruminants in developing countries. Agricola Experimental Station, Utha State University, Logan, Utha, USA. P. 11.

Krysl LJ, Galyean ML, Judkins MB, Branine ME, Estell RE (1987). Digestive physiology of steers grazing fertilized and nonfertilized blue grama rangeland. J. Range Manage. 40(6):493-501.

Langlands JP, Sanson J (1976). Factors affecting the nutritive value of the diet and composition of rumen fluid grazing sheep and cattle. Aust. J. Agric. Res. 27(5):691-698.

Littell RC, Milliken GA, Stroup WW, Wlfinger RD, Schanberger O (2006). SAS for Mixed Models (2nd Edn), SAS Institute, Care, NC. P. 42.

McCollum FT, Galyean ML, Krysl LJ, Wallace JD (1985). Cattle grazing blue grama rangeland. I. Seasonal diets and rumen fermentation. J. Range Manage. 38(6):539-542.

McCracken BA, Krysl LJ, Park KK, Holcombe DW, Judkins MB (1993). Steers grazing endophyte-free tall fescue: Seasonal changes in nutrient quality forage intake. digesta kinetics, ruminal fermentation and serum hormones and metabolites. J. Anim. Sci. 71(6):1588-1595.

McDowell LR (1997). Feeding minerals to cattle on pasture. Anim. Feed Sci. Technol. 60(3-4):247-271.

McDowell LR (2003). Mineral in Animal and Human Nutrition, 2th Edn, Elsevier Science. P. 103.

Mehrez AZ, Orskov ER, McDonald I (1977). Rates of rumen fermentation in relation to ammonia concentration. Brit. J. Nutr. 38(3):437-441.

Minson DJ (1990). The chemical composition and nutritive values of tropical grasses. In: Skerman PJ, Riveros F. (eds.) Tropical Grasses, Rome FAO, pp. 163-180.

Navarro JN, Galt D, Holechek J, McCormick J, Molinar F (2002). Long-term impacts of livestock grazing on Chihuahua desert rangelands. J. Range Manage. 55(4):400-405.

NRC (2000). Nutrient Requirements of Beef Cattle (7th ed). National Academic Press, Washington D.C., USA.

Obeidat BS, Thomas MG, Hallford DM, Keisler DH, Petersen MK, Bryant WD, Garcia MD, Narro L, Lopez R (2002). Metabolic characteristics of multiparous Angus and Brahman cows grazing in the Chihuahua Desert. J. Anim. Sci. 80(9):2223-2233.

Orskov ER (1982). Protein Nutrition in Ruminants, Academic Press, London. 38 P.

Park KK, Krysl LJ, McCracken BA, Judkins MB, Holcombe W (1994). Steers grazing intermediate wheatgrass at various stages of maturity: Effects on nutrient quality, forage intake, digesta kinetics, ruminal fermentation and serum hormones and metabolites. J. Anim. Sci. 72(2):478-486.

Paterson J, Cohran R, Klopfenstein T (1996). Degradable and undegradable protein response of cattle consuming forage-based diets. In: Proceedings of the Western Section American Society Animal Science. P. 103.

Petersen MK (1987). Nitrogen supplementation of grazing livestock. In: Proceedings of the Grazing Livestock. University of Wyoming. P. 115.

Russell JB, O`Connor JD, Fox DG, Van Soest PJ, Sniffen CJ (1992). A net carbohydrate and protein system for evaluating cattle diets: I. Ruminal fermentation. J. Anim. Sci. 70(11):3551-3561.

SAS (2003). SAS/STAT Users Guide. Version 9.1, SAS Institute Inc., Cary, N.C., USA.

Satter LD, Slyter LL (1974). Effect of ammonia concentration on rumen microbial protein production in vitro. Brit. J. Nutr. 32(1):199-208.

Van Soest PJ (1982). Nutritional Ecology of the Ruminant, Corvallis, O and Books. 182 P.

Van Soest PJ (1994). Nutritional Ecology of the Ruminant, 2nd Edn, Corvallis, O and Books. P. 92.

Van Soest PJ, Robertson JB, Lewis BA (1991). Methods for dietary fiber, neutral detergent fiber, and non-starch polysaccharides in relation to animal nutrition. J. Dairy Sci. 74(10):3583-3597.

Waterman RC, Grings EE, Geary TW, Roberts AL, Alexander LJ, MacNeil MD (2007). Influence of seasonal forage quality on glucose kinetics of young beef cows. J. Anim. Sci. 85(10):2582-2595.

18

Bacterial assessment and keeping quality of milk obtained from savanna brown doe

Olatunji, A. E.[1], Ahmed R[2] and Njidda, A. A.[3]

[1]Department of Animals Science, University of Abuja, Nigeria.
[2]Niger State Agricultural Development Programme (ADP) Minna, Nigeria.
[3]Department of Animal Science, Bayero University Kano, Nigeria.

Bacterial activities of milk obtained from Savanna brown doe, were chemically assessed before and after pasteurization. A total of 60 L of milk was collected from a randomly selected doe in 10 different herds within Minna, and was stratified into 3 treatments (T_1 - T_3), with 5 replicates (R_1 - R_5), in a completely randomized design (CRD). After collection one quarter of it was homogenously pooled and immediately taken to the laboratory for analysis (T_1), the other portion was left on the laboratory table to ferment (T_2) .The last quarter was pasteurized using the145°F (63°C) for 30 min (LTLT) (T_3). The biochemical results revealed an uneven disparity in all the treatments with high protein for fresh milk while fat was highest for pasteurized milk, this could be attributed to low activity of proteolytic and spoilage microorganism in fresh milk and the multiplication of fat splitting microorganism in the unpasteurized milk, the bacterial count (*Pseudomonas, Lactobacillus, Bacillus, Staphylococcus* and *streptococcus)* and frequently occurrence in treatments T_1 - T_2 indicate that these treatments was heavily loaded with different types of bacteria (proteolytic, lipolytic, coliform and lactic acid) when compared with T_3 (pasteurized), this could be due to lack of proper hygienic measure at all stages of collection and storage and/or pasteurization and diseased udder at the time of milk collection. Producer of milk and milk products should be pasteurized immediately after collection and should observe absolute aseptic measures when handling milk and milk products.

Key words: Savanna brown doe, milk keeping quality.

INTRODUCTION

Milk is a well-recognized high quality nutritional food elaborated by nature to foster the young ones and also good to maintain balance diet by the adult, its production and consumption had increased because of this knowledge, especially in most developing countries of the world. Unfortunately milk is the most easily perishable food. As a result attempts have been made to keep the quality of fresh milk in its original form as long as possible.

Milk is susceptible to contamination from sources like vessels, equipment used for milk and storage. Secreted milk by healthy udder is sterile but may become contaminated by the bacterial present in the tubules from where milk flows, in storage space and the cisterns (Uraih and Izuagbe, 1990). Milk is considered spoilt and unsafe for human consumption when it thickens. Some microorganisms spoil milk by impacting colour to it (Olatunji, 1997). *Serratia marcesceus* infection result in a bluish gray to brownish colour of milk, while *Pseudomonas synxanth* causes a yellow colour.

Table 1. Experimental procedure.

Replicate	T_1	T_2	T_3
R_1	T_1R,	T_2R_1	$T_3 R_1$
R_2	T_1R_2	$T_2 R_2$	$T_3 R_3$
R_3	T_1R_3	$T_2 R_3$	$T_3 R_3$
R_4	T_1R_4	T_2R_4	T_3R_4
R_5	T_1R_5	T_2R_5	T_3R_5

Olatunji (1997) observed that milk curdled or coagulated without acid production when the casein content is coagulated and will result in proteolysis by either *Bacillus subtilis*, *Bacillus cereus*, *Pseudomonas* spp and *Streptococcus lignitaciens*; given bitter flavour because of the release of peptides, while *Pseudomonas fluorescence* and *Candida lipolytica* will split fat in milk to produce glycerol and fatty acids which result in rancid taste and sourness.

Olatunji (1997) observed that pasteurization, especially the longtime low temperature (LTLT): 145°F (63°C) for 30 min is another means of preserving the quality of fresh milk.

The study was conducted to assess the effect of pasteurization on keeping quality of milk obtained from savanna brown doe.

MATERIALS AND METHODS

A total of sixty (60) L of milk was collected from 10 locally breed Savanna brown doe in Minna for the study. The collection period which lasted for 1 month 7 days with 2 weeks interval was collected from animal that are on management and feeding regime that are purely the traditional Fulani husbandry type, where animals graze from place to place in search of green pasture and towards the evening small quantity of sorghum bran is provided to supplement energy intake from forage.

Also milk-handling system conformed to the traditional system in that the kids are tied close to the dam to foster milk-let-down before milking which is done by any member of the family especially women and young once after the udder teat has being clean, using water from nearby stream. Animals were milked early morning time (8.00 h) and about 20 L of milk was collected from thoes that were chosen randomly within the same herd in Minna for the 3 collected periods making a total collection of 60 L of milk.

Milk samples collected were divided into 3 equal parts. One part was taken to laboratory fresh daily for analysis while another was kept on the laboratory table to sour, while the last portion was pasteurized under the long time low temperature pasteurization(145° for 63°C for 30 min). Samples were taken from the soured and pasteurized for laboratory analysis after the first, third, fourth, and fifth day of storage for biochemical and microbial analysis. The experimental procedure composed of 3 treatments that is, fresh milk (T_1), soured milk (T_2) and pasteurized milk (T_3) and 5 replicates (R_1 - R_5) assigned randomly in a completely randomized design (CRD) (Table 1).

Samples taken for microbial assay and biochemical analysis (Crude protein, fat and total solid) was done according to AOAC (1980). The routine laboratory procedure for standard microbial culture and plate count employed for evaluation are as follows:

Microbial analysis

The milk sampled was assessed for bacteriological quantity using the standard plate count. Total bacterial count, proteolytic, lipolytic, lactic acid bacteria (LAB) and Coliform counts were carried out by inoculating serially diluted sample in nutrient agar, milk agar, tributyrin agar, De Man Rogosa Sharpe agar and Ma'Conkey agar respectively and incubating them at 37°C for 48 h. The counts were expressed as colony forming units per milliliter of samples (cfu/ml).

Characterization and identification of isolates

Characterization of bacterial isolate was carried out using colonial morphology, microscopic techniques and biochemical test including gram staining, production of coagulase, oxidase, catalase and urease, methyl red-voges proskauer test, starch and gelatin hydroysis, spore stain, nitrate reduction and utilization of carbohydrates such as glucose, sucrose, mannitol, fructose, inositol, maltose and arabinose. The organisms were identified by comparing their characteristics with those of known taxa using the schemes of Cowan (1974) and Cruickshank et al. (1975).

RESULTS

Results of the biochemical analysis in percentage (%) ranged from 6.28 to 6.56 (pH), 14.00 to 25.42 (TS), 9.06 to 11.32 (CP) and 8.00 to 16.21 (EE) (Table 2).

Bacteria count per milliliter ranged from 9.8×10^7 to 1.4×10^8 in unpasteurized milk and from 1.2×10^8 to 1.7×10^8 in pasteurized milk, from day one to fifth day respectively. Value for fresh milk was 1.9×10^7 (cfu/ml) (Table 3), While values for Coliform count ranged from 1.2×10^4 to 1.8×10^5 cfu/ml, (pasteurized milk) and from 4.8×10^4 to 1.8×10^5 (cfu/ml) for unpasteurized milk (Table 4).

The bacterial isolated include species of *Bacillus*, *Micrococcus*, *Streptococcus*, *Staphylococcus* and *Escherichia* others include *Proteus*, *Pseudomonas*, *Lactobacillus*, *Achromobacter* and *Areobacter*. Percentage number of isolate was higher for T2 (15%) Unpasteurized fermented milk, compared to the pasteurized milk (7%) in T3 (Table 5).

DISCUSSION

Proximate analysis of the milk sample revealed that protein content was highest for fresh milk followed by unpasteurized milk and least for pasteurized milk, this could be as a result of low activity of proteolytic and spoilage micro-organism as indicated by Talaro and Talaro (1996). The fat content was highest for pasteurized milk followed by fresh and least for unpasteurized probably due to multiplication of fat splitting micro-organism in the unpasteurized milk. Frazier and Westofff (1988) reported that fat are subjected more often to chemical and microbial spoilage. The acidity nature of the unpasteurized milk may be due to the inferiority of lactic acid bacteria, which metabolized

Table 2. Chemical analysis of sampled Milk (DM Basis) (%).

Variable	PH	TS	CP	FAT
Fresh milk	6.56	14.00	11.32	11.00
UPGM	6.28	18.00	10.62	8.00
PGM	6.34	25.42	9.06	16.21

UPGM = Unpasteurized Goat milk FM = fresh milk, UGM = pasteurized goat milk.

Table 3. Bacterial counts of milk sample (cfu/ml).

Storage days	Pasteurized	Unpasteurized milk
1	9.8×10^7	1.2×10^8
2	1.0×10^8	1.3×10^8
3	1.1×10^8	1.4×10^8
4	1.2×10^8	1.5×10^8
5	1.4×10^8	1.7×10^8

Values for fresh milk = 1 Goat = 1.9×10^7 cfu.ml.

Table 4. Counts of coliform bacteria in milk samples (cfu/ml).

Storage days	Pasteurized	Unpasteurized milk
1.	6.4×10^4	7.6×10^4
2.	4.8×10^4	9.3×10^4
3.	1.1×10^5	1.2×10^4
4.	1.7×10^5	1.5×10^5
5.	1.8×10^5	1.8×10^5

Values for fresh milk = 1 Goat 5.9×104 cfc/ml.

Table 5. Frequency of occurrence of bacterial in milk sample (Cfu/ml).

Bacteria	Unpasteurized milk	Pasteurized milk
Bacillis spp	2(13.3)	2(28.6)
Micrococcus spp	2(20.0)	1.(14.3)
Streptococcus spp	2(13.3)	0(0.0)
Staphylococcus spp	4(26.7)	1(14.3)
Escherichia coil	1(6.7)	1(14.3)
Proteus spp	1(6.7)	0(14.3)
Pseudomonas	0(0.0)	0(0.0)
Lactobacillus spp	2(13.3)	0(14.3)
Achromobacter spp	0(0.0)	1(0.0)
Areobacter spp	0(0.0)	0(0.0)
Total number of isolates obtained	15%	7%

Value in parenthesis represents percentage isolates.

sugar to lactic acid (Talaro and Talaro, 1996). This also might help to check the proliferation of spoilage bacteria in the milk samples.

The result obtained for bacteria indicated that the samples were heavily contaminated. However, lower count was obtained in the pasteurized milk than unpasteurized. This agreed with Frazier and Westhoff (1988) who recorded low pathogenic bacteria count after pasteurization. Coliform count in the samples analyzed were quite high, especially after the fourth day in both pasteurized and unpasteurized milk. The presence of coliform in the sample could be due to the fact that after defecation, the local milk handler did not clean their hands properly or use water contaminated with facial matter from nearby stream. Coliform bacteria are undesirable in milk and milk product (Prescott et al., 1990; Umoh et al., 1990).

The proliferation of *Bacillus* spp. and *Staphylococcus* spp. in the sample reflects the abundance of the organisms in nature. *Bacillus* spp. produce spores, which help the organisms to withstand harsh conditions and germinate when the conditions become fovourable (Umoh et al., 1990). *Staphylococcus aureus* inhabits the skin and nostrils of a man and animals from where they could be shed on foods through coughing and sneezing (Ado and Wong, 2000). *Lactobacillus* and *Streptococcus* species are desirable, as these organism are responsible for the aromas and flavours of milk and milk products (Bryan, 1980).

The presence of pigment producing bacteria like *Pseudomonas* causes discolouration in milk under storage (Sale, 1967). Other undesirable organisms in the milk samples obtained are *Staphylococcus* and *Coliform* (*Escherichia coli* and *Proteus* spp.). Talaro and Talaro (1996) reported that coliform, bacillus and pseudomonas can spoil milk and cheese by proteolysis because of gas production, sliminess and off-flavour.

CONCLUSION AND RECOMMENDATION

The bacterial isolates observed in this study are suspected to contaminate the sample from various sources, which could be due to poor handling and storage after milk collection. The environment, utensils used, the state of hygiene of the animal from which the milk was collected and the sanitary condition of the milk collectors are all possible source of contamination.

It is recommended that the milk collection should be done with utmost hygienic measure and that milk should be pasteurized immediately after collection to reduce the load of bacteria especially the pathogenic ones. Government should endeavor to assist the poor fulani milk producer, in buying and getting these product into a collection centers were proper equipment for pasteurization are provided before the products get to the consumer, in view of the danger inherent in this product.

ACKNOWLEDGEMENTS

The author acknowledges the contribution of Prof U. J. Ijah of the Department of Biological Science and staff of Biological Science, Microbiology Unit Federal University of Technology, Minna for analyzing this product and Malam Tahid, of the Department of Animal Science, University of Abuja who did the type-setting of the whole work.

REFERENCES

AOAC (1980). Association Of Official Analytical hemists Official Methods Of Analysis Washington D.C. U.S.A.

Ado SA, Wong M (2000). Isolation and Characterization Of Bacillus Cereus From 'Nono' (A Fermented Milk) Spectrum J. 2:164-167.

Cowan ST (1974). Manual For The Identification Of Medical Bacteria, University of Cambridge Press Cambridge.

Cruickshank KR, Duguid JR, Marmion BP, Swain PH (1975). Medical Microbiology (4th ed) Churchill Livingstone, New York. pp. 356-357.

Frazier WC, Westhoff DT (1988). Food Microbiology (4th ed) Mcgraw-Hill Book Company, Singapore. pp. 419-428.

Olatunji EA (1997). Effect Of Different Methods On Keeping quality Of Cheese. Proceeding of 10th Annual Conference of The Biotechnology Society of Nigeria. pp. 54-63.

Prescott ML, Harley SP, Klein AD (1990). Microbiology W.M.C. Brown Publisher, Subuque, pp. 850-851.

Sale AJ (1967). Fundamental Principles Of Bacteriology (6th ed). M.C. Graw. Hill Company New York St Louis San Francisco. 22:17-22.

Talaro K, Talaro A (1996). Foundations In Microbiology, W.M.C. Brown Publisher (7th ed), Dubuque.

Umoh VJ. Adesiyun AA, Gomwalk, NE (1990). Seasonal Rarfation, Characteristic and Entero-toxin Production By *Staphylococcal* Isolate From Ferment, Milk Product. J. Food Microbiol. 3:468-545.

Uraih N, Izuagbe Y (1990). Dairy Microbiology Public Food. Indust. Microbiol. 12:49-54.

19

The one-humped camel in Southern Africa: Unusual and new records for seven countries in the Southern African Development Community

R. Trevor Wilson

Bartridge Partners, Bartridge House, Umberleigh EX37 9AS, UK.

The one-humped camel (Camelus dromedarius) was introduced to colonial southern Africa in the late 19th and early 20th centuries for military and police work to maintain law and order, for use in the postal services and for experiments in connection with rinderpest. Remnants of these (originally very small) populations survive in Botswana, Namibia and the Republic of South Africa but there are no surviving camels in Zimbabwe to where they were also introduced. This paper documents unusual and new records of camels in seven southern African countries. In several countries these are the first national introductions and they arrived as result of gifts to Heads of State by the then President of Libya.

Key words: Livestock introductions, exotic livestock, *Camelus dromedarius*.

INTRODUCTION

The total world population of the one-humped camel (*Camelus dromedarius*) in 2011 has been estimated at about 20.2 million (FAOStat, 2012). This species is native to the Near East, South Asia and North Africa including the Horn of Africa. There have been, however, many introductions and attempted introductions to areas outside its native range, including Europe, North and South America, the Caribbean and Australia.

The concept of using camels in southern Africa originated as early as 1861 when a memorandum of 27 May from Robert Moffat Jr., a trader on the southeastern border of the Khalagare Wilderness, was sent to Richard Southey, the Colonial Secretary (Cape Town Archives Repository, Source CO, Volume 4120, Reference M46). It was, however, to be almost 30 years before the first one-humped camels arrived in the region. These arrived at Walvis Bay in German South West Africa (Namibia) from the Canary Islands in 1889 for use by the Schutztruppe (Protection Force) in a military role (Grunow,

1961; Wilson, 2012a). The next imports were to the Cape of Good Hope (Republic of South Africa) where 10 camels arrived at Cape Town from Tenerife (Canary Islands) on 27 March, 1897 for "experimental purposes in connection with the rinderpest epidemic" that swept through eastern and southern Africa in 1895/1896 and killed up to 90% of all cattle (Wilson, 2009). A total of 34 camels having sailed from Karachi (now in Pakistan) to Beira in Portuguese East Africa (Mozambique) and from there by train - in coal trucks covered with tarpaulin - arrived in Salisbury (Harare) in Rhodesia (Zimbabwe) on 9 May, 1903 (Flint, 1903; Wilson, 2007). These last were subsequently used for general transport and by the police and postal services and it was from these that the only record of foot and mouth disease in camels arises (Anon, 1904; Wilson, 2008)). There were no direct imports to British Bechuanaland (Botswana) which was administered by the Cape of Good Hope Government in Cape Town but camels were used there as part of the

Figure 1. Camels at Tete in Portuguese East Africa (Mozambique) (Source: Clay, 1962).

Cape Postal Services and for police work (Wilson, 2013). Descendants of the early imports survive in the 21st century in small numbers in the Republic of South Africa and Namibia. None of the original population is known in Zimbabwe where those that did not die in service were turned out to fend for themselves, at least one of these was shot by professional hunters near Masvingo close to the Great Zimbabwe ruins (Louw Hoffman, personal communication) and its seems likely that others also suffered this fate. Ironically, the last original camels in official service were in Botswana, where they were in police use until the early 1980s and where 200 plus animals have recently been transferred to village groups for use in ecotourism.

MATERIALS AD METHODS

This paper documents unusual and new records from Mozambique, Zambia, Malawi, Swaziland, Lesotho, Zimbabwe and Tanzania. Information was gathered from a search of the literature and from internet searches. This was complemented by personal correspondence and intercourse with relevant people in the various countries and from the author's own studies and experience.

RESULTS

A small number of unusual records date from the late 19th and the early 20th centuries. In the first years of the 21st century camels have arrived by various routes in several southern African countries. By far, the most bizarre of these were the animals that arrived in at least

four countries by air as gifts from the late Brother Leader and Guide of the Revolution of the former Great Socialist People's Libyan Arab Jamahiriya (or more simply the President of Libya). At least four countries - Lesotho Swaziland, Zimbabwe and Mozambique, received camels in such a way. Botswana, Madagascar and Namibia were also to receive camels from Libya but there is no evidence that they did so.

Portuguese East Africa/Mozambique

David Livingstone, the missionary-explorer, was the first to fail to introduce camels to Portuguese East Africa when he sailed into what would now be considered territorial waters at the mouth of the Ruvuma River on 22 March, 1866. Tide and wind were against him however, and after two days he sailed back north and landed six camels at Mikindani in what is now Tanzania (Waller, 1874; Wilson, 2012b). The so far undisputed first camels in Mozambique were thus the 34 brought by Colonel Flint from Karachi that arrived at Beira towards the end of April, 1903 (Flint, 1903). This group was loaded onto a train immediately on arrival at Beira and transported directly to Salisbury (Harare) where they arrived on 9 May, 1903. None of the group was lost or died on the long journey from South Asia to Southern Africa. The presence of camels at Tete in northern Portuguese East Africa (located at 16° 17' S, 33° 58' E) on the banks of the Zambezi in 1904 has been established by a photograph taken there at the time (Figure 1) but nothing further is known about these animals.

President Guebuza of Mozambique received a present of an adult male camel, two adult females and two calves from President Gaddafi, in 2009. The animals arrived by air after a week of travel (suggesting they were the last of one of Gaddafi's camel deliveries to southern Africa), tightly restrained by ropes and in a totally exhausted state (Samuel Bila, personal communication). It was 24 h before the animals were able to stand when they were then transferred to a private wildlife sanctuary at Matola 30 miles north of Maputo where they proved extremely difficult to handle. Heavy infestations by ticks resulted in severe disease problems and in spite of treatment with tetracycline and penicillin four of the animals died (Antonio Rocha, personal communication). At about the same time as the arrival of Gaddafi's camels a private shipment of five camels arrived at Beira port from Kenya (apparently with a legal import certificate) whence they were transferred to various tourist establishments at Vilanculos Bay, on the coast between Beira and Maputo (Figure 3) (Cuniffe, 2011). In mid 2012, the number of camels had been reduced to three of which one was apparently owned by the President of Mozambique (the sole survivor of the Gaddafi lot) and two by the Anchor tourist establishment (Dale Fraser, personal communication).

Northern Rhodesia/Zambia

The first of only two references to camels in Northern Rhodesia (Zambia) dates from 1896. At that time, Hugh Marshall, the first Magistrate and Postmaster at Abercorn (now Mbala), wrote to his brother in England "Fancy a Traction Engine for the Lakes Coy. For the Katunga-Blantyre Road! Another white elephant! Oh, yes - some talk of camels as well" (Clay, 1962). The other record is of an imprecise location but somewhere along the northern bank of the Zambezi. This was in 1915 when eight Germans and one rebel Dutchman with five camels and one horse were captured by a mobile unit of the Northern Rhodesia Rifles. It is believed the group was escaping from Southwest Africa (Namibia) in an attempt to join the forces of Colonel von Lettow-Vorbeck in German East Africa (Tanzania). They were prevented from escaping by the single guard in charge of them by the simple expedient of confiscating their trousers every night because "it is impossible to ride a camel without trousers" (Brelsford, 1954).

Southern Rhodesia/Zimbabwe

The most singular reference to camels in "South Central Africa" is in a work of fiction rather than it being a fact (Haggard, 1900). The ancient ruins of Zimbabwe provide the setting for the novel in which camels are used as both pack and riding animals. The considerate author even provides an illustration of a camel that is a very good likeness (Figure 2).

At the ZANU-PF conference held in Bulawayo on 8 December, 2011, the Zimbabwe president took time off from imperialist bashing to lambast his ex-bosom friend the late Libyan leader Muammar Gaddafi as ignorant and as someone who failed to invest in Africa but splashed his money in Europe to please the West, who killed him at the end. Gaddafi had made an undertaking to invest in Africa but "we saw him dishing out camels and we got four which are at the farm" (that is Mugabe's farm) (ZimEye, 2011). As for Lesotho, however, "many people now wonder when the camels will be returned to the NTC (that is, the National Transitional Council of Libya) (Mashiri, 2011).

Lesotho

On 10 January, 2008, Lesotho's foreign minister and another top government official were at the airport to meet a huge Libyan transport aircraft and receive two adult camels and two calves which were a gift from Gaddafi to the Lesotho Prime Minister. On arrival, the animals were whisked away to a secret destination to be kept under the supervision of the Ministry of Agriculture and Food Security (Thakalekoala, 2008). According to another source, the Kingdom of Lesotho received "a precious gift of five camels - two females, one male and two babies - from the leader of the Libyan revolution, Colonel Muammar Kadhafi" (NetNews Publisher, 2008). Not everyone, however, was overjoyed at the arrival of the camels. Questions were asked by a local agricultural newspaper about the procedure of the Government of Lesotho in importing live animals and in particular with respect to the control of dangerous diseases that might be transmitted by imported animals. The Lesotho animal health authorities claimed to have conducted all necessary test measures before permitting the import of the camels and had also put them in quarantine on arrival. By 2011, nonetheless, one camel had died from pneumonia and two from ingestion of foreign bodies. Libya apparently supplied a further three camels to replace those which had died but one of these also died. By mid 2011, three camels were still on the premises of the Ministry of Agriculture and Food Security and were being maintained at public expense. The livestock services claimed that this was legitimate as their staff had benefited in acquiring more skills in the welfare, animal health, and feeding of camels. According to the newspaper questions were asked in parliament about the

Metem noted that there hung the body of a black dwarf

Figure 2. Camel on the beach at Vilanculos bay, Mozambique
(note double Indian saddle) (Source: Teagan Cunliffe).

costs related to the camels and what use they were. The Minister of Agriculture replied that the owners (presumably the Prime Minister) regularly visit them as they recognize and acknowledge that such fragile animals with their soft feet need love and to be well taken care of and that it was also to be noted that since the camels had arrived many schools and parents visited the department of live stock and concluded that the more they stay, the more the department would learn more about the welfare of this species. The newspaper remained sceptical and asked when they would go home (Anon., 2011).

Swaziland

In a situation analogous to that of Lesotho, King Mswati III of Swaziland was given a present of six camels by Libyan president Muammar al Gaddafi as a token of friendship on 1 July, 2009 (Royal Forums, 2009; Ndlovu, 2009). The two females, one male and three calves (two female and one male) arrived on board a cargo aircraft and were met by the royal veterinarians and the King's private secretary. The camels were to stay in quarantine for 30 days before being moved to a game reserve with the King having the final say in their location.

The Principal Secretary in the Ministry of Agriculture said that "it is a great pleasure for the country to be blessed with these rare beasts, especially because most people here are not familiar with these animals". Prior to Gaddafi's gift, there had been serious suggestions that dairy camels should be introduced to the dry southeast lowlands of Swaziland as these animals had done well in Kenya and Israel (Fayolle, 2008).

This proposal has yet to be given further consideration. A camel breeding association of Swaziland has a site on Facebook but there is no information other than the name.

Figure 3. Illustration of a camel in the fictional work "Elissa" by Rider Haggard.

Malawi

Malawi has at least two lots of camels. One is at the Kuti Community Wildlife Park (at the southern end of Lake Malawi) where there are three camels whose names are Mphatso, Nguva and Diesel. These were brought into the park in May 2009 (but again the provenance is unknown) to provide visitors with a unique safari opportunity (Figure 4) but this was later stopped in the interests of the camels as it proved stressful to them (Kuti Project, 2012). The other site is in the low-lying hot and humid Shire lowlands in southeast of Malawi. The area is cultivated to sugar and is patrolled by security personnel mounted on camels. Although the sharp sugar roots cut the camels' feet they wear specially designed shoes when on patrol. The camels are said to be thriving and three calves have been born since their arrival in Malawi "a few years ago". The provenance of these animals is given as Tanzania (Bemyguest, 2012).

Tanzania

Irrespective of whether or not the Malawi camels arrived from Tanzania there are some unusual records for this latter country which is not within the normal range of the species. (Tanzania has always been considered to be an East African country but as a member of the Southern African Development Community it is legitimate to include it in this paper.) Historical records relating to Livingstone (Wilson, 2012b) and current records for northern Tanzania (Wilson, 2011) have already been published. There are now two camels at Kigamboni, 10 km south of Dar es Salaam city centre, that were imported from Somalia by a private individual travelling through Kenya by lorry in 2009 (Figure 5). Some 25 km farther south west at Dar es Salaam Zoo there are 48 camels that have increased from a small group imported from Ethiopia in 2006: these are typical Afar camels and much smaller physically than the Somali camels. The provenance of three female camels at the Sokoine University of Agriculture about 160 km southwest of Dar es Salaam in a hot subhumid climate is not exactly known but anecdotal evidence puts them in the same group as those presented by Gadaffi to the several Presidents of SADC countries to which reference has already been made.

DISCUSSION

There have been many attempts, over many hundreds of years, to introduce the one-humped camel to areas outside its normal range. Few of these have been successful. In the high period of the Roman empire some

Figure 4. Camel with local handler at Kuti Community Safari Park, Malawi (Source: Kuti Project).

Figure 5. A camel south of Dar es Salaam, Tanzania imported from Somalia by a private individual (Photograph by Trevor Wilson).

2000 years ago camels were used for transport in many of the northern provinces including what are now Belgium, Germany and France as well as in Italy itself (de Grosso-Mazzorin, 2006; Pigière and Henrotay, 2012) but these never became naturalized or truly domestic animals. There were many introductions to Italy (mainly from nearby Libya and Tunisia) over the centuries but these were always limited, the toys of the rich families (Cochi, 1858) and again never became common farm animals although one herd of about 200 head survived through to the 1940s when they were slaughtered by troops fighting the Second World War who used them as food.

The Moors took camels to Spain and there was a royal stud at Aranjuez during the eighteenth century and there were later imports at least until the 1830s (Graells, 1854) but, as in Italy, the animals failed to naturalize. There were imports to the USA in the 1850s for use by the military in the southwestern deserts but these efforts were terminated by the Civil War (Lesley, 1929) although a few camels are still used as tourist animals in parts of Texas (Doug Baum, personal communication). Several attempts have been made to introduce camels to South and Central America, including Peru, Bolivia and Brazil (Legge, 1936), Venezuela (Dareste, 1857), Barbados and Jamaica (Legge, 1936) and Cuba (Dareste, 1857). In all these cases, numbers were small and the camels failed to survive for a number of reasons including disease, climate and inability or incompetence on the part of the owners and handlers.

Australia provides the only example of successful long term introduction of camels outside their natural range. For many years they were the principal means of transport in the interior, being largely instrumental in the completion of a railway and then of transporting wool to the railhead (McKnight, 1969).

The history of introductions to the Republic of South Africa (Cape Colony), Namibia (German South West Africa), Botswana (British Bechuanaland) and Zimbabwe (Southern Rhodesia) has been briefly touched upon in this paper. In none of these countries was the introduction "successful" beyond a short period although a few camels remain in each country where they are regarded as curiosities and find limited use for tourism. Most recent introductions to the SADC countries have been political ploys by an eccentric and now defunct Arab dictator. Each of these introductions has proved to be biologically unsuccessful and has usually resulted in internal problems and ridicule for the recipient government.

The one-humped camel is supremely adapted to hot dry environments with limited feed resources. As such it has some potential, if introduced, for mitigating some effects of climate change in the arid and marginal areas of southern Africa. Before serious attempts at large scale introduction are made, however, there should be serious longer term studies on the biological, economic, environmental and social aspects of such an action.

ACKNOWLEDGEMENTS

My interest in camels in southern Africa was aroused when Anna Mupawaenda introduced me to an academic at the University of Zimbabwe. His office had a photograph of cattle that had died of rinderpest. Looking for this photograph in the Zimbabwe national archives, I accidentally came across photographs of "Colonel Flint's camels" and I carried on from there. With the exception of the section on camels in Tanzania (the two camels south of Dar es Salaam were brought to my attention by Jeffrey Lewis) of the information in this paper on camels in the 21st century has been gleaned from social networking, newspaper and (semi)official sites on the world wide web. This does not invalidate what has been learned but usually leaves it incomplete. No thanks are due to the many e-mail addresses connected to these sites, including several whose owners are nationally and internationally responsible for the collection and dissemination of data relevant to their own domesticated animal genetic resources. The number of e-mail addresses and telephone contacts of many officials in the various countries that do not exist when attempts are made to contact them is astonishing. I am extremely grateful to three referees whose comments assisted greatly in improving the original draft of the paper.

REFERENCES

Anon (1904). Foot and mouth disease in camels. Rhodesia Agric. J. 1:84.

Anon (2011). When are the camels going home? The Silo, 2(9): June 2011. http://www.thesilo.co.ls/home/ src/news/when-are-the-camels-going.html (accessed 14 July 2012).

Bemyguest (2012). King of the antelopes Lower Shire Valley. http://www.bemyguestmagazine.com/wildlife/lowershire.html (accessed 01 July 2012)..

Brelsford WV (Ed.) (1954). The Story of the Northern Rhodesia Regiment. Government Printer: Lusaka. P. 29.

Clay G (1962). Camels in Northern Rhodesia. Northern Rhodesia J. 5:176-177.

Cochi I (1858). Sur la naturalisation du dromadaire en Toscane. Bull. Soc. Imp. Zool. Acclimatation 5:479-483.

Cuniffe T (2011). An oddity: camels in Vilanculos, Mozambique. http://blog.getaway.co.za/travel-stories/an-oddity-camels-in-vilanculos-mozambique/ 25 January 2011 (accessed 19 July 2012).

Dareste A (1857). Rapport sur l'introduction projetée du dromadaire au Brésil. Bull. Soc. Imp. Zool. Acclimatation 4:190-215.

de Grosso Mazzorin J (2006). Cammelli nell'antichità: le presenze in Italia. In: Archaeozoological studies in honour of Alfredo Riedel. Bolzano, Italy. pp. 231-242.

FAOStat (2012). FAO Statistical Year Book 2012. Food and Agriculture

Organization: Rome. http://faostat3.fao.org/home/index.html (accessed 10 June 2013).

Fayolle A (2008). Final Evaluation of the Multi-annual Micro-Projects Programme (MPP) (Framework Contract Benef – Lot No 1 Letter of Contract N° 2007/144232. Italtrend: Rome.

Flint J (1903). Camels in Rhodesia. Rhodesia Agric. J. 1:5-6.

Graells MP (1854). Sur l'acclimatation des animaux en Espagne. Bull. Soc. Imp. Zool. Acclimatation 2:109-116.

Grunow W (1961). Entstehung und Einsatz des Kamelreiterkorps der kaiserlichen Schutztruppe für Deutsch-Südwest Afrika (Origins and deployment of the Kaiser's Camel Corps Defence Force in German Southwest Africa). Mitteilungsblatt des Traditionsverbandes ehemaliger Schutz-und-Überseetruppen 61:44-49.

Haggard HR (1900). Elissa or The Doom of Zimbabwe. Longmans, Green, and Co.: London. pp. 1, 3, 115, 120, 157, 231, 236, 241, 243.

Kuti Project (2012). Kuti Project: The animals. http://www.kuti-malawi.org/animals.html (accessed 1 July 2012).

Legge CM (1936). The Arabian and the Bactrian camel. J. Manchester Geog. Soc. 46:21-48.

Lesley LB (1929). Uncle Sam's camels. Harvard University Press: Cambridge, USA.

Mashiri CC (2011). When will Mugabe return Libya's 4 camels to the NTC? http://www.thezimbabwean.co.uk/comment/blogs/clifford-mashiri/55384/when-will-mugabe-return-libyas.html 12 December 2011 (accessed 14 July 2012).

McKnight TL (1969). The camel in Australia. Melbourne University Press: Melbourne, Australia.

Ndlovu H (2009). King presented with six camels. Swazi Observer, 2 July 2009. http://www.observer.org.sz/index.php?news=5462 (accessed 10 July 2012).

NetNews Publisher (2008). Lesotho Welcomes Five Camels From Libya. http://www.netnewspublisher.com/ lesotho-welcomes-five-camels-from-libya/ 11 January 2008 (accessed 14 July 2012).

Pigière F, Henrotay D (2012). Camels in the northern provinces of the Roman Empire. J. Archaeol. Sci. 39:1531-1539.

Royal Forums (2009). Libyan President Honors King of Swaziland with Camels. http://www.theroyalforums. com/9198-libyan-president-honors-king-of-swaziland-with-camels 2 July 2009 (accessed 10 July 2012).

Thakalekoala T (2008). Gaddafi sends camels for prime minister. http://www.iol.co.za/news/africa/gaddafi-sends-camels-for-prime-minister-1.385397 11 January 2008 (accessed 14 July 2012).

Waller H (1874). The last journals of David Livingstone, in Central Africa, from 1865 to his death. Continued by a narrative of his last moments and sufferings, obtained from his faithful servants Chuma and Dusi, in Two Volumes. – Vol. I [1866–1868]. John Murray: London.

Wilson RT (2007). The one-humped camel in Southern Africa: Their presence and use in Rhodesia (Zimbabwe) in the early twentieth century. South Afr. Anim. Sci. 8:21-28. http://www.sasas.co.za/Popular/Popular.html.

Wilson RT (2008). Perceptions and problems of disease in the one-humped camel in southern Africa in the late nineteenth and early twentieth centuries. J. South Afr. Vet. Assoc. 79:58-61.

Wilson RT (2009). The one-humped camel in Southern Africa: imports to and use in the Cape of Good Hope in the late nineteenth and early twentieth centuries. J. Camel Pract. Res. 16:1-17.

Wilson RT (2011). The one-humped camel and the environment in northern Tanzania. J. Camel Pract. Res. 18:25-29.

Wilson RT (2012a). The one-humped camel in Southern Africa: imports to and use in South West Africa/Namibia. J. Camel Pract. Res. 19:1-6.

Wilson RT (2012b). The one-humped camel in Tanzania: attempted introduction by the missionary-explorer David Livingstone in 1866. J. Camel Pract. Res. 19:111-113.

Wilson RT (2013). The one-humped camel in Southern Africa: use in police and postal services and as a tourist attraction in British Bechuanaland/Botswana. Botswana Notes Rec. 45: in press.

ZimEye (2011). Gaddafi only gave me "four Camels" – Mugabe. http://www.zimeye.org/?p=41448.9 December 2011 (accessed 10 July 2012).

20

The health status of turkeys against the microclimatic conditions

Katarzyna Ognik[1], Anna Czech[1], Bożena Nowakowicz-Dębek[2] and Łukasz Wlazło[2]

[1]Department of Biochemistry and Toxicology, Faculty of Biology and Animal Breeding,
University of Life Sciences in Lublin, Poland.
[2]Department of Animal Hygiene and Environment, Faculty of Biology and Animal Breeding,
University of Life Sciences in Lublin, Poland

The objective of this study was to evaluate the sanitary microbiological status of a poultry house building as well as to determine morphological and biochemical blood markers of turkey hens with a deteriorated health status (diarrheas, numerous deaths, drowsiness, deformed tarsal joint). Microorganisms isolated from litter, buildings and turkey hens constituted the natural environmental flora detected in these types of buildings, hence the sanitary and hygienic status of the turkey rearing house was found satisfactory. The turkeys' hens with the poorer health status were characterized by worst production performance, decreased levels of hemoglobin, protein, alkaline phosphatase and minerals (Ca, Mg and Fe). In addition, increased levels were noted for lipid peroxidation products (H_2O_2, MDA), catalase, glucose, cholesterol, high-density lipoprotein (HDL) cholesterol fraction, uric acid, and lactic dehydrogenase.

Key words: Turkey, health status, microclimate, blood parameters.

INTRODUCTION

Optimal parameters of a poultry house environment are one of the key elements that determine the welfare of reared birds. Environmental factors, that is, hygiene, microclimate, feeding as well as various types of contaminations substantially affect the health status and production performance of animals (Kocaman et al., 2006). Amongst a variety of contaminations, great significance is ascribed to those of microbiological origin. In the case of a poultry house, they may originate from, among other things, feedstuff material, litter, walls of the hen house, floor cracks, drinkers, etc., and their number is determined by microbiological parameters. Apart from saprophytes, the microflora of poultry houses may as well include pathogenic microorganisms (Cencek et al., 2000). Hence, it is of great importance to provide the animals

appropriate feeding and rearing conditions, which will enable eradication of diseases and thereby eliminate death of the animals. Bearing in mind the above, the objective of this study was to evaluate the sanitary microbiological status of a poultry rearing house as well as to determine the production, morphological and biochemical blood markers of turkey hens with deteriorated health status.

MATERIALS AND METHODS

The experiment was carried out on Big 5 line turkey hens reared from the 1st to the 16th week of life. The turkey hens of all groups were fed *ad libitum* based on all-mash feed mixtures with a constant access to drinking water. The feed mixtures were

Table 1. Nutrient content of the standard diets.

Component (Feeding period)	Starter (1-2 week of age)	Grower I (3-5 week of age)	Grower II (6-9 week of age)	Grower III (10-12 week of age)	Finisher I (14-16 week of age)
Maize meal (%)	25.6	27.4	23.8	35.2	47.4
Wheat bran (%)	20.0	25.0	30.0	25.0	25.0
Wheat bran (%)	3.0	-	-	-	-
Soybean meal, 46% protein (%)	41.0	41.7	38.8	32.7	20.4
Soybean meal 45% protein (%)	2.0	-	-	-	-
Fish meal 60% (%)	3.5	-	-	-	-
Fodder chalk (%)	1.2	1.7	1.7	1.4	1.5
Soybean oil (%)	0.5	1.0	2.5	3.0	3.0
Cytromix Plus[1] (%)	0.2	0.2	0.2	0.2	0.2
Farmix [1] (%)	3.0	3.0	3.0	2.5	2.5
Nutrient composition					
CP (%)	27.1	25.5	24.5	22.0	17.5
ME kcal kg^{-1}	2736	2803	2913	3007	3129
Crude fibre (%)	2.86	2.77	2.72	2.71	2.7
Lysine %	1.81	1.71	1.57	1.38	1.17
Methionine +Cysteine (%)	0.98	0.90	0.88	0.79	0.70
Tryptophan (%)	0.34	0.28	0.27	0.23	0.19
Arginine (%)	1.77	1.57	1.50	1.32	0.98
Calcium (%)	1.39	1.23	1.17	1.06	0.94
Phosphorus available (%)	0.77	0.67	0.59	0.57	0.47
Sodium (%)	0.15	0.16	0.15	0.15	0.15

[1]Cytromix Plus - citric acid. fumaric acid, phosphoric acid (62%); [2]Farmix— mineral and vitamin premix provided the following per kilogram of diet – 433333,0 IU of vitamin A; 133333,0 IU of vitamin D_3; 73,3 mg of vitamin K_3; 100,0 mg of vitamin B_1; 291,7 mg of riboflavin; 175,0 mg of vitamin B_6; 0,9 mg of vitamin B_{12}; 58,3 mg of folic acid; 10,5 mg of biotin; 2182,0 mg of niacin; 13333,0 mg of choline; 4 200 mg of calcium pantothenicum; 4 000 mg of Mn; 2 666 mg of Zn; 1 666 mg of Fe; 833 mg of Cu; 26 mg of I; 10 mg of Se; 6,7 mg of Co; 13 g of Ca; 15,5 g of P.

produced at the poultry farm based on recipes and premixes of the Polsanders-Poland company. All mixtures were composed based on wheat, maize meal, post-extraction soybean bean, soybean oil, with the iso-protein and iso-energetic balance maintained (Table 1). In addition, use was made of a Farmix type premix. Group I (n=12,000) included control birds with the normal health status, not displaying any pathological symptoms (proper body weight gains and feed intake). Birds belonging to group II (n=12,000) were turkey hens in which poorer health status was diagnosed on the 9th week of life (including: diarrheas, numerous deaths, drowsiness, deformed tarsal joint causing impaired walking), which resulted in the observed smaller body weight gains and feed intake. Birds of the experimental groups (I and II) were located in two separate turkey rearing houses: Building 1 (control birds) and building 2 (birds with a poorer health status). Owing to the fact that up to the 5th week of life the poults were located in the turkey rearing house, and since the 6th week of life they were fattened in different rearing buildings, the weekly weighing of turkey hens (100 turkey hens per building) was began since the 6th week of their life, and repeated after each completed week of life. Feed intake was monitored as well. Production performance results noted for particular groups were then used to calculate the European Production Performance Index WEO (Faruga and Pudyszak, 1999).

$$WEO = \frac{\text{mean body weight after rearing (kg) x liveabilit y (\%) x 100}}{\text{day of rearing x feed conversion (kg kg - 1)}}$$

On the 12th week of birds life, blood was sampled (by a veterinarian) from the wing vein of 100 birds from each group for morphological and biochemical analyses. The collected blood samples were determined for: hematocrit (Ht) number– with the microhematocrit method, hemoglobin (Hb) level – with the colorimetric method accounted to Drabkin, as well as for numbers of white blood cells (WBC) and erythrocytes (RBC) – with the chamber method (Bomski, 1989). The percentage composition of leucocytes (leucogram) was also determined by staining blood smears with the Pappenheim's method (Bomski 1989). Using monotests by Cormay company, blood plasma was assayed spectrophotometrically for levels of the selected biochemical markers, that is, total protein (TP), glucose (Glu), uric acid (UA), total cholesterol (CHOL), triacylglycerols (TG), high-density lipoprotein (HDL) and low-density lipoprotein (LDL) fractions of cholesterol. Samples of blood plasma were additionally analyzed for levels of: Zinc, iron, magnesium, calcium, and phosphorus – with the method of atomic absorption spectrometry (AAS). Further on, monotests by Cormey company were also used to determine the activity of alkaline phosphatase (ALP). Spectrophotometric analyses were also carried out in blood plasma to determine parameters of blood redox status: Superoxide dismutase (SOD) – with the adrenaline method accounted to Misra, in: Greenwald (1985), modified in respect of the wavelength of 320 nm. The method was modified to achieve higher selectivity of transitory reaction products at this wavelength (Bartosz, 1995). Catalase (CAT) was determined according to Bartosz (1995). In terms of the parameters of the antioxidative system, analyses were also carried out for the total antioxidant potential of plasma (FRAP) accounted

Table 2. Microorganisms identified in the rearing environment of turkeys and intensity of their occurrence.

Sample collection site	Microorganisms /occurrence intensity			
Feeder	CNS (+)	*Trichosporon asahii* (+)	-	-
Drinker	*Candida crusei/inconspicua* (+)	*Trichosporon asahii* (++)	CNS (+)	*E. coli* (+)
Partitions/walls	CNS (++)	*Streptococcus fecalis* (+)	*Candida rusei/inconspicua* (+)	-
Litter	*Candida crusei/inconspicua* (+++)	CNS (+)	*Streptococcus fecalis* (+)	*E. coli* (++)
Cloaca	*Candida famata* (++)	*Candida crusei /inconspicua* (++)	*Streptococcus fecalis* (+)	*E. coli* (++)
Cloacal area	*Candida crusei /inconspicua* (+++)	*Trichosporon asahii* (+)	*Streptococcus fecalis* (+)	*E. coli* (+)

+ Few, + + increase in the number, + + + abundant growth.

to Benzie and Strain (1996). Blood plasma of the experimental birds was also analyzed for the level of lipid peroxidation products, that is, concentration of peroxides (H_2O_2) according to Gay and Gębicki (2000, 2002), and concentration of malondialdehyde (MDA) as the end product of tissue lipids oxidation according to Ledwożyw et al. (1986).

In view of the fact that the problem of birds rearing referred exclusively to the turkey hens kept in one of the rearing buildings (building 2), it was stated that the problem was due to the poorer conditions of a turkey house microenvironment. For this reason, on the 12th week of birds rearing samples were collected from particular poultry houses for microbiological analyses. The material to be analyzed was litter collected into sterile containers as well as swabs collected with sterile cotton swabs from the birds (n=80) and from the walls and partitions inside the building. The collected material was transported in portable refrigerators to the laboratory and subjected to microbiological analyses. The samples were analyzed for microbiological contamination and for the occurrence of pathogenic bacteria of the genus *Salmonella*. The litter was fixed in a solution of physiological salt with the addition of *Tween 80*; next dilutions were prepared and inoculated onto McConkey's and Sabouraud's culture media. Swabs collected in the turkey rearing houses were inoculated from the cotton swabs onto the following media: Agar with blood and SS agar with pre-proliferation. The inoculates were incubated for 24 h at 37°C. Pure cultures were isolated from grown colonies by means of multiple reducing inoculations. Colonies were identified microscopically after staining accounted to Gram, by means of the following tests: Catalase, coagulase and biochemical tests API 20 C AUX by bioMerieux (bioMerieux Polska Ltd.).

Numerical data obtained were subjected to a statistical analysis using STATISTICA ver.5 software, with the one-way analysis of variance ANOVA, at a significance level of 0.05 and 0.01.

RESULTS

The microbiological analyses of litter samples collected from both poultry rearing houses and of swabs collected from the birds demonstrated intensified occurrence of yeast-like fungi of the following genera: *Candida crusei, Candida famata* and *Trichosporon asahi*. The presence of *C. crusei* was detected in litter, on drinkers and in the cloacal area of the birds from both groups examined. *C. famata* was isolated mainly from the material collected with cotton swabs from the cloaca of turkey hens, whereas *T. asahi* – from drinkers and cloaca of the birds. The analyses did not demonstrate the presence of *Salmonella* genus bacilli in any of the samples collected from both turkey rearing houses. The other microbial species isolated from litter, buildings and animals represented the environmental microflora of this type of buildings that is *E. coli*, fecal streptococci – *S. faecalis,* and coagulase-negative staphylococci – CNS, usually having no impact on the pathological symptoms observed in the turkey hens from group II (Table 2). The main parameters of the rearing performance of turkey hens are presented in Table 3. As shown by data collated therein, in Group II the survivability of turkey hens was lower by 10.5% than in the control group. The observed (especially in the 10th week of life) disease symptoms in the turkey hens from Group II affected lower (than in the control group) body weight gains, that is, by 23.6% in 9 to 12 weeks of life, and by 15.4% in 13 to 16 weeks of life. In turn, in the period of 6 to 16 weeks of life, the body weight gains of the turkey hens from Group II were lower by 12% than those of the control birds. Data achieved were also reflected in values of the European Production Performance Index – WEO. A lower by 75.7 point value of that index was noted in Group II (372.5 point) when compared to the control group (448.2 point).

The main parameters of the hematological of blood of turkey hens are presented in Table 4. As shown by results achieved in the reported study, no differences were noted between the groups in the value of hematocrit and in the number of erythrocytes. In turn, the level of hemoglobin in the turkey hens with lower body weight gains and poorer health status turned out to be lower ($p \leq 0.05$) when compared to the control birds. The analysis of results achieved in the reported experiment demonstrated also a significant increase ($p \leq 0.05$) in the

Table 3. Production performance of turkey hens.

Specification	Week of life	Experimental groups	
		I – control	II
Weekly body weight gains (kg/bird)	6-9	2.11	2.08
	9-12	2.79	2.13
	13-16	2.59	2.19
	6-16	7.58[a]	6.68[b]
Feed intake (kg/kg)	6-16	2.51	2.43
Survivability (%)		95	85
WEO, pts		448.2[a]	372.5[b]

[a,b] Values in rows denoted with different letter differ significantly at p≤0.05.

Table 4. Level of hematological blood markers of turkey hens.

Marker	Feeding groups				SEM
	I		II		
	M	SD	M	SD	
Ht (L^{-1})	0.29	0.023	0.28	0.012	0.006
Hb (g L^{-1})	7.82[a]	0.35	6.58[b]	0.30	0.090
RBC (10^{12} L^{-1})	3.41	0.79	3.40	0.30	0.012
WBC (10^{9} L^{-1})	12.10[b]	1.47	19.5[a]	5.94	1.23

M, The arithmetic average; SD - standard deviation; [a, b] values in rows denoted with different letter differ significantly at p≤0.05. [A, B,] values in rows denoted with different letter differ significantly at p≤0.01.

Table 5. Level of biochemical blood markers of turkey hens.

Marker	Feeding groups				SEM
	I		II		
	M	SD	M	SD	
TP (g L^{-1})	78.9[A]	9.41	45.8[B]	4.34	1.12
GLU (mmol l^{-1})	14.71[b]	0.93	17.08[a]	1.25	0.26
CHOL (mmol l^{-1})	1.78[b]	0.52	2.77[a]	0.84	0.023
HDL (mmol l^{-1})	1.83[b]	0.38	2.25[a]	0.38	0.082
TG (mmol l^{-1})	0.29	0.05	0.32	0.05	0.024
UA (µmol l^{-1})	234.1[b]	30.1	297.5[a]	37.6	3.45
ALP (U l^{-1})	891.1[A]	79.5	476.94[B]	68.7	5.79
LDH (U l^{-1})	858.7[b]	37.4	1082.3[a]	49.2	1.83

M, The arithmetic average; SD, standard deviation; [a, b,] values in rows denoted with different letter differ significantly at p≤0.05 A, B, values in rows denoted with different letter differ significantly at p≤0.01.

level of white blood cells in the group of turkey hens with pathological signs (group II) compared to the control birds. The parameters of the biochemical of blood of turkey hens are presented in Table 5. The reported study did not demonstrate any significant differences between the groups in terms of the levels of triglycerides. Furthermore, a significant (p≤0.01) decrease was observed in the levels of total protein and alkaline phosphatase in the group of turkey hens displaying poorer health status, compared to the control birds. In the case of turkey hens exhibiting the poorer health status also levels of glucose, uric acid, cholesterol with HDL fraction, and the activity of lactic dehydrogenase appeared to be significantly (p≤0.05) higher than in the control birds. The results obtained in our study point also to a significant, compared to the control group, increase

Table 6. Level of mineral components in blood of turkey hens.

| Mineral component | Feeding groups | | | | SEM |
| | I | | II | | |
	M	SD	M	SD	
Ca (mmol L^{-1})	3.52[a]	0.30	2.91[b]	0.51	0.013
Mg (mmol L^{-1})	0.95[a]	0.08	0.80[b]	0.08	0.010
P (mmol L^{-1})	1.79	0.14	1.95	0.21	0.071
Zn (µmol L^{-1})	10.88[b]	3.50	18.65[a]	8.78	1.14
Fe (µmol L^{-1})	72.5[a]	3.59	59.6[b]	5.32	1.04
Mn (µmol L^{-1})	0.542	0.15	0.536	0.14	0.021

M, The arithmetic average; SD - standard deviation; [a, b], values in rows denoted with different letter differ significantly at p≤0.05.

Table 7. Level of pro- and antioxidative markers in blood of turkey hens.

| Marker | Feeding groups | | | | SEM |
| | I | | II | | |
	M	SD	M	SD	
H$_2$O$_2$ (µmol L^{-1})	2.34[b]	0.83	3.85[a]	0.07	0.19
MDA (µmol L^{-1})	6.49[b]	1.82	7.7[a]	1.08	0.058
SOD (U ml^{-1})	25.93	0.11	26.19	0.24	0.24
CAT (U ml^{-1})	6.88[b]	1.98	9.03[a]	1.90	0.27
FRAP (µmol L^{-1})	63.7	9.77	68.9	3.21	1.50

M, The arithmetic average; SD, standard deviation; [a, b,] values in rows denoted with different letters differ significantly at p≤0.05.

in the level of uric acid and urea in blood plasma. As shown by data presented in Table 6, there were no significant differences between the groups in the contents of phosphorus and manganese. The values reported for contents of calcium, magnesium and iron were significantly (p≤0.05) lower in Group II when compared to control birds. Values of the parameters of the redox system in blood plasma of turkey hens are presented in Table 7. No significant differences were noted between the groups in the activity neither of SOD nor in the level of FRAP. The analysis of results achieved in the reported experiment demonstrated also increase level of hydrogen peroxide and MDA in Group II by 39.2 and 15.7%, respectively compared to the control birds. Also the activity of CAT in the group of birds exhibiting pathological symptoms was higher (by 23.8%) than in the control group.

DISCUSSION

Worthy of notice is the fact that the microflora of litter displays some dynamics of changes closely linked with microclimatic conditions occurring in farm buildings. Amongst the saprophytic microorganisms the pathogenic ones are also likely to occur. All isolated fungi species

are often detected in the farm environment and represent opportunistic microorganisms that may become causative agents of a disease only under specified conditions. The microclimatic conditions in a building, feces composition and, consequently, litter parameters are changing over the entire rearing cycle of turkeys, which affects the sanitary status of litter. Such changes trigger the modification of bioaerosol in the air of the turkey rearing house, and with the air falling down changes are also proceeding in the microbiological composition of the surface of walls and partitions (Trawińska et al., 2008; Kołacz, 2000). As reported by Kołacz (2000), the hygienic status of the microenvironment animals are living in exerts a significant effect on their health status and rearing performance.

Hematological analyses are among the methods which may contribute to the detection of some changes in health status and can be a useful aid for diagnosis diseases in birds (Moreira dos Santos Schmidt et al., 2009). Levels of the hematological and biochemical blood markers of birds are affected by a number of factors, including age, sex, species, breed, feeding, physiological status, and rearing technology. Birds are additionally characterized by considerable individual diversification (Koncicki and Krasnodębska-Depta, 2005). As shown by results achieved in the reported study, no differences were noted between the groups in the value of hematocrit and in the number of erythrocytes. It should be emphasized, however, that the Ht values of turkey hens from both groups turned out to be lower than the reference values for turkeys (0.30 – 0.33) (Koncicki and Krasnodębska–Depta, 2005) and then the values reported elsewhere (Krasnodębska-Depta et al., 2003). In turn, the level of hemoglobin in the turkey hens with lower body weight gains and poorer health status turned out to be lower when compared to the control birds Despite the observed differences in hemoglobin level, the values reported were corresponding with results achieved for turkeys hens (without pathological symptoms) by Czech et al. (2010) and by Krauze et al. (2007). Results of the increased levels of hemoglobin and hematocrit in turkeys in the course of histomonadosis were noted by Koncicki et al. (1999).

In contrast, diminished levels of the erythrocytic system markers in turkeys infected with a pathogenic strain of E. coli bacilli were noted by Krasnodębska–Depta et al. (2003). The decreased level of hemoglobin is a symptom of anemia. It also occurs, though to a lesser extent, after hemorrhages as well as in the course of infectious, parasitic diseases and intoxications (Stankiewicz, 1973b). According to Konwicki and Krasnodębska–Depta (2005), it may also be due to pathogenic conditions, that is, colibacteriosis and coccidiosis. The increased levels of the erythrocytic system markers may also be explained by the condensation of blood due to body dehydration resulting from diarrhea. Hence, there is no simple elucidation of the reduced level of hemoglobin noted in our study owing to the observed symptoms of diarrhea occurring in the birds. The analysis of results achieved in the reported experiment demonstrated also a significant increase in the level of WBC in the group of turkey hens with pathological signs. The increased level of WBC in the course of histomonadosis in turkey hens was noted by Konicicki et al. (1999). Likewise, an increase in leucocytes level in the pathogenic condition of birds was observed by Krasnodębska–Depta et al. (2003).

The increased level of WBC in turkeys was also reported as affected by the administration of biostimulants (Czech et al., 2010; Truchliński et al., 2005a). In contrast, the diminished level of leucocytes resulting from the administration of synthetic antioxidants to turkey hens. Leucocytosis, that is, an increased number of leucocytes, occurs in the case of coccidiosis, in the acute course of salmonellosis and colibacteriosis in chickens and turkeys, as well as in histomonadosis and intoxication of turkeys with ionophore coccidiostats (Koncicki et al., 2000). As reported by Winnicka (2008), leucocytosis may appear in neoplasmic diseases, leukemias, anemias and different inflammatory states in the body. While elucidating the effect of biostimulants on WBC level increase, other researchers emphasize that leucocytosis may also indicate the stimulation of the immune system (Sembratowicz et al., 2004;

Wagner, 1996). The reported study did not demonstrate any significant differences between the groups in terms of the levels of TG. Nevertheless, worthy of notice is the noted TG level which in both groups turned out to be significantly higher than the reference values stipulated for turkeys (1.06 - 1.57) (Koncicki and Krasnodębska–Depta, 2005; Winnicka, 2008), and then the values reported by Makarski and Zadura (2006). The reduced concentration of triacylglycerols is observed in the case of afla- and ochratoxin as well as in monoensin intoxication (Rostek, 2010). In spite of that, the reported values of the above-mentioned markers were corresponding with results of a study conducted on turkeys by Rostek (2010). The increased level of total protein in the course of histomonadosis was noted by Koncicki et al. (1999). Also Krasnodębska–Depta et al. (2003) observed an increase in total protein level in plasma of turkeys infected with a pathogenic strain of E. coli bacilli. The same authors reported on a significant suppression of alkaline phosphatase activity in the study on turkeys infected with a pathogenic strain of E. coli bacilli. The suppressed activity of this enzyme may be indicative of turkeys anemia. It is also reported in the case of the inflammatory state of bones and in green liver disease (Bayyari et al., 1997; Krasnodębska–Depta et al., 2003).

According to Koncicki and Krasnodębska–Depta (2005), the reduced level of this marker was also noted as a result of heat stress in ducks, IBD (Gumboro disease), colibacteriosis of turkeys, and turkeys intoxication with salinomycin and furazolidone. In contrast, an increased level of this marker is observed in the course of rickets and osteomalacia, aflatoxins intoxication in hens and ducks, and neoplasmic processes in bones. Hypoproteinemia, namely a condition characterized by a decreased level of total protein, occurs in the case of protein losses to exudates (round heart disease of turkeys), in the course of viral diseases, that is, Gumboro disease, hydropericardium hepatitis syndrome and Newcastle disease, ochratoxicosis in hens, monoensin intoxication in turkeys, and under the influence of heat stress in hens and turkeys (Koncicki and Krasnodębska – Depta, 2005). According to Stankiewicz (1973a), it may also be due to enteric ailments, e.g. diarrhea, which corresponds with results of our study on turkeys with the observed cases of diarrhea. In turn, an increase in the total protein level (hyperproteinemia) is observed in conditions of rehydration, histomonadosis and colibacteriosis of turkeys (Koncicki and Krasnodębska–Depta, 2005).

The increased levels of glucose and uric acid were also observed by Truchliński et al. (2005a, b) in their experiments with turkey hens (without pathological symptoms). Stankiewicz (1973a) reports that hyperglycemia occurs also as a result of hypoxia and acute infectious diseases. The increased level of cholesterol in turkeys administered biostimulating preparations was also observed by Dmoch and Polonis (2007) as well as by Makarski and Zadura (2006). Similar results of the increased cholesterol level in the course of histomonadosis were noted by Koncicki et al. (1999) and Krasnodębska–Depta et al. (2003) in turkeys infected with a pathogenic *E. coli* strain. According to Koncicki and Krasnodębska–Depta (2005), hypercholesterolemia – being the condition characterized by an increased level of cholesterol, occurs in the course of histomonadosis in turkeys and MD (Marek diseases), as well as IBD in hens. The increased concentration of glucose (hyperglycemia) was observed in the course of amyloidosis in geese and ducks, ochrotoxin intoxication in chickens and upon heat stress in ducks. The increased values of those markers are likely to be due to hepatitis and kidney failure, as well as due to accelerated catabolism of proteins which usually occurs in bacterial diseases (Krasnodębska–Depta et al., 2003). The increased level of uric acid is also reported as a result of infectious diseases – ND, IBD, HHS (hydropericardium hepatitis syndrome) and colibacteriosis of turkeys, in the course of gouty diathesis and ochratoxin intoxication in hens, and ionophore coccidiostatics intoxication in turkeys (Koncicki and Krasnodębska–Depta, 2005). As indicated by literature data, increased levels of LDH enzyme may be observed in the course of histomonadosis (Koncicki et al., 1999) and in turkeys infected with a pathogenic strain of *E. coli* bacilli (Krasnodębska–Depta et al., 2003).

The enhanced activity of LDH may result from damage of liver and cardiac muscle (Makarski and Zadura, 2006; Truchliński et al., 2005a). Similar observations were made by Koncicki and Krasnodębska–Depta (2005) and Koncicki et al. (1999) who have, additionally, claimed that this condition is maintaining in all diseases proceeding with tissue necrosis, in the course of histomonadosis in turkeys, in intoxications with aflatoxin and monoensin, and as a result of many infectious diseases, including IBD, HHS and colibacteriosis. In the case of the turkey hens characterized by lower body weight gains and poorer health status, the study demonstrated also reduced levels of calcium, magnesium and iron. A diminished level of calcium in the case of turkeys exposed to a short-term heat stress was reported by Krasnodębska–Depta and Koncicki (2002). In turn, a reduced level of iron in blood of turkey hens receiving an additive of synthetic antioxidants was noted by Czech et al. (2010), whereas decreased levels of calcium and iron in turkeys administered copper chelate – by Dmoch and Polonis (2007).

According to Konwicki and Krasnodębska–Depta (2005), both an increase and a decrease in calcium level are observed in the case of rickets and osteomalacia, and as a result of many infectious diseases and intoxications. The observed decrease in the level of iron may be elucidated by the increased utilization of this element for heme synthesis, which may in a consequence affect an increase in hemoglobin level (Brodacki et al.,

2006). Hence, the diminished level of iron recorded in our study may be the reason of the reduced hemoglobin level in blood of the turkey hens from group II. In turn, the reported decrease in the level of magnesium may contribute to disorders in calcium and potassium metabolism as well as in water-electrolyte balance (Kłopocki and Winnicka, 1987). A significant ($p \leq 0.05$) increase in the blood level of zinc, when compared to the control birds, was observed in the group of turkey hens with a poorer health status (Group II). The increased level of this element was also reported as a result of biostimulants administration to turkeys (Ognik et al., 2004; Ognik and Sembratowicz, 2007). The excessive content of zinc may inhibit the absorption of copper and iron and accelerate the excretion of iron from the body. The high level of zinc additionally hinders iron and copper complexation into a heme molecule and may be the causative agent of anemia (Murray et al., 1994; Kleczkowski et al. 2004).

The maintenance of the balance between neutralization and production of reactive oxygen species is indicated by the enhancement in the activity of catalase as well as by increased levels of hydrogen peroxide and MDA in Group II. A lack of works addressing the effect of the pathological condition in poultry, turkeys in particular, on blood levels of redox parameters makes the comparison and confrontation as well as in-depth reference of own results with findings of other authors difficult. The higher level of MDA at the suppressed activity of SOD and CAT was noted by Truchliński et al. (2007) in turkey hens exposed to cooling and crowding stress. The observed increase in the concentration of lipid peroxidation products, that a significant increase in peroxides and malondialdehyde, indicates the oxidative stress occurring in the body of the birds (Truchliński et al. 2007). As shown in literature data (Kleczkowski et al., 2004; Ogryczak et al., 2001), the diminished activity of catalase is usually observed at the onset of the pathological process, whereas its enhancement appears after disease termination. The reported decrease in the level of iron might have also be the reason of the suppressed activity of catalase (Czech et al., 2010; Wieleba and Pasternak, 2001), for iron is the activator of this enzyme.

REFERENCES

Bartosz G (1995). Oxygens second face. PWN Warszawa [in Polish].
Bayyari GR, Huff WE, Rath NC, Balog JM, Newberry LA, Villines JD Skeels JK (1997) Immune and physiological responses of turkeys with green–liver osteomyelitis complex. Poult. Sci. 76:280-288.
Benzie IFF, Strain JJ (1996). The ferric reducing ability of plasma (FRAP) as a measure of "antioxidant power" the FRAP assay. Anal Biochem. 239:70-76.
Bomski J (1989). Basic hematological laboratory research. PZWL, Warszawa [in Polish].
Brodacki A, Batkowska J, Zadura A (2006). Influence of Different Poultry Production Systems on Hematological and Biochemical Indices in Blood of Turkey Broiler. Ann UMCS sec EE 24:335-342.
encek T, Ziomko I, Majdański R (2000). Acarinosis of laying hens induced by *Dermanyssus gallinae* (Redi, 1674). Medycyna

Wet. 56:14-116.

Czech A, Ognik K, Chachaj R (2010) Efficiency of synthetic antioxidant mixture additive in turkey hens' diet. Ann UMCS sec EE 4:1-7.

Dmoch M, Polonis A (2007). Influence of copper biopelx on selected hematological and biochemical indicators and content of mineral components in broilers blood. Acta Sci. Pol. Zoot. 6:11-18.

Faruga A, Pudyszak K (1999). Rearing efficiency and meat quality of slaughter turkey-hens fed with fodder supplemented with herbs. Zesz Nauk Prz Hod. 45:349-357.

Gay C, Gębicki JM (2000). A critical evaluation of the effect of sorbitol on the ferric-xylenol orange hydroperoxide assay. Anal. Biochem. 284:217-220.

Gay C, Gębicki JM (2002). Perchloric acid enhances sensitivity and reproducibility of the ferric-xylenol orange peroxide assay. Anal Biochem. 304:42-46.

Greenwald RA (1985). CRC Handbook of methods for oxygen radical research. CRC Press Boca Raton.

Kleczkowski M, Kluciński W, Sikora J, Kasztelan R (2004). The role of chosen mineral components in antioxidant processes in living organism. Medycyna Wet. 60:242-245.

Kłopocki T, Winnicka A (1987). Reference values of basic tests in veterinary medicine. Published by the Warsaw Agricultural University, Warsaw [in Polish].

Kocaman B, Esenbuga N, Yildiz A, Laçin E, Macit M (2006). Effect of Environmental Conditions in Poultry Houses on the Performance of Laying Hens, Int. J. Poult. Sci. 5:26-30.

Koncicki A, Krasnodębska-Depta A (2005). The possibility of exploit of hematological and biochemical research data in poultry diseases diagnostic. Magazyn Wet. Supl. Drób. 5:20-22.

Koncicki A, Krasnodębska-Depta A, Souleymane G (1999). Hematological and biochemical indicators of blood in turkey histomonasis. Medycyna Wet 55(10):674-676.

Koncicki A, Krasnodębska-Depta A, Szweda W, Rumińska-Groda E, Guiro S (2000). Turkey infections of Salmonella in Poland. Medycyna Wet 56:524-527.

Kołacz R (2000). Standards of hygiene, animal welfare and environmental protection in animal production in the light of the EU. Foundation "FUNDAR", Wroclaw.

Krasnodębska–Depta A, Koncicki A (2002). Effect of short-term heat stress on some blood biochemical indices in turkeys. Medycyna Wet 58:223-226.

Krasnodębska-Depta A, Koncicki A, Mazur-Gonkowska B (2003). Hematological and biochemical indicators in blood stream of turkeys infected pathogenic strain of E. coli. Medycyna Wet. 59:623-625.

Krauze M, Truchliński J, Adamczyk M, Modzelewska–Banachiewicz B (2007). A note on the effect of garlic, coneflower preparation and 1,2,4 – triasole derivative on the hematological and immunological indices of blood of Big – 6 turkey – hens. Ann UMCS sec DD 62:1–7.

Ledwożyw A, Michalak J, Stępień A, Kędziołka A (1986). The relationship between plasma triglicerydes, cholesterol, total lipids and lipid peroxidation products during human atherosclerosis. Clin. Chim. Acta 155:275-284.

Makarski B, Zadura A (2006). Influence of copper and lysine chelate on hematological and biochemical component levels in turkey blood. Ann UMCS sec. EE 24:357-363.

Moreira dos Santos Schmidt E, Paulillo EC, Regina Vieira Martins G, Moura Lapera I, Pereira Testi AJ, Nardi Junior L, Denadai J, Fagliari J (2009). Hematology of the Bronze Turkey (Meleagris gallopavo) Variations with Age and Gender. Int. J. Poult. Sci. 8:752-754.

MurrayRK, Granner DK, Mayes PA, Rodwell VW (1994). Harper's Biochemistry. PZWL Warszawa [in Polish].

Ognik K, Sembratowicz I (2007). Influence of Biostymina and Bioaron C on Some Anti – Oxidation Indices of Turkey – Hen's Blood. Pol. J. Environ. Stud. 16:209-212.

Ognik K, Sembratowicz I, Modzelewska–Banachiewicz B (2004). Concentration of chosen elements and activity of antioxidant enzymes in turkey hens' blood taking Echnovit C and 1,2,4-triazole derivate. J. Elementol. 9:445-449.

Ogryczak D, Jurek A, Kleczkowski M, Kluciński W, Sitarska E, Sikora J, Kasztelan R, Shaktur A (2001). The characteristic of SuperOxide Dismutase activity in blood of cows with a different mineral demand. Biul Magnezol. 6:616-622.

Rostek K (2010). The influence of vitamin E and trolox on biochemical and hematological indices in turkey hen's blood. Ann UMCS sec EE 28:26-33.

Sembratowicz I, Ognik K, Truchliński J (2004). Influence of Biostymina and Bioaron C on some anti-oxidation and immune indices of turkey-hens' blood. Ann UMCS sec EE 22:318-323.

Stankiewicz W (1973a). Laboratory research in veterinary diagnosis, PWN Warszawa [in Polish].

Stankiewicz W (1973b). Veterinary hematology, PWRiL Warszawa [in Polish].

Trawińska B, Saba L, Wdowiak L, Ondrasovicova O, Nowakowicz-Dębek B (2008). Evaluation of salmonella rod incidence in poultry in the Lublin Province over the years 2001–2005. Ann. Agric. Environ. Med. 15:173-167.

Truchliński J, Ognik K, Sembratowicz I (2007). Influence of prolonged and interrupted stress in a form of crowding and cooling of turkey-hens on anti-oxidation indices of blood. Medycyna Wet 63:95-98.

Truchliński J, Ognik K, Sembratowicz I, Szelkowska B (2005a). Influence of plant preparation Ginsengin 200 on chosen blood hematological and biochemical indicators and turkey breeding effects. Ann UMCS sec EE 23:343-348.

Truchliński J, Szelkowska B, Ognik K, Sembratowicz I (2005b). Influence of beta-glucan and nettles herb on chosen turkey blood indicators. Ann. UMCS sec EE 23:335-341.

Wagner H (1996). Immunstimulantien und Phytkoterapeutika. Zeitschrift Phyto. 7:91-96.

Wieleba E, Pasternak K (2001). Trace elements in animal antioxidant system. Medycyna Wet 57:788-791.

Winnicka A (2008). Referential value of basic laboratory research in veterinary. SGGW Warszawa [in Polish].

21

Turkey farming: Welfare and husbandry issues

Phil Glatz and Belinda Rodda

SARDI Pig and Poultry Production Institute, J. S. Davies Building, Roseworthy Campus, Roseworthy, South Australia 5371, Australia.

A review was undertaken to obtain information on the major welfare issues associated with turkey farming. In the hatchery there are some negative effects of long term storage of turkey fertile eggs on post-hatch growth and quality of chicks. There is a view that free range turkeys housed on deep litter in naturally ventilated sheds with natural light and access to forage and shelter belts is beneficial to bird welfare. However, an increase in mortality in the last few weeks of growth can be caused by very hot or cold environmental temperatures. Turkey welfare can be compromised at high stocking density. The selection of fast growing strains of turkeys has resulted in leg and locomotory problems. Mortality rates in turkeys caused by gait problems range from 2 to 4%. However, intermittent lighting improves bird activity and a decrease in locomotory problems. Under commercial conditions, domestic turkeys are often aggressive towards other birds. Beak treatment is used to prevent injuries caused by cannibalism, bullying, and feather and vent pecking with infrared beak treatment the most common trimming method used. However birds that have been severely beak treated can develop chronic pain. The barren environment of turkey houses has been identified as a major cause of poor animal welfare and responsible for cannibalism. Use of straw bales in the shed and elevated platforms gives the bird the chance to explore the environment and reduce pecking. Foot pad dermatitis (FPD) is a common condition in turkeys and is largely caused by wet litter. Apart from bird flu, Blackhead is one of the most serious poultry diseases in turkeys. Mortality can reach 70% in some flocks. Good management is essential to maintain turkey health and welfare including taking action to minimise contact of turkeys with wild birds and other animals. Pick-up of turkeys from sheds for transport to processing plant can result in welfare concerns. Mortality has long been a concern in relation to turkey transport. During this procedure the heads or wings of the birds can be injured against the solid sides of the crates, birds are exposed to temperature extremes, sudden acceleration and braking of the vehicle, vibration, fasting, injuries, social disruption and noise.

Key words: Turkey, welfare, cannibalism, beak treatment, stocking density, cannibalism, transport.

INTRODUCTION

For turkey meat birds to meet their genetic potential in growth there is a need for farmers to use best practice husbandry and management (Case et al., 2010). Good nutrition is a key factor in achieving high growth rates and good meat yield but management factors such as shed temperature and lighting also influence growth. Turkeys are raised on the floor in modern intensive barn systems and on free range farms and at the end of the growing period are transported to an abattoir where they are stunned and slaughtered (Hartung et al., 2009).

Welfare considerations associated with turkey production are becoming increasingly important (Kijowski et al., 2005). Genetic selection for rapid growth and higher body weight has resulted in health problems. Natural mating by commercial turkeys is difficult due to their high body weight and poor ability to mate. High stocking density on farms causes poor air quality and contributes to cannibalism as well as difficulty in inspecting all the birds. Farm procedures like catching, transport and slaughter can cause welfare problems for turkeys. Stress associated with transport especially in inappropriate containers on long journeys results in poor carcass quality. Stunning and slaughter can also have a high impact on animal welfare. To prevent cannibalism, turkeys are raised under low light intensity or are beak treated which may also result in a welfare concern (Ostovic et al., 2009). Poor welfare of breeding birds results from restricted feeding programs, and use of forced molting to control their body weight. There has been an increasing move toward organic poultry production in free range systems to overcome the issue of raising turkeys in sheds with high stocking density.

Housing systems can have a high influence on animal welfare. Turkeys grown in a traditional farming system are preferred by consumers (Kijowski et al., 2005) and there is a strong move toward producing organic poultry and the selection of slow growing lines which have better meat quality. Slaughter of such lines is usually carried out at about 25 weeks. Organically produced turkeys are kept mainly in Germany and the UK (Zeltner and Maurer, 2009). The organic diet must be sourced from ingredients grown without fertilisers, herbicides and insecticides. In some organic diets no synthetic additives are permitted in the diet. Organic is a term defined by law and all organic meat producers are governed by a strict set of guidelines, including registration and certification, production, permitted and non-permitted ingredients, the environment and conservation, processing and packing. This paper identifies the major welfare issues in turkey industry and the extent to which welfare is compromised.

Definition of welfare

There has been considerable debate about how animal welfare should be assessed (Fraser, 2003; Sandøe et al., 2004) with many definitions provided for animal welfare. The variation in both the definition and methods of assessing welfare has resulted in considerable debate and disagreement on how welfare should be assessed and interpreted in most farmed species (Hemsworth and Coleman, 1998). In this review we use the following definition: the provision of good welfare for turkey means meeting high standards of husbandry which includes care of animals, good housing, protection from the environment, maintaining good health, preventing disease, recognising and treating disease, providing good nutrition and good stockpersonship.

Assessment of turkey welfare

The Brambell Committee (Brambell et al., 1965) recommended all animals are entitled to good welfare and defined the basic freedoms. In Europe a project (LayWel, 2006) developed a series of welfare assessment protocols for laying hens. However no such protocols have been specifically developed for assessing turkeys although all aspects of the welfare of turkeys from placement on the farm and transport to the processing plant are summarised in a report on the welfare of turkeys by the Farm Animal Welfare Council (FAWC, 1995).

The layer welfare assessment system in LayWel (2006) has an emphasis on scoring animals according to their health status, feather cover, injuries and behaviour. The welfare scores reflect how the bird is interacting with its environment. Current methods of assessing welfare in the turkey industry have concentrated on assessing the impacts of housing and husbandry on production, behaviour and physiology. To assess bird welfare it is important that observations and measurements are made on individual animals. For example the poultry welfare quality project (LayWel, 2006) used the welfare freedoms as the basis for assessing bird animal welfare and focused on 4 welfare categories. These included; 1) injury, disease and pain; 2) hunger, thirst and productivity; 3) behaviour and 4) fear, stress and discomfort. Scores were given for a range of welfare risks in the various poultry production systems. The findings suggest that birds housed in more intensive systems are at a greater welfare risk and this may also apply to turkeys. Beak treatment is a controversial husbandry practice in turkeys. Beak treatment methods used have included using a cold blade, hot blade, biobeaker and the infrared method. One approach has been to assess welfare of beak treated birds on the basis of their performance. The results from these studies showed that the husbandry practice of beak treatment has a positive impact on bird welfare by reducing injuries to other birds, but the initial impact from the treatment was to cause a reduction in growth.

Another approach that has been used to determine the welfare impact of beak trimming was to assess the potential chronic pain in the beak stump by making an anatomical assessment of the incidence of neuromas. The presence of neuromas is an indicator that the bird may be feeling chronic pain in the beak stump and is a negative emotion that birds may experience as discussed by Duncan and Fraser (1997).

The welfare state of birds has also been assessed using preference and behavioural demand tests (Dawkins, 1980). These tests determined if preferences are influenced by the animal's feelings and place a value on the bird's choice (Dawkins, 1983) particularly when an evaluation is made of how hard the animal works to obtain their preference. In the case of turkeys preference testing of beak trimmed versus control birds has not been evaluated in terms of the resources they will select and if there is a decline in the strength of their demand to access a facility or resource because they may be feeling pain from beak trimming.

A further approach that has been used to assess welfare of turkeys has been to assess the effect of various stocking densities and lighting regimes on bird behaviour. Dawkins (2003) indicated it is difficult to attribute poor welfare in birds if certain behaviours are absent. More recently there has been a greater emphasis on behavioural indicators of poor coping such as fearfulness, aggression and stereotypies.

Embryo and chick mortality

Adoption of best practice fertile egg handling, storage, incubation and hatching conditions ensures hatched chicks have optimum health and welfare. There are some negative effects of long term storage of turkey fertile eggs. Storing fertile eggs for more than 1 week increases embryonic abnormalities and chick mortality which is a welfare issue. For example when eggs are stored for 4 days there are fewer chicks that die at hatch, better hatchability is achieved and there is reduced incubation time compared with eggs that are stored for 14 days. Post-hatch growth and quality of chicks and poults from fertile eggs stored for long periods also suffers (Fasenko, 1997).

Housing system

There are 3 types of turkey production systems; conventional, barn and free-range. In conventional housing, commercial turkeys are kept in enclosed houses (some with side curtains) with environmental control of heating, ventilation and lighting. Sheds can house up to 20,000 birds and on larger farms the turkeys are usually distributed across a number of sheds. Turkeys are grown to a variety of ages from 3 to 5 months depending on the strain used. Stocking density is adjusted by moving birds to other sheds or having a pickup for early slaughter.

In the free range system turkeys are housed on deep litter in naturally ventilated sheds with natural light and have access to forage and shelter belts. Kijowski et al. (2005) indicates that the free range system is beneficial to bird welfare. However, Herendy et al. (2004) found that carcass yield decreased in male and female birds housed in the extensive system. Burs and Faruga (2006) also noted problems in the outdoor system and found an increase in mortality in the last two weeks of growth (20 to 22 weeks) due to ground frost at night. Burs et al. (2007) reported no significant changes in blood plasma biochemistry of turkey-toms kept in a shelter with access to open-air runs compared to those raised traditionally in a brooder house. Birds kept in a free range system have a lower incidence of Foot Pad Dermatitis (FPD) compared to intensive systems (Sarica and Yamak, 2009, 2010) presumably because birds have greater exposure to damp litter while indoors.

Selection for high growth rate

The merging of the genetic stocks within a small number of individual companies has increased the demands on breeders to minimise disease risk, to maintain diverse genetic pools and to have appropriate breeding goals (Wood et al., 2006). However, the selection of fast growing strains has resulted in leg and locomotory problems for turkeys (Nestor, 1984; Nestor et al., 1985), higher mortality rates particularly under intensive housing conditions (Martrenchar et al., 1999). Fast growth has been accompanied by internal organs (such as the heart and lungs) lacking the capacity to meet the demands of metabolism. Birds have greater difficulty coping with heat stress (Yahav, 2007). Injuries on the back of the female from the claws of the male during mating are another issue that has emerged as a result of genetic selection for fast growing strains (Rauw et al., 1998). As a consequence, natural mating by breeder turkeys is very difficult due to their size and weight and artificial insemination is normally practiced by breeder farms.

ENVIRONMENT

Brooding

Supplementary heat is required for five weeks when brooding turkey poults in winter. Tom turkey poults require supplementary heat also during late summer because they are not yet fully feathered and the digestive system has not yet fully developed (Gencoglan et al., 2009).

Heat stress

Heat stress is major welfare problem in the turkey industry. Huge economic losses can occur because of mortality and decreased production due to high environmental temperatures. The utilization of food additives to improve poultry welfare during the hot summer months is essential under farm production conditions. Ascorbic acid has often been provided as a supplement to minimize the impact of heat stress. However, studies on ascorbic acid supplementation in the diet during summer showed that there was no improvement in body weight, feed intake, feed conversion ratio, slaughter weight, carcass yield, composition and thigh and breast pigmentation and shank and tibia bone characteristics (Konca et al., 2008). On the other hand addition of 1% of arginine to the feed contributed to a significant improvement in turkey breeders' welfare during summer. Birds showed more frequent dust-bathing, improved egg-laying and sexual behaviour and there was less aggression among birds (Bozakova et al., 2009). The heat stress for young turkeys (assessed by body temperature and surface temperature) when exposed to environmental temperatures from 25 to 35°C can be alleviated using mechanical ventilation from 1.5 to 2.5 m/s (Yahav et al., 2008).

Air quality

Ammonia and other toxic gases need to be kept below certain levels for bird health and also to maintain good production. If excessive, they reflect inadequate ventilation or poor litter management. Ammonia levels greater than 10 ppm can reduce feed intake with effects on body weight and production. It can cause lesions of the air sacs and cause inflammation of the cornea and conjunctiva (Carlile, 1984). Also, because stock people find high ammonia concentrations aversive, they are likely to give birds only a cursory examination during routine inspections and this could delay diagnosis of health problems. While there is some argument over whether birds can smell ammonia, the evidence of adverse effects on birds exposed to ammonia would suggest they are at least as sensitive as people (Wathes, 1998).

Ammonia concentrations increase with wet litter. Corrective action needs to be taken if concentrations exceed 20 ppm at the level of the bird. Patches of wet litter need to be removed from the shed and replaced with dry litter. Humans can smell ammonia at a level of 10 to 15 ppm; it irritates eyes and nasal mucous membranes at concentrations of 25 to 35 ppm. Thus, if there is an ammonia smell there is a potential air quality problem. In free range paddocks birds may create boggy patches in water which can also result in odour problems (albeit not ammonia). Ammonia concentrations (and other gaseous odours) are considered critical to bird welfare. Hydrogen sulphide is highly toxic to humans (and animals) with adverse clinical symptoms occurring above 10 ppm. If it can be smelt there is a serious ventilation problem. Its main source is from the anaerobic decomposition of faeces/manure and this is more likely to occur if litter becomes wet and caked. CO_2 cannot be detected by smell. At normal concentrations (0.3% or 3,000 ppm) CO_2 is involved in the regulation of respiration. Thus, if CO_2 in the atmosphere increases, this results in an increased respiration rate which functions to minimize the increase in body CO_2. Another noxious gas that is also odourless and colourless is carbon monoxide; concentrations should be below 50 ppm. Carbon monoxide can bind with haemoglobin in the blood much more easily than oxygen (210 times faster) and this drastically reduces the amount of haemoglobin available to carry oxygen. Exposure to all these odours may affect the respiratory system leading to health problems for turkeys (Fallschissel et al., 2009).

In domestic poultry there is a strong relationship between production and welfare (Al Homidan et al., 1998;

Feddes et al., 1995; Hayter and Besch, 1974; Kristensen et al., 2000). The immunological challenges often associated with poor air quality can lead to a reduction in feed intake and production (Kelley et al., 1987; Kemeny, 2000). Airborne particles could also increase the susceptibility of birds to diseases through irritant action or via allergic reactions (Harry, 1978). It is likely that improving air quality in turkey houses could improve production and provide a better working environment for stockpersons. Wathes (1998) indicated that the minimum ventilation rate required to provide acceptable levels of atmospheric dust should be 3.66 m^3/h/kg. High NH_3 concentrations in broiler and turkey houses can adversely affect bird performance. Acidifiers have been used in poultry houses to reduce NH_3 levels. The metabolic biostimulant, Bio-Kat reduced NH_3 concentration by 61% in the exhaust air of treated litter compared with untreated litter. The Bio-Kat treatment was most effective during the first 10 to 12 days, and its efficacy decreased over time (Shah et al., 2007). Reuse of litter also increases NH_3 emission rates by 130% compared to fresh litter while variability of NH_3 emission rates between houses can occur due to differences in the way litter is managed (Gay et al., 2006).

Air quality and environmental enrichment

There has been some work on providing an enriched environment for turkeys to reduce boredom and encourage locomotory behaviours. The behaviour of turkeys provided with elevated levels, straw bales and access to an open area were assessed in terms of air quality inside the barn as well as health and welfare of birds and humans and the resulting emissions into the environment (Hinz and Berk, 2002). Mean values of dust concentration were mostly below accepted limits (dust 3 mg/m^3, NH_3 20 ppm, CO_2 3000 ppm). The concentration of endotoxins ranged from a few hundred to 12 000 ng/m^3 with an overall average of 3000 ng/m^3. It was concluded that enriching the environment in turkey sheds by providing structures had no negative effects on animals' or farmers' health and welfare.

Lighting and turkey welfare

Photoperiod

A survey of the turkey breeder industry indicated that farmers have a good understanding of light management (Grimes and Siopes, 1999). Most farmers provide turkeys with 14 to 15 h of light daily with some farmers using day lengths of 16 to 18 h. Sodium lights were the most common light source, followed by incandescent and fluorescent light. However the EU has set new energy efficiency requirements for lamps. Incandescent light and halogen bulbs are being replaced by more energy-efficient bulbs.

There is a welfare view that there should be at least 6 h period of continuous darkness every 24 h and that intermittent lighting program should be avoided due to the increased incidence of leg abnormalities (Clarke et al., 1993). Most farmers do not use intermittent or step-down/step-up lighting programs although research indicates these programs improve gait (Classen et al., 1994) and reduce leg problems (Hester et al., 1986) due to an increase in bird activity. However, there has been some use of intermittent programs, because when the light is switched on there is an increase in bird activity which results in a higher feed intake and a decrease in locomotory problems (Hester and Kohl, 1989) and reduced pecking (Lewis et al., 1998). However a light regime of 8 periods of IL:2D reduced injuries caused by wing and tail pecking but increased injuries due to head pecking (Sherwin et al., 1999).

Light intensity

To prevent outbreaks of feather pecking, bullying and cannibalism the light intensity in sheds is usually maintained at low levels with turkeys preferring intensities of 5-25 lux (Sherwin, 1998). At low light levels turkeys find it difficult to explore the environment and they develop eye problems (Siopes et al., 1984). In addition stockpersons cannot detect birds that are being pecked, sick or need culling. No behavioral differences were observed between turkeys provided 10-80 lux (Denbow et al., 1990) although Barber et al. (2004) showed that turkeys have some preference for temporal variation in the lighting. Turkeys spent most of their time in the brightest light at 2 weeks of age, but in 20 and 200 lux at 6 weeks. At 2 weeks of age, all behaviours were observed to occur most often in 200 lux. At 6 weeks, resting and perching were observed least often in <1 lux, whereas all other activities were observed more in the two brightest light environments.

Yahav et al. (2000) observed that body weight of 18 week-old turkeys, was highest under the lowest light intensity of 10 lux and coincided with higher weight gain and lower food intake and better food conversion efficiency. Light intensity affected heart muscle weight but not the weight of breast muscle or abdominal fat. It was suggested that differences in feed conversion were related to differential investment of energy expenditure for maintenance.

Beak treatment

Under commercial rearing conditions, domestic turkeys are often aggressive towards other bird which can lead to serious injuries or even death. Aggressive encounters and injuries due to head, feather and tissue pecking seriously threaten the welfare of domestic turkeys and also result in economic losses for the turkey industry. Turkeys will exhibit an increase in aggression when unfamiliar birds are housed together but a marked drop in aggression occurs as they become familiar with each other (Buchwalder and Huber-Eicher, 2003, 2005). At present beak trimming is unavoidable if turkeys are reared in naturally ventilated sheds where light intensity cannot be controlled.

Methods of beak treatment

Following the development of the hot blade beak trimming machine in 1943 there have been refinements to the machine including some control of cutting and cauterisation and control of blade temperature. The Lyon Electric Company in San Diego, California has been manufacturing hot blade machines for beak-trimming turkeys for over 60 years. The Lyon Company (1982) suggest that precision beak-trimming of 6 to10 day-old turkey poults is one of the most accurate methods available. The machines have a timed cauterisation of 2 s and Lyon suggests that properly done, this method of beak-trimming will suffice for the productive life of the bird (Glatz, 2000).

The electric arc beak trimming machine uses a high voltage electrical current to burn a small hole in the upper beak of turkeys. In the 1980's the Bio-Beaker (Sterwin Laboratories, Millsboro, Delaware, USA) was developed which used a high voltage arc (1500 Volt AC electric current) across two electrodes to burn a small hole in the upper beak of turkeys. The primary advantage of the electric arc trimmer is that an adequate beak-trim is achieved during the first day of life, making the unit ideal for use in the hatchery. This allows treated birds to eat and drink normally for the first few days with their beaks intact. In turkeys, (Grigor et al., 1995) the beak tip fell off in 5 to 7 days and the wound healed by 3 weeks (Grigor et al., 1995; Noble and Kestor, 1997).

Since 2002, infrared beak treatment has been introduced and is by far the most popular methods used worldwide. It is an innovative procedure and uses an infrared (IR) energy source to treat the beak (Glatz, 2005). Immediately following treatment, the beak looks physically the same as it did before treatment and the bird is able to continue to use its beak normally. The IR method has proven to be safe, effective and the most welfare friendly method currently available of controlling cannibalism and feather pecking in poultry. Birds are restrained on a circular carousel using a head restraint and infrared energy is focused on the area of the beak being treated. The heat generated by an IR lamp penetrates the beak's outer layer to the epidermis. Damage to the epidermal layer, inhibits further germ layer growth. Immediately after treatment, the beak remains intact with the beak tip adjacent to the treatment line appearing lighter in colour due to reduction in blood flow to the treated area of the beak. There is no blood loss or open wound exposure. The equipment can be adjusted for various exposure times, levels of IR energy, and amount of beak treated. The beak tissues exposed to IR energy generally slough off after a few weeks giving the bird a blunted beak (Glatz, 2005).

Welfare issues associated with beak treatment

Jongman et al. (2008) indicated that the objections to the use of beak trimming in domestic poultry include the removal of sensory receptors, with a subsequent reduction in feed intake, pecking efficiency, pecking preferences, permanent loss of temperature and touch responses, behavioural evidence (hyperalgesia and guarding behaviour) for persistent pain and the potential for loss of magnetoreception (Mora et al., 2004). The adverse effects of beak trimming can be divided into; 1) acute pain while the procedure is performed (Grigor et al., 1995) until several days later (Lee and Craig, 1990), 2) sensory deprivation during a large part of the animal's life (Hughes and Michie, 1982; Gentle et al., 1997), and 3) chronic pain as a consequence of the formation of neuromas (Breward and Gentle, 1985; Gentle, 1986a; Lunam et al., 1996). Traumatic neuromas in the beak stump after trimming have been implicated as a cause of chronic pain in commercial hens (Breward and Gentle, 1985; Gentle, 1986b; Lunam et al., 1996).

In turkeys, a study on male turkey poults compared the effects of beak trimming (IR, hot blade and the electric arc methods) versus non-trimmed controls. Birds were fed either mash or crumbles. Beak trimming method did not affect time spent in feeding, foraging, drinking, preening, standing, or walking, nor did it affect body weight (Kassube et al., 2006; Noll and Xin, 2007). In the poults fed mash, feed efficiency was improved in all beak trimmed groups compared to controls. Up to 6 weeks of age, there was greater mortality in the group trimmed by the hot blade method compared to the control and the hot blade trimmed group showed the most beak re-growth (Noll and Xin, 2007). These results suggest that the amount of beak that was trimmed was insufficient to control feather pecking as indicated by the number of birds that had to be removed due to damaging pecking: 19% of controls and 21% of hot blade treated birds, compared to 7% of electric arc and 11% of infrared treated birds (Noll and Xin, 2006; Noll and Xin, 2007). This work would suggest that, in turkeys, infrared and electric arc beak trimming were preferable to either hot blade or no beak treatment although this depends on the amount of beak that is removed using the hot blade or treated using the IR and arc method. A study of the histology and pathology of infrared treated beaks showed that both upper and lower mandibles were damaged by

IR indicating the method was no better for bird welfare than the electric methods (Fiedler and Konig, 2006). However, as suggested by Gentle and McKeegan (2007), the IR process is automatic and standardized and can be performed with a greater degree of control than mechanical methods. Ruszler et al. (2004) also noted that because the process is performed at the hatchery, birds undergo less handling and have a reduced risk of injury.

A recent study indicates that IR beak treatment does not result in chronic pain or other adverse consequences for sensory function (McKeegan and Philbe, 2012). By looking at the long term effects of IR beak treatment on birds up to the age of 50 weeks they found that re-innervation and scarring was visible, but no neuromas or abnormal proliferations of nerve fibres were observed at any age. However, Glatz and Hinch (2008) reported the presence of persistent traumatic neuromas in birds that had been IR beak treated. This was unexpected leading to the conclusion that the application of excess heat or excess tissue removal was responsible for neuroma formation which may not be typical of routine IR beak treatment (McKeegan and Philbe, 2012). The poultry companies that are using the IR technology are putting emphasis on optimizing the beak treatment process and retaining the maximum amount of upper and lower beak tissue that is adequate to control feather pecking and cannibalistic behaviours (pers. comm. Andrew Gomer, Novatech).

Alternatives to beak treatment

Beak-trimming has been used for many years as a standard method to prevent cannibalism (Glatz and Bourke, 2006) but the technique is coming under increasing scrutiny. Currently a number of European countries are working towards the EU Welfare Directive by legislating for a ban on beak trimming. The EU had earlier indicated this must be achieved by January 1, 2011 but a number of countries have delayed implementation. The aim was to eliminate welfare effects of beak trimming on birds namely; reduction in feed intake, pecking efficiency, and pecking preferences, loss of temperature and touch responses and magneto-reception and overcoming persistent pain. The costs associated with cannibalism are significant. Mortality from cannibalism can be greater than 20% depending on the production system and management strategies (Glatz, 2005). Alternative methods have not been fully evaluated but what is known is discussed.

Lighting

As reported above low light intensity can be used to prevent feather pecking and cannibalism in turkeys.

However, the poultry house must be light proof to use the low light intensity strategy. In some turkey houses natural daylight can pass into the shed through fan cowlings and stimulate pecking in birds. It has been shown that, in small groups of intact male domestic turkeys, supplementary ultraviolet radiation, visual barriers, and added straw (environmental enrichment) minimize the incidence of injurious pecking for birds housed under incandescent light at an intensity of 5 lux.

Groups of non beak-trimmed birds up to 5 weeks of age were assessed at higher light intensities when provided fluorescent light and incandescent light (Moinard et al., 2001). Fluorescent light significantly reduced the incidence of tail injuries, and tended to reduce injuries to the wings, compared with incandescent light. No difference was observed between 5 and 10 lux for either tail or wing injuries. The incidence of tail and wing injuries was significantly and positively correlated with light intensity. Injuries to the head were minimal in all treatments. These results suggest that turkey poults may be kept with minimal injurious pecking, under fluorescent light at an intensity of 10 lux, with appropriate environmental enrichment.

Environmental enrichment

A study investigated how intensively housed turkey hens used different elements of environmental enrichment (elevated plateaus with ramp, straw bales, racks with perches and batches of pallets) under practical rearing conditions on a farm. The animals preferred the enrichment structures early in their life but as they aged, the use of the environmental elements was reduced. The elevated plateau was significantly preferred to straw bales, batches of pallets and racks with perches. Environmental enrichment using elevated resting places such as plateaus and straw bales were preferred by turkeys and seem to have a potential to improve their welfare (Spindler and Hartung, 2009).

Two open-sided houses with male turkeys were enriched with raised platforms, round and square bales of straw and wire baskets filled with hay. One separate turkey house was left unenriched as the control. The enrichment structures influenced the resting behaviour. In both enriched houses, the total time of locomotor activity was significantly lower on square bales of straw and on raised platforms as compared to the non enriched groups (Letzgub and Bessei, 2009). In both enriched houses, animals showed more locomotion in the unenriched area compared to the raised platforms and square bales. In enriched houses even less locomotor activity was observed than in the free space, because the turkeys preferred the raised platforms and square bales of straw for resting.

Additionally, Berk et al. (2002) found that the activity of turkeys was related to the spatial distribution of their faeces. Accumulation of faeces was highest in the veranda section of the poultry house adjacent the range. Enrichment of the range with bushes or trees is required to encourage activities in the range to improve animal health and welfare and achieve an even distribution of faeces. It is suggested that free range areas be provided forage, shade and protection to the hens. In commercial poultry, claims are made that the provision of string enrichment devices will eradicate the propensity to feather peck and thereby eliminate the need for beak trimming (Jones and Ruschak, 2002). Likewise Renz and Walkden-Brown (2007) found that a string enrichment device reduced pecking in chicks. This form of enrichment is likely to sustain the birds' interest, to promote desirable 'natural behaviours' like exploration and foraging, to potentially reduce boredom, and to significantly reduce the expression of feather pecking as well as the amount of pecking-related feather damage. String has the added advantages of low cost, durability and ready availability. Its beneficial effects are considered unlikely to be constrained by genotype or housing system. String enrichment devices have been on a number of layer farms in Europe (Jones, 2005).

Surveys in Europe have shown that increases in pecking is related to poor litter condition. A number of authors have suggested that feather pecking and cannibalism in domestic poultry may be considered as redirected ground pecking, based on strong similarities in the performance of both behaviours. Blokhuis and Wiepkema (1998) report the main strategy to prevent feather pecking is to provide an adequate substrate. Substrate conditions during the rearing period affect the development of feather pecking and the use of scratch grain is recommended. During the rearing period Gleaves (1999) recommended the location of semi sold milk or whey blocks around the house, hanging green leafy vegetables and spreading grass clippings to prevent feather pecking. An alternative approach is to use scratching trays in the shed and provide high fibre grain to encourage more forage related activities in birds.

There is potential to improve the ranging ability of laying birds in free-range systems and get the birds out of the shed (where they tend to feather peck) by using shelterbelts, crop rotations (Miao et al., 2006), shade and sand baths. Improving the attractiveness of the range for birds is therefore an important aspect to investigate. Currently many range areas are just fenced open fields with hardly any cover. This does not allow birds the opportunity to seek shelter from weather or predators, or make the free range area stimulating for the birds to use (Hegelund et al., 2002).

Studies have shown that there is a positive relationship between the availability of cover and the percentage of laying birds in the range (Zeltner and Hirt, 2003; Hegelund et al., 2002; Nicol et al., 2003; Bestman and Wagenaar, 2003). Enrichment with shade and shelter and providing a variety of these facilities enables birds to meet their behavioral needs. Trees provide an area where birds can dust bathe (Dawkins, 2003), and seek shade and protection from predators. More birds use the range area when cloud cover is prevalent (Hegelund et al., 2002) and when man made shade areas are provided. The use of the range decreases as the flock size increases. A greater percentage of the birds use the range in small flocks compared to larger flocks (Hegelund et al., 2002; Hirt et al., 2000). Hens in the range usually remain close to the poultry house (Furmetz et al., 2005) and leave the area denuded of forage. However, when trees or shrubs or shaded areas are provided about 75% of hens in larger flocks will use the range (Bestman and Wagenaar, 2003). Nevertheless poor use of the range by hens remains a major issue in all free range systems. Birds are unable to hide from predators if there is no overhead protection provided by trees or other shaded areas. Even though feather pecking is reduced when the hens use the free range frequently, feather pecking remains a serious problem on free range farms (Bestman and Wagenaar, 2003, 2006). Reduced feather pecking occurs when birds are reared in the same facility, stocking density is low, high quality litter is used and perches are provided (Bestman and Wagenaar, 2003, 2006; Knierim et al., 2008). Further work on the use of the above facilities on free range turkey farms is required.

Nutrition

In laying hens an adequate amount of insoluble fibre in the diet appears to be important for minimising the outbreak of cannibalism in laying hens and may have relevance to turkeys. Millrun, oat hulls, rice hulls and lucerne meal are effective sources of fibre. It has been suggested that the physical properties of the fibre, modulate the function of the gizzard, giving the birds a calm feeling. In addition it has been suggested that the increased rate of digesta passage, increases hunger and results in laying birds spending more time eating and less time pecking (Choct and Hartini, 2005).

Genetics

Aviagen Turkeys is the premier supplier of turkey breeding stock worldwide, supporting the brands of B.U.T. and Nicholas. Aviagen Turkeys has pedigree breeding programs in the USA and Europe. In domestic poultry, Kjaer (2005, 2009) reported on the considerable interest in the genetics of feather pecking and cannibalism. It was considered that a genetic solution might be more sustainable, efficacious and cost effective than beak-trimming. Differences in the rate of feather pecking, quality of plumage and mortality from cannibalism between populations of domestic fowl are well documented. The nature of the genetic background

of these differences is less well known.

Selection lines differing in the propensity to perform feather pecking or cannibalistic pecking have been developed. Realised heritability of 0.1 to 0.7 has been reported.

Stocking density

Stocking density is an important issue in turkey welfare with high stocking density being a major animal welfare concern. Currently there are a wide range of recommendations for stocking densities for growing turkeys. Gunthner and Bessei (2006) showed that the behavioural effects of stocking density are only observed for sitting/lying, preening and feather pecking, with more sitting/lying, preening and feather pecking at the lower stocking densities.

The behavioural responses to the different stocking densities were generally small in magnitude. Bessei and Gunthner (2006) determined the water consumption of male and female turkeys under different stocking densities throughout the growing period and the influence of disease, vaccination and medical treatment on water intake. There was no significant effect of stocking density on water and feed intake and water: feed ratio. Abdel-Rahman (2005) observed that turkey welfare was poorer at higher stocking densities with impacts on behaviour, higher blood corticosterone levels, reduced body weight and poorer health.

Majumdar et al. (2003) indicated that poults reared in floor spaces of <0.30 m^2/bird had lower feed intake and better feed conversion ratio (FCR) during the prestarter period compared with poults reared in a larger floor area of 0.46 m^2/bird. However, during the starter period, poults reared in a smaller floor area consumed less feed but there was no significant difference in the FCR. Buchwalder and Huber-Eicher (2004) found that in small groups of turkeys, an increase in floor space reduces the number of aggressive pecks and threats aimed at introduced unfamiliar birds. Additionally, they found evidence that there might be a critical distance below which retreating from an opponent is not successful in avoiding aggressive encounters.

Martrenchar et al. (1999) compared different animal welfare traits at three different stocking densities of 8, 6.5 and 5 birds/m^2. No decrease in locomotory activity was observed at the highest density, contrary to results reported for broiler chickens (Blokhuis and van der Haar, 1990; Lewis and Hurnik, 1990; Martrenchar et al., 1997). However, resting birds were more distracted by other birds as the stocking density increased. The birds' gait appeared to be worst at the highest density. Birds reared at a density of 8 birds/m^2 showed a higher incidence of hip lesions (scabs and scratches) and of FPD than those reared at 6.5 or 5 birds/m^2 indicating that bird welfare is compromised at the highest density.

Group size

Few authors have studied the influence of group size on welfare. In laying hens the incidence of cannibalism increases with groups less than 12 birds (Hughes and Wood-Gush, 1977). Conversely, in large groups (more than 100 birds) the difficulty of establishing a stable social hierarchy makes feather pecking behaviour independent of group size (Hughes et al., 1997). It has been demonstrated that it is possible to keep male turkey broilers at a high light intensity (60 lux) without the occurrence of severe feather pecking if group size and stocking density are low (animals housed in pairs at a density of 0.2 birds/m^2) (Sherwin and Kelland, 1998). Further studies to determine the optimal group size in turkeys are required.

Litter

Foot pad dermatitis (FPD)

FPD is a common condition amongst commercially grown turkey poults and is largely caused by litter quality. In turkeys 48% of female and 46% of male flocks have noticeable signs of FPD (Martrenchar et al., 2001). The skin of the footpad becomes hard and scaly, often developing horn-like pegs of abnormal keratin. The footpad can become swollen and frequently splits. In the centre of the lesion the epidermis separates, and is often totally necrotic. The cause of FPD is complex, but many contributing factors have been suggested, such as diet (Clark et al., 2002), skin structure, bird weight and sex, litter moisture, litter type (Mayne, 2005) and ventilation (Martrenchar et al., 2001). It may not be possible under high commercial stocking densities to have flocks with a low prevalence of FPD (Martrenchar et al., 2001).

Litter type

The litter types used are mainly straw and wood shavings. Litter quality is affected by factors such as stocking density, air temperature and moisture, season, consistency and amount of faeces and drinker design. Wet litter is one of the key factors affecting FPD, followed by biotin deficiency. Turkey poults reared on wet litter have an increased incidence and severity of FPD lesions, but the problem is alleviated by replacing the wet litter with dry litter. Recent research has demonstrated the association between biotin levels and FPD. There are some indications that increased stocking density is associated with an increase in FPD. Supplementation of the diet with biotin has been shown to reduce the severity and incidence of lesions if birds are reared on dry litter, but if on wet litter, lesions may still occur (Mayne, 2005).

Due to differences in water adsorption capacity and the rate of its further release, litter moisture in turkey sheds can be significantly affected by the type of material used. Turkeys are housed for longer periods than broilers and the impact of litter quality on FPD is greater. Kuczynski and Sobodzian-Ksenicz (2002) compared; 1) long rye straw and softwood shavings in a summer-autumn flock, 2) long rye straw and chopped rye straw in an autumn-winter flock and 3) softwood shavings and chopped straw in a spring-summer flock. The amount of litter caking was increased on long straw and the resulting incidence and severity of FPD caused birds to suffer poor health, welfare and production. Youssef et al. (2009) housed birds on dry, clean wood shavings and replaced it daily with fresh, clean and dry litter. There was no effect of uric acid or NH$_4$Cl in the litter, but the FPD severity was increased markedly by litter with higher water content. Dairy compost was evaluated as a possible bedding substrate for turkeys compared to shavings. There were no significant differences in livability, but body weight was lower for birds housed on the dairy compost and there was a greater incidence of FPD (Frame et al., 2004).

Sobodzian-Ksenicz et al. (2008) investigated the effect of applying two different additives (brown fine coal and microbe vaccine solution) to the straw litter on the physical and chemical characteristics of bedding and on turkey performance. Both additives led to a significant rise in litter temperature, which positively affected its physical parameters and contributed to the improvement of the birds' welfare and performance (lower mortality, higher final weight). Pintaric and Dobeic (2000) showed that the addition of a bioenzymatic additive resulted in a 12% drop in ammonia release from the litter. By adding the bioenzymatic additive to bedding more frequently, it was possible to achieve a larger drop in ammonia and unpleasant odour emissions.

Productivity

The productivity of turkeys raised on deep litter was compared to those raised on a slatted floor (Oblakova et al., 2004). Significantly higher live weight and better FCR was found for birds raised on a grid floor compared to birds on deep litter. However, Wojcik et al. (2004) found that turkey cocks kept on a slatted floor made of metal mesh had lower final body weight and higher body weight losses during transportation and a higher number of birds with damaged carcasses in comparison with the turkey cocks kept on a litter floor.

Disease

The introduction of Codes of Practice for Poultry in a number of countries (eg. Standing Committee on Agriculture and Resource Management, 2003) has meant that persons who are responsible for turkeys must ensure that the bird's health and welfare are maintained. The basic requirements of the Codes of Practice that relate to health include provision of sufficient food and water to sustain health, protection from disease (including those diseases that are caused by poor management) and avoidance of pain, distress, suffering and injury in birds.

Good management is essential to maintain turkey welfare including taking action to minimize contact of turkeys with wild birds and other animals. Appropriate hygiene, proper housing, and brooding and appropriate stocking density are essential when welfare of turkeys is being judged. The housing facilities and equipment used in turkey farming need to be cleaned and disinfected as much as is practicable before restocking to prevent the carry-over of disease-causing organisms to incoming birds. Free range turkeys should not be kept on land which has become contaminated with organisms which cause or carry disease to an extent which could seriously prejudice the health of turkeys (Standing Committee on Agriculture and Resource Management, 2003). Preventative health programs and performance targeting can greatly contribute to the efficiency and ultimate viability of turkey farms. Good management requires that sick and injured turkeys are treated without delay and isolated if necessary. Records of sick animals, deaths, treatment given and response to treatment need to be kept to assist disease investigations. Each turkey shed should include, whenever necessary, a hospital pen where sick birds should be placed (European Council, 1998). Turkeys which have an incurable disease, irreparable injury or painful deformity that create unacceptable levels of suffering should be humanely euthanased. The euthanasia of animals raises welfare problems. Regulation proposals state that the method used should not cause pain or distress. Drowning or suffocating by methods such as putting live birds into tied-up bags is forbidden (European Council, 1998). In practice, farmers use cervical dislocation.

Biosecurity

A survey was conducted to determine the potential disease/pathogen risk pathways on commercial turkey farms. A questionnaire was sent to the farms which related to domestic and wild birds on the farm, proximity to waterfowl habitats, water sources and treatment, biosecurity practices, personnel, vehicles and equipment movement and disposal methods for dead birds, litter and manure. It was shown that drinking water, movement of personnel between farms and contact with wild birds were the main potential pathways for pathogen transfer to domestic birds (Rawdon et al., 2008).

Salmonella and Campylobacter

The bacteria Salmonella and Campylobacter are bacteria which cause the industry significant concern particular

with food safety. A study to estimate prevalence and risk factors for Salmonella spp. and Campylobacter spp. caecal colonization (Arsenault et al., 2007) found that in turkeys the odds of Salmonella colonization were 5-8 times greater for flocks which allowed visitors to enter the premises especially staff who came from the hatchery. The prevalence of Campylobacter-positive flocks was 46% for turkeys. For turkeys the odds of Campylobacter flock colonization were 3 times greater in flocks having a manure heap 200 m from poultry house and 4 times greater in flocks drinking un-chlorinated water. Uncertainty exists concerning the key factors contributing to Campylobacter colonization of poultry, especially the possible role of vertical transmission from breeder hens to young birds. A longitudinal study of Campylobacter colonization was performed in turkey flocks (Smith et al., 2004). Management practices such as proper litter maintenance, control of people movement between the farm and other turkey flocks, were likely responsible for the absence of Campylobacter in the flocks before processing.

Food borne salmonella outbreaks in humans have been associated with consumption of foods of animal origin, including turkey meat from processing plants. Trampel et al. (2000) found that to reduce Salmonella on turkey carcasses may require removal of litter and faeces from feathers before turkeys enter a processing plant. Preslaughter practices of feed withdrawal, catching, loading, transport, and holding do not significantly alter the prevalence of Salmonella in market-age turkeys. It may be possible to monitor the Salmonella status of turkey farms based on samples collected at the abattoir (Rostagno et al., 2006). The ability of 2 probiotic cultures (P1 and P2) to reduce environmental Salmonella in commercial turkey flocks 2 weeks prior to processing with or without the use of a commercial organic acid was evaluated (Vicente et al., 2007). The administration of selected probiotic candidate bacteria in combination with organic acids can reduce environmental Salmonella in turkey houses prior to transport, and that this practice could help to reduce the risk of Salmonella cross-contamination in the processing plant. Intestinal tracts of turkeys from 10 conventional turkey farms, where antimicrobials were routinely used, and 5 organic turkey farms, where antimicrobials had never been used, were collected and cultured for Campylobacter species (Luangtongkum et al., 2006). None of the Campylobacter isolates obtained were resistant to gentamicin, while a large number of the isolates from both conventional and organic poultry operations were resistant to tetracycline. Multidrug resistance was observed mainly among Campylobacter strains isolated from the conventional turkey operation (81%).

Shed disinfection

Mueller-Doblies et al. (2010) showed that disinfectants

containing a mixture of formaldehyde, glutaraldehyde and quaternary ammonium compounds perform better under field conditions than oxidising products and should therefore be the first choice for disinfection of turkey premises to eliminate *Salmonella* contamination.

Field study

A field study was conducted to estimate the sanitary condemnation proportion in male turkey broiler flocks, to describe the reasons for condemnation and the related macroscopic lesions, and to investigate whether primary production information would predict the risk of condemnation (Lupo et al., 2010). Emaciation, arthritis were the main reasons for condemnation, representing 76% of the condemned carcasses. Three variables were significantly associated with increased risk of condemnation: locomotor disorders on the farm, high cumulative mortality 2 weeks before slaughter, and clinical signs observed during the *ante mortem* inspection at the slaughterhouse.

Role of probiotics

The effects of selected probiotic bacteria or antibiotics on performance of poults suffering mild idiopathic diarrhoea and stunting (Higgins et al., 2005) were compared. Poults receiving antibiotics followed by a probiotic culture had significantly higher weight gain than non treated or probiotic-treated poults.

Control of meal worms

To control the lesser mealworm, *Alphitobius diaperinus* (Panzer), in turkey houses (Salin et al., 2003) a combined treatment included an adulticidal compound (pyrethroid: cyfluthrin) and a larvicidal compound (insect growth regulator (IGR): triflumuron). The combined insecticide treatment greatly reduced the adult and larval stocks throughout the different growing periods, and control of *A. diaperinus* populations was achieved by the end of the second treatment.

Blackhead

Blackhead may be the most serious poultry disease in turkeys (Beyer and Moritz, 2000). Mortality has been reported to reach up to 70% in some flocks. Early signs of this disease include drowsiness, drooping of the head and wings, walking with an unusual gait, soiled vent feathers due to diarrhoea and bright yellow faeces resulting from the infection of the liver. The bird also may become anorexic leading to a considerable loss of weight and a depressed, weak appearance. Sometimes, the head of the bird appears to be cyanotic, which is a bluish

or black discoloration of the skin due to deficient oxygenation of the blood-hence the name, Blackhead. Once infected, it often is difficult to eliminate Blackhead from a flock. Therefore, prevention is the best strategy. One medication that can be used as a preventative is Nitarsone, which also is known as Histostat-50 (Alpharma, Fort Lee, New Jersey). Histostat-50 is a premix that can be added to the feed on a continuous basis up to 5 days prior to processing or marketing.

Turkey pick up and transport

Pick-up of turkeys from farms and transport can result in welfare concerns. Catchers are often required to carry birds upside down through a shed to a truck outside especially when the containers are not able to be taken inside the shed. Birds are usually caught by one or both legs and then placed into the crate. During this procedure the heads or wings of the birds can be injured against the solid sides of the crates. These methods are criticized by the European Council (1998). It has been shown in broiler chickens that, although corticosterone levels were higher in birds handled in an inverted position than in those handled in an upright position, stress due to the crating was greater than stress due to handling before crating (Kannan and Mench, 1996).

Surveys on meat chicken have identified a high prevalence of heart failure and dislocation of the femur at the hip, probably due to the stress of catching, loading and transporting and to catching and carrying birds by one leg, respectively (Gregory and Austin, 1992). Wing injuries may occur when the wings protrude out of the crates and become trapped between containers during loading and unloading.

Four designs of a modular turkey transport systems were compared for carcass damage and heart rate of turkeys during loading. Three systems required turkeys to be manually loaded. Another system was loaded by herding turkeys into it (Prescott et al., 2000). Birds in the manually loaded systems had similar levels of fractures and bruising. Birds which were herded into the module had less damage and heart rate was lower for birds.

Wichman et al. (2009) investigated how different crate heights affected the turkey's ability to alter their body position while being transported and what effects this might have on their welfare using behavioural observations. The main findings from the study were that the degree of physical confinement in the cages influenced the bird's behaviour and low height crates decreased the bird's ability to move and change their position.

Transport

Mortality has long been a concern in relation to poultry transport (Bayliss and Hinton, 1990). When birds are being transported they are exposed to a number of stressors including temperature extremes, sudden acceleration and braking of the vehicle, vibration, abrasion on the crates, fasting, withdrawal of water, social disruption, noise and high temperature. Kowalski et al. (2001) showed that transport of turkeys caused a considerable increase in the levels of creatine kinase, triglycerides, corticosterone, adrenaline and noradrenaline, as well as a decrease in the total lipid content. Crowding and overheating resulted in a significant decrease in the level of glucose, values of humoral and cellular immunity indices, as well as an increase in the concentration of triglycerides (overheating) and corticosterone (crowding). Activation of the sympathetic system via increased plasma levels of adrenaline and noradrenaline were observed. The transport of live animals has important economic and welfare implications. A commercially-available organic acid product (Optimizer™) was added to the drinking water of commercial turkeys during preslaughter feed withdrawal (Pixley et al., 2010). A significant reduction in rate of weight loss during holding at the processing plant was observed in the treated turkeys.

Young poult transport

A study done on turkey flocks identified hatchery- and transportation-associated risk factors for poult mortality in the first 14 d after placement (Carver et al., 2002). Hatchery and transportation-related risk factors for flock mortality included desnooding, truck, truck temperature, shipping time, and weather conditions at placement.

Carcass lesions

The duration of transport between farm and slaughterhouse has been positively correlated with the prevalence of some carcass lesions (McEwen and Barbut, 1992).

Stunning

Sometimes, the birds are not sacrificed properly, which makes manual sacrifice necessary (Mota-Rojas et al., 2008). The main point of concern regarding the slaughtering procedure itself is the intensity of the stunning current in water bath stunners. In the EU the recommended minimum current for a turkey is 150 mA per bird applied for a minimum of four seconds when delivered as 50 Hz sinusoidal alternating current. These parameters have been shown to induce cardiac arrest in >90% of turkeys and so eliminate chances of recovery (Gregory and Wilkins, 1989). The effect of stunning method (gas vs. electrical) on some turkey breast meat quality traits was evaluated. Turkey breast meat from gas stunned birds seems to have more favourable quality characteristics in comparison to breast meat of electrical stunned birds.

A new humane slaughter method for broilers using low atmospheric pressure was aimed at developing an alternative method of slaughtering broilers adapted to existing plant equipment. This method could be adopted in turkeys. Insensibility via electroencephalogram (EEG) and electrocardiogram (ECG) and loss of posture was recorded in birds. A 90% reduction in the EEG and ECG signal occurred within 32 s after a pressure of 21.4 kPa was reached; within 35 seconds the chickens' heart exhibited complete fibrillation of both the atria and ventricles. Finally, at 37 seconds after attaining the desired pressure, loss of posture was recorded indicating death (Thaxton et al., 2009).

CONCLUSION

The welfare of turkeys can be affected during most processes from the hatchery, rearing on the farm, transport and processing. Genetic selection, housing conditions, transport and slaughter can all be causes of poor welfare. The major issues are concerns associated with disease, poor locomotion due to high growth rate, chronic pain from beak treatment, behavioral problems caused by high stocking density, lack of enrichment in the turkey house and in the free range and poor air quality in turkey sheds. Depopulation of sheds and transport and slaughter can also result in poor welfare for birds. Changes in the current practices may lead to higher costs which cannot be sustained by producers.

REFERENCES

Abdel-rahman MA (2005). Study on the effect of stocking density and floor space allowance on behaviour, health and productivity of turkey broilers. Assiut. Vet. Med. J. 51(104):1-13.

Al Homidan A, Robertson JF, Petchey AM (1998). Effect of environmental factors on ammonia and dust production and broiler performance. Brit. Poult. Sci. 39 (Supplement): S10.

Arsenault J, Letellier A, Quessy S, Normand V, Boulianne M (2007). Prevalence and risk factors for Salmonella spp. and Campylobacter spp. caecal colonization in broiler chicken and turkey flocks slaughtered in Quebec, Canada. Prev. Vet. Med. 81(4):250-264.

Barber CL, Prescott NB, Wathes CM, Sueur C le, Perry GC (2004). Preferences of growing ducklings and turkey poults for illuminance. Anim. Welf. 13(2):211-224.

Bayliss PA, Hinton MH (1990). Transportation of broilers with special reference to mortality rates. App. Anim. Behav. Sci. 28:93-118.

Berk J, Haneklaus S, Schnug E (2002). Free range behaviour of male turkeys and its effect on the spatial variability of phosphorous in soil. In: Koene, P. (Ed) Proceedings of the 36th international congress of the ISAE, the Netherlands, P. 133.

Bessei W, Gunthner P (2006). Water intake in growing turkeys. World Poult. 22(Special):10-12.

Bestman MWP, Wagenaar JP (2003). Farm level factors associated with feather pecking in organic laying hens. Livest. Prod. Sci. 80:133-140.

Bestman MWP, Wagenaar JP (2006). Feather pecking in organic rearing hens. Joint Organic Congress, Odense.

Beyer RS, Moritz JS (2000). Preventing Blackhead Disease in Turkeys and Game Birds, Kansas State University.

Blokhuis HJ, Van der Haar JW (1990). The effect of stocking density on the behaviour of broilers. Archiv. für Gef. 54:74-77.

Blokhuis HJ, Wiepkema PR (1998). Studies of feather pecking in poultry. Vet. Quart. 20:6-9.

Bozakova N, Oblakova M, Stoyanchev K, Yotova I, Lalev M (2009). Ethological aspects of improving the welfare of turkey breeders in the hot summer period by dietary L-arginine supplementation. Bulg. J. Vet. Med. 12(3):185-191.

Brambell FWR, Barbour DS, Barnett MB, Ewer TK, Hobson A, Pitchforth H, Smith WR, Thorpe WH, Winship FJW (1965). Report of the Technical Committee to Enquire into the Welfare of Animals Kept Under Intensive Husbandry Systems. Her Majesty's Stationery Office, London.

Breward J, Gentle MJ (1985). Neuroma formation and abnormal afferent nerve discharges after partial beak amputation (beak-trimming) in poultry. Experientia 41:1132-1135.

Buchwalder T, Huber-Eicher B (2003). A brief report on aggressive interactions within and between groups of domestic turkeys (Meleagris gallopavo). App. Anim. Behav. Sci. 84(1):75-80.

Buchwalder T, Huber-Eicher B (2004). Effect of increased floor space on aggressive behaviour in male turkeys (Meleagris gallopavo). Appl. Anim. Behav. Sci. 89(3/4):207-214.

Buchwalder T, Huber-Eicher B (2005). Effect of group size on aggressive reactions to an introduced conspecific in groups of domestic turkeys (Meleagris gallopavo). Appl. Anim. Behav. Sci. 93(3/4):251-258.

Burs M, Faruga A (2006). Effect of husbandry conditions and genotype of young slaughter turkeys on production and slaughter traits. Pol. J. Nat. Sci. 21(2):647-657.

Burs M, Glogowski J, Faruga A (2007). Some plasma biochemical parameters of young slaughter turkeys raised under varied environmental conditions. Pol. J. Nat. Sci. 22(2):214-227.

Carlile FS (1984) Ammonia in poultry houses: a literature review. World's Poult. Sci. J. 40:99-113.

Carver DK, Fetrow J, Gerig T, Krueger KK, Barnes HJ (2002). Hatchery and transportation factors associated with early poult mortality in commercial turkey flocks. Poult. Sci. 81(12):1818-1825.

Case LA, Miller SP, Wood BJ (2010). Factors affecting breast meat yield in turkeys. World's Poult. Sci. J. 66(2):189-202.

Choct M, Hartini S (2005). Interaction between nutrition and cannibalism in laying hens. In Poultry Welfare Issues, Beak Trimming (ed. PC Glatz), Nottingham University Press, pp. 111-115.

Clark S, Hansen G, McLean P, Bond P Jr, Wakeman W, Meadows R, Buda S (2002). Pododermatitis in turkeys. Avian Dis. 46(4):1038-1044.

Clarke JP, Ferket ER, Elkin RG, McDaniel CD, McMurtry JP, Freed M, Krueger KK, Watkins BA, Hester PY (1993). Early dietary protein restriction and intermittent lighting. 1. Effects on lameness and performance of male turkeys. Poult. Sci. 72:2131-2143.

Classen HL, Riddell C, Robinson FE, Shand PJ, McCurdy AR (1994). Effect of lighting treatment on the productivity, health, behaviour and sexual maturity of heavy male turkeys. Brit. Poult. Sci. 35:215-225.

Dawkins MS (1980). Animal Suffering: The Science of Animal Welfare. Chapman and Hall, London.

Dawkins MS (1983). Battery hens name their price: consumer demand theory and the measurement of animal needs. Anim. Behav. 31:1195-1205.

Dawkins MS (2003). Behaviour as a tool in the assessment of animal welfare. Zoology 106:383-387.

Denbow DM, Leighton AT, Hulet RM (1990). Effect of light sources and light intensity on growth performance and behaviour of female turkeys. Brit. Poult. Sci. 31:439-443.

Duncan IJH, Fraser D (1997). Understanding animal welfare. In: Appleby MC, Hughes BO (eds) Animal Welfare, CAB International, Oxon, pp. 19-31.

European Council (1998). Recommendation Project on Turkey Broilers (7th draft). Permanent Committee of the European Convention on Farm Animal Welfare, Strasbourg, France. P. 16.

Fallschissel K, Klug K, Kaempfer P, Jaeckel U (2009). Detection and identification of airborne bacteria in a German turkey stable. In: Sustainable animal husbandry: prevention is better than cure, Volume 2. Proceedings of the 14th International Congress of the International Society for Animal Hygiene (ISAH), (Eds: Briese, A., Clauss, M., Springorum, A. and Hartung, J.) Vechta, Germany. pp. 593-596.

Fasenko GM (1997). Turkey Egg Storage: Effects on Embryo and Poult

Viability. University of Alberta. Poult. Res. Centre News 6:1.

FAWC (1995). Report on the welfare of turkeys. www.fawc.or.uk/reports/turkeys. Accessed on 11 November, 2013.

Feddes JJR, Taschuk K, Robinson FE, Riddell C (1995). Effect of litter oiling and ventilation rate on air quality, health and performance of turkeys. Can. Agric. Eng. 37(1):57-62.

Fiedler HH, Konig K (2006). Assessment of beak trimming in day-old turkey chicks by infrared irradiation in view of animal welfare. Archiv. fur Gef. 70:241-249.

Frame DD, Kelly EJ, Fields N, Bagley LG (2004). Dairy compost as a source of turkey brooder bedding. J. App. Poult. Res. 13(4):614-618.

Fraser D (2003). Assessing animal welfare at the farm and group level: the interplay of science and values. Anim. Welf. 12:433-443.

Furmetz A, Keppler C, Knierim U, Deerberg F, Hess J (2005). Laying hens in a mobile housing system – Use and condition of the free-range area. In: Hess, J., Rahmann, G. (Eds) Ende der Nische, Beiträge zur 8. Wissenschaftstagung Ökologischer Landbau, Kassel P. 313.

Gay SW, Wheeler EF, Zajaczkowski JL, Toppe RPA (2006). Ammonia emissions from U.S. tom turkey growout and brooder houses under cold weather minimum ventilation. App. Eng. Agric. 22(1):127-134.

Gencoglan S, Gencoglan C, Akyuz A (2009). Supplementary heat requirements when brooding tom turkey poults, South. Afr. J. Anim. Sci. 39(1):1-9.

Gentle MJ (1986a). Neuroma formation following partial beak amputation (beak-trimming) in the chicken. Res. Vet. Sci. 41:383-385.

Gentle MJ (1986b). Beak-trimming in poultry. World's Poult. Sci. J. 42:268-275.

Gentle MJ, McKeegan DEF (2007). Evaluation of the effects of infrared beak trimming in broiler breeder chicks. Vet. Rec. 160:145-148.

Gentle MJ, Hughes BO, Fox A, Waddington D (1997). Behavioural and anatomical consequences of two beak-trimming methods in 1- and 10-d-old chicks. Brit. Poult. Sci. 38:453-463.

Glatz PC (2000). Beak-trimming methods – A review. Asian-Aust. J. Anim. Sci. 13:1619-1637.

Glatz PC (2005). Poultry Welfare Issues - Beak Trimming. pp. 1-174 (ed. Glatz P.C.) Nottingham University Press: Nottingham, UK.

Glatz PC, Bourke M (2006). Beak Trimming Handbook for Egg Producers. Best Practice for Minimising Cannibalism in Poultry. Landlinks Press, Melbourne, ISBN: 0643092560. P. 88.

Glatz PC, Hinch G (2008). Minimise cannibalism using innovative beak trimming methods. Final report to Australian Poultry CRC Pty Limited, ISBN 1 921010282. P. 81.

Gleaves JW (1999). Cannibalism. Cause and prevention in poultry. WWW article published by Cooperative Extension Institute of Agriculture and Natural Resources. University of Nebraska. Lincoln.

Gregory NG, Wilkins LJ (1989). Effect of stunning current on downgrading in turkeys. Brit. Poult. Sci. 30:761-764.

Gregory NC, Austin SD (1992). Causes of trauma in broilers arriving dead at poultry processing plants. Vet. Rec.131:501-503.

Grigor PN, Hughes BO, Gentle MJ (1995). An experimental investigation of the costs and benefits of beak trimming in turkeys. Vet. Rec. 136:257-265.

Grimes JL, Siopes TD (1999). A survey and overview of lighting practices in the U.S. turkey breeder industry. J. Appl. Poult. Res. 8(4):493-498.

Gunthner P, Bessei W (2006). Behavioural reactions of growing turkeys to different stocking densities. EPC 2006 - 12th European Poultry Conference, Verona, Italy, 10-14 September, 2006: paper 316.

Harry EG (1978). Air pollution in farm building and methods of control: a review. Avian Pathol. 7:441-454.

Hartung J, Nowak B, Springorum AC (2009). Animal welfare and meat quality. In: Improving the sensory and nutritional quality of fresh meat. (Eds: Kerry, J. P. and Ledward, D.), pp. 628-646.

Hayter RB, Besch EL (1974). Airborne-particle deposition in the respiratory tract of chickens. Poult. Sci. 53:1507-1511.

Hegelund L, Kjaer J, Kristensen IS, Sorresen JT (2002). Use of the outdoor area by hens in commercial organic egg production systems. Effect of climate factors and cover. In: 11th European Poultry Conference – Abstracts. Archiv für Gef. 66:141-142.

Hemsworth PH, Coleman GJ (1998). Human-livestock interactions. The stockperson and the productivity and welfare of intensively farmed animals. CAB International, Wallingford, Oxon, UK.

Herendy V, Suto Z, Horn P, Szalay I (2004). Effect of the housing system on the meat production of turkey. Acta Agric. Slov. (Suppl. 1):209-213.

Hester PY, Kohl H (1989). Effect of intermittent lighting and time of hatch on large broad-breasted white turkeys. Poult. Sci. 68:528-538.

Hester PY, Peng IC, Adams RL, Furumoto EJ, Larsen JE, Klingensmith PM, Pike OA, Stadelman WJ (1986) Comparison of two lighting regimens and drinker cleaning programmes on the performance and. incidence of leg abnormalities in turkey males. Brit. Poult. Sci. 27:63-73.

Higgins SE, Torres-Rodriguez A, Vicente JL, Sartor CD, Pixley CM, Nava GM, Tellez G, Barton JT, Hargis BM (2005). Evaluation of intervention strategies for idiopathic diarrhea in commercial turkey brooding houses. J. Appl. Poult. Res. 14(2):345-348.

Hinz T, Berk J (2002). Airborne contaminants in tom turkey production under enriched husbandry conditions. Ann. Anim. Sci. 2(1):39-41.

Hirt H, Hordegen P, Zeltner E (2000). Laying hen husbandry: group size and use of hen-runs. In: Alföldi, T. Lockeretz, W. and Niggli, U. (Eds) Proceedings 13th International IFOAM Scientific Conference, Basel. P. 363.

Hughes BO, Michie W (1982). Plumage loss in medium bodied hybrid hens: The effect of beak trimming and cage design. Poult. Sci. 23:59-64.

Hughes BO, Wood-gush DGM (1977). Agonistic behaviour in domestic hens: the influence of housing method and group size. Anim. Behav. 25:1056-1062.

Hughes BO, Carmichael NL, Walker AW, Grigor PN (1997). Low incidence of aggression in large flocks of laying hens. Appl. Anim. Behav. Sci. 54:215-234.

Jones RB (2005). Environmental Enrichment can reduce feather pecking. In Poultry Welfare Issues, Beak Trimming (ed. P.C Glatz), Nottingham University Press, pp. 97-100.

Jones RB, Ruschak C (2002). Domestic chick's responses to PECKABLOCKS and string enrichment devices. Proc. Br. Soc. Anim. Sci. P. 215.

Jongman EC, Glatz PC, Barnett JL (2008). Changes in behaviour of laying hens following beak trimming at hatch and re-trimming at 14 weeks. Asian-Aust. J. Anim. Sci. 21(2):291-298.

Kannan C, Mench JA (1996). Influence of different handling methods and crating periods on plasma corticosterone concentrations in broilers. Brit. Poult. Sci. 37:21-31.

Kassube H E, Hoerl Leone E, Estevez I, Xin H, Noll SL (2006). Turkey beak trim and feed form. 2. Effect on turkey behaviour. Poultry Science Association Annual Meeting. P. 17.

Kelley KW, Brief S, Westly HJ, Novakofski J, Bechtel, PJ, Simon J, Walker ER (1987). Hormonal regulation of the age-associated decline in immune function. Ann. New York Acad. Sci. 496:91-97.

Kemeny DM (2000). The effects of pollutants on the allergic immune response. Toxicology 152(1-3):3-12.

Kijowski J, Mikoajczak A, Kwitowski Nencki J, Sliga M (2005) Traditional rearing and slaughter of christmas Turkeys in England. Pol. J. Food. Nut. Sci. 14(1):75-78.

Kjaer JB (2005). Genetics. In Poultry Welfare Issues: Beak Trimming. PC Glatz, ed. Nottingham Univ. Press, Nottingham, UK. Pp.101-109.

Kjaer JB (2009). Feather pecking in domestic fowl is genetically related to locomotor activity levels: implications for a hyperactivity disorder model of feather pecking. Behav. Gen. 39(5):564-570.

Knierim U, Staack M, Gruber B, Keppler C, Zaludik K, Niebuhr K (2008). Risk factors for feather pecking in organic laying hens –starting points for prevention in the housing environment. 16th IFOAM Organic World Congress, Modena, Italy, June 16-20, 2008. Archived at http://orgprints.org/view/projects/conference.html.

Konca Y, Krkpnar F, Yaylak E, Mert S (2008). Effects of dietary ascorbic acid on performance, carcass composition and bone characteristics of turkeys during high summer temperature. Asian-Aust. J. Anim. Sci. 21(3):426-433.

Kowalski A, Mormede P, Jakubowski K, Jedlinska-Krakowska M (2001). Susceptibility to stress of Big-6 type turkeys subjected to different kinds of stress in light of assorted hormonal, biochemical, immunological and behavioural indices. Pol. J. Vet. Sci. 4(2):65-69.

Kristensen HH, Burgess LR, Demmers TGH, Wathes CM (2000). The preferences of laying hens for different concentrations of atmospheric ammonia. App. Anim. Behav. Sci. 68:307-318.

Kuczynski T, Sobodzian-Ksenicz O (2002). Physical properties of different types of litter and their effect on animal health and welfare in turkey housing. Ann. Anim. Sci. 2(1):31-33.

LayWel ? (2006). Welfare implication of changes in production systems for laying hens. Report on the FP6 European Research Program. www.laywel.eu/web/ date accessed 11 November 2013.

Lee HY, Craig JV (1990). Beak-trimming effects on the behaviour and weight gain of floor-reared, egg strain pullets from three genetic stocks during the rearing period. Poult. Sci. 69:568-575.

Letzbug H, Bessei W (2009). Effects of environmental enrichment on the locomotor activity of turkeys. 2009 Poultry Welfare Symposium Cervia, Italy. P. 46.

Lewis NJ, Hurnik FJ (1990). Locomotion of broiler chickens in floor pens. Poult. Sci. 69:1087-1093.

Lewis PD, Perry GC, Sherwin CM (1998). Effect of intermittent light regimens on the performance of intact male turkeys. Anim. Sci. 67:627-636.

Luangtongkum T, Morishita TY, Ison AJ, Huang SX, McDermott PF, Zhand QJ (2006). Effect of conventional and organic production practices on the prevalence and antimicrobial resistance of Campylobacter spp. in poultry. Appl. Environ. Microbiol. 72(5):3600-3607.

Lunam CA, Glatz PC, Hsu YJ (1996). The absence of neuromas in beaks of adult hens after conservative trimming at hatch. Aust. Vet. J. 74:46-49.

Lupo C, Bouquin S le, Allain V, Balaine L, Michel V, Petetin I, Colin P, Chauvin C (2010). Risk and indicators of condemnation of male turkey broilers in western France, February-July 2006. Prev. Vet. Med. 94(3/4):240-250.

Lyon Electric Company Inc. (1982). A general guide to beak-trimming. Bulletin No. 144.

Majumdar S, Bhanja SK, Singh RP, Agarwal SK (2003). Performance of turkey poults at different cage density during summer. Ind. J. Vet. Res. 12(2):26-33.

Martrenchar A, Boilletot E, Huonnic D, Polf (2001). Risk factors for foot-pad dermatitis in chicken and turkey broilers in France. Prev. Vet. Med. 52(3/4):213-226.

Martrenchar A, Huonnic D, Cotte JE, Boilletot E, Morisse JR (1999). The influence of stocking density on different behavioural, health, and productivity traits of turkey broilers kept in large flocks. Brit. Poult. Sci. 40:323-331.

Martrenchar A, Morisse JE, Huonnic D, Cotte JP (1997). The influence of stocking density on some behavioural, physiological and productivity traits of broilers. Vet. Res. 28:473-480.

Mayne RK (2005). A review of the aetiology and possible causative factors of foot pad dermatitis in growing turkeys and broilers. World's Poult. Sci. J. 61(2):256-267, 319, 324-325, 330, 336, 342.

McKeegan DEF, Philbey AW (2012). Chronic neurophysiological and anatomical changes associated with infra-red beak treatment and their implications for laying hen welfare. Anim. Welf. 21:207-217.

McEwen SA, Barbut S (1992). Survey of turkey downgrading at slaughter carcass defects and associations with transport, toenail trimming, and type of bird. Poult. Sci. 71:1107-1115.

Miao ZH, Glatz PC, Ru YJ, Wyatt SK, Rodda BK (2006). Integration of hens into a crop and pasture rotation system in Australia-production and agronomic aspects. Aust. Poult. Sci. Symp. 18:94-99.

Moinard C, Lewis PD, Perry GC, Sherwin CM (2001). The effects of light intensity and light source on injuries due to pecking of male domestic turkeys (Meleagris gallopavo). Anim. Welf. 10(2):131-139.

Mora CV, Davidson M, Wild JM, Walker MM (2004). Magnetoreception and its trigeminal mediation in the homing pigeon. Nature 291:433-434.

Mota-Rojas D, Maldonado MJ, Becerril MH, Flores SCP, Gonzalez-Lozano M, Alonso-Spilsbury M, Camacho-Morfin D, Ramirez RN, Cardona AL, Morfin-Loyden L (2008). Welfare at slaughter of broiler chickens: a review. Int. J. Poult. Sci. 7(1):1-5.

Mueller-Doblies D, Carrique-Mas JJ, Sayers AR, Davies RH (2010). A comparison of the efficacy of different disinfection methods in eliminating Salmonella contamination from turkey houses. J. Appl.

Micro. 109(2):471-479.

Nestor KE (1984). Genetics of growth and reproduction in the turkey. 9. Long-term selection for increased 16-week body weight. Poult. Sci. 63:2114-2122.

Nestor KE, Bacon WL, Saif YM, Renner PA (1985). The influence of genetic increases in shank width on bodyweight, walking ability and reproduction in turkeys. Poult. Sci. 64:2248-2255.

Nicol CJ, Potzsch C, Lewis K, Green LE (2003). Matched concurrent case-control study of risk factors for feather pecking in hens on free-range commercial farms in the UK. Brit. Poult. Sci. 44:515-523.

Noble DO, Kestor KE (1997). Beak-trimming of turkeys. 2. Effects of arc trimming on weight gain, feed intake, feed wastage, and feed conversion. Poult. Sci. 76:668-670.

Noll S, Xin H (2007). Performance and behavior of market tom turkeys as influenced by beak trim and form of feed. Midwest Poult. Fed. Conv. pp. 1-6.

Noll SL, Xin H (2006). Turkey beak trim and feed form. 1. Effect on turkey performance. Poult. Sci. Assoc. Ann. Meeti. P. 17.

Oblakova M, Bozakova N, Lalev M, Yotova I, Stoyanchev T (2004). Effect of type of floor on productivity of turkey broilers. Bulg. J. Agric. Sci. 10(5):623-627.

Ostovic M, Pavicic Z, Tofant A, Balenovic T, Kabalin AE, Mencik S (2009). Welfare of turkeys in intensive production. In: VIII Symposium of Poultry Days 2009 (Ed: Balenovic, M.) Porec, Croatia, pp. 31-34

Pintaric S, Dobeic M (2000). Effects of bioenzymatic and mineral additives in bedding material on decreasing ammonia emissions in turkey breeding. Slov. Vet. Res. 37(3):133-144.

Pixley C, Barton J, Vicente JL, Wolfenden AD, Hargis BM, Tellez G (2010). Evaluation of a commercially available organic acid product during Feed Withdrawal and its relation to carcass shrink in commercial turkeys. Int. J. Poult. Sci. 9(6):508-510.

Prescott NB, Berry PS, Haslam S, Tinker DB (2000). Catching and crating turkeys: effects on carcass damage, heart rate, and other welfare parameters. J. Appl. Poult. Res. 9 (3):424-432.

Rauw WM, Kanis E, Noordhuizen-Stassen EN. Grommers FJ (1998). Undesirable effects of selection for high production efficiency in farm animals: a review. Live. Prod. Sci. 56:15-33.

Rawdon T, Tana T, Frazer J, Thornton R, Chrystal N (2008). Biosecurity risk pathways in the commercial poultry industry: free-range layers, pullet-rearers and turkey broilers. Surveillance (Wellington) 35 4):4-9.

Renz KG, Walkden-Brown SW (2007). Environmental enrichment strategies. Aust. Poult. Sci. Symp. 19:94-99.

Rostagno MH, Wesley IV, Trampel DW, HURD HS (2006). Salmonella prevalence in market-age turkeys on-farm and at slaughter. Poult. Sci. 85(10):1838-1842.

Ruszler PL, Novak CL, McElroy AP, Denbow DM (2004). Stress determination in pullets beak trimmed at one day versus seven days versus no beak trimming. US Poultry and Egg Association, pp. 1-2.

Salin C, Delettre YR, Vernon P (2003). Controlling the mealworm Alphitobius diaperinus (Coleoptera: Tenebrionidae) in broiler and turkey houses: field trials with a combined insecticide treatment: insect growth regulator and pyrethroid. J. Econ. Entom. 96(1):126-130.

Sandøe P, Forkman F, Christiansen SB (2004). Scientific uncertainty - how should it be handled in relation to scientific advice regarding animal welfare issues? Anim. Welf. 13:121-126.

Sarica M, Yamak US (2009). The effects of production systems (barn and free-range) on foot pad dermatitis and body defects of white turkeys. World Poult. Sci. Association (WPSA), Proceedings of the 8th European Symposium on Poultry Welfare, Cervia, Italy. P. 52.

Sarica M, Yamak US (2010). The effects of production systems (barn and free-range) on foot pad dermatitis and body defects of white turkeys. J. Anim. Vet. Adv. 9(5):958-961.

Shah SB, Baird CL, Rica JM (2007). Effect of a metabolic stimulant on ammonia volatilization from broiler litter. J. Appl. Poult. Res. 16(2):240-247.

Sherwin CM (1998). Light intensity preferences of domestic male turkeys. Appl. Anim. Behav. Sci. 58:121-130.

Sherwin CM, Kelland A (1998). Time-budgets, comfort behaviours and injurious pecking of turkeys housed in pairs. Brit. Poult. Sci. 39:325-332.

Sherwin CM, Lewis PD, Perry GC (1999). The effects of environmental enrichment and intermittent lighting on the behaviour and welfare of male domestic turkeys. Appl. Anim. Behav. Sci. 62:319-333.

Siopes TD, Timmons MB, Baughman GR, Parkhurst CR (1984). The effects of light intensity on turkey poult performance, eye morphology, and adrenal weight. Poult. Sci. 63:904-909.

Smith K, Reimers N, Barnes HJ, Lee BC, Siletzky R, Kathariou S (2004). Campylobacter colonization of sibling turkey flocks reared under different management conditions. J. Food Prot. 67(7):1463-1468.

Sobodzian-Ksenicz O, Houszka H, Michalski A (2008). Effect of addition of brown coal and microbe vaccine to litter on bedding quality and production results in turkey farming. Anim. Sci. Pap. Rep. 26(4):317-329.

Spindler B, Hartung J (2009). Influence of environmental enrichment on the behaviour of female Big 6 turkeys reared on an ecological farm. In: Sustainable animal husbandry: prevention is better than cure, Volume 1. Proceedings of the 14th International Congress of the International Society for Animal Hygiene (ISAH) (Eds: Briese, A., Clauss, M., Springorum, A. and Hartung, J.) Vechta, Germany. pp. 359-362.

Standing Committee on Agriculture and Resource Management (2003). Model Code of Practice for the Welfare of Animals. Domestic Poultry, 4th Edition. CSIRO Publications, East Melbourne, Australia.

Thaxton P Jr, Vizzier-Thaxton W, Schilling K, Christensen R, Stuckey JL, Purswell SL, Branton, White P (2009). A New Humane Slaughter Methods for Broilers - Low Atmospheric Pressure (LAPS). World Poult. Sci. Association (WPSA), Proceedings of the 8th European Symposium on Poultry Welfare 18-22 May, 2009 Cervia, Italy. P. 42.

Trampel DW, Hasiak RJ, Hoffman LJ, Debey MC (2000). Recovery of Salmonella from water, equipment, and carcasses in turkey processing plants. J. Appl. Poult. Res. 9(1):29-34.

Vicente J, Higgins S, Bielke L, Tellez G, Donoghue D, Donoghue A, Hargis B (2007). Effect of probiotic culture candidates on Salmonella prevalence in commercial turkey houses. J. Appl. Poult. Res. 16(3):471-476.

Wathes CM (1998). Aerial emissions from poultry production. World's Poult. Sci. J. 54:241-251.

Wichman M, Norring M, Pestell M, Hanninen L (2009). Being able to stand up - consequences for the welfare of turkeys in transit, Proceedings of the 8th European Symposium on Poultry Welfare, organised by the World's Poult. Sci. Association (WPSA), and held in Cervia, Italy. P. 101.

Wojcik A, Sowinska J, Iwanczuk-Czernik K, Mituniewicz T (2004). The effect of a housing system in slaughter turkeys on mechanical damage to carcass and meat quality. Czech J. Anim. Sci. 49(2):80-85.

Wood BJ, Wojcinski H, Buddiger N (2006). Company consolidation and the responsibility of the primary turkey breeders. Proceedings of the 8th World Congress on Genetics Applied to Livestock Production, Belo Horizonte, Minas Gerais, Brazil. pp. 07-09.

Yahav S (2007). The crucial role of ventilation in performance and thermoregulation of the domestic fowl. Aust. Poult. Sci. Symp. 19:14-18.

Yahav S, Hurwitz S, Rozenboim I (2000). The effect of light intensity on growth and development of turkey toms. Brit. Poult. Sci. 41(1):101-106.

Yahav S, Rusal M, Shinder D (2008). The effect of ventilation on performance body and surface temperature of young turkeys. Poult. Sci. 87(1):133-137.

Youssef I, Beineke A, Kamphues J (2009). Effects of diet composition and litter quality on development and severity of foot pad dermatitis (FPD) in young fattening turkeys. In: Sustainable animal husbandry: prevention is better than cure, Volume 1. Proceedings of the 14th International Congress of the International Society for Animal Hygiene (ISAH), (Eds: Briese, A., Clauss, M., Springorum, A. and Hartung, J.) Vechta, Germany, pp. 393-396.

Zeltner E, Hirth H (2003). Effect of artificial structuring on the use of laying hen runs. Brit. Poult. Sci. 44:533-537.

Zeltner E, Maurer V (2009). Welfare of organic poultry. WPSA Proceedings. 8th European Symposium on Poultry Welfare. Cervia, Italy. pp. 104-112.

22

How far do egg markets in India conform to the law of one price?

Sendhil R.[1], D. Babu[2], Ranjit Kumar[2] and K. Srinivas[2]

[1]Directorate of Wheat Research, Karnal (Haryana), India.
[2]National Academy of Agricultural Research Management, Hyderabad, India.

Growing demand for eggs in India is also accompanied by its production at six per cent compound annual growth rate. The law of one price states that in an efficient market, all identical goods or commodities should have a single price. Johansen's cointegration test was done to identify whether spatially separated egg markets in India share a common linear deterministic trend and the law of one price holds true. Daily wholesale prices from January 2011 to November 2012 in major egg markets across the country were collected and analysed. Post checking for unit root employing the Augmented Dickey Fuller test statistic, cointegration results indicated a strong spatial integration between regional egg prices in the long-run implying the price co-ordination despite production in multiple regions. However, empirical results indicated that the law of one price does not hold true in Indian egg market.

Key words: Egg, cointegration, law of one price.

INTRODUCTION

Eggs are one of the nature's most perfectly balanced foods, containing proteins, vitamins and minerals essential for good human health. Besides nutritional value and culinary purpose, it has immense export potential. India ranks fifth in the world producing 66.45 billion eggs with an average growth rate of 6 per cent per annum (Karthikeyan and Nedunchezhian, 2013). The country stands fourth in exporting eggs to the rest of the world with a turnover of 40.76 US$ million by exporting 33.92 million tonnes of eggs. However, the country has imported 315 tonnes of eggs valued at 1.25 US$ million (FAO Stat, 2011).Among Indian states, Tamil Nadu accounts for the maximum egg production followed by Andhra Pradesh, Karnataka, Maharashtra, Gujarat, Madhya Pradesh, Odisha and North Eastern States (GoI, 2012; APEDA, 2012).

Poultry eggs are considered to be identical goods, which are marketed across the regions through well connected infrastructure facilities. The industry witnessed rapid strides in egg production till late 70's. However, rising primary input costs *viz.,* medicines, feed, electricity, taxes, etc., coupled with domination of middlemen had led to the crisis in 1981-1982 when egg prices fell drastically. Consequent to this, over 20,000 marginal poultry farmers lost their only source of livelihood in India. In order to prevent the ailing poultry sector, the National Egg Coordination Committee (NECC) was formed in 1982 as an institutional support to the poultry farmers. Since then, the NECC has been performing its designated functions, including declaration of market prices across various markets on daily basis, in order to enhance transparency in the egg marketing system (Saran and Gangwar, 2008). However, in the recent past, the soaring price of eggs across the country set a serious concern on the welfare of consumers. The difference in prices prevailed across egg markets in India raised the question of price integrity among the spatially separated markets.

Economic theory states that price variables should have a long-run equilibrium relationship and even if they drift away from equilibrium for a while, economic forces will bring these back to its equilibrium position (Kaur et al., 2010). Technically, the co-movement or long-run relationship between the spatial prices has been conceived as market integration (Fackler, 1996). The concept of cointegration was originally proposed by Granger (1981) which recognised that even though several price series have unit root, a linear combination of them could exist which would not have such a property. Integration between markets and price transmission largely depends on the dynamic relationship that arises due to trade distortions. In the case of eggs too, there is a likelihood of market integration and price transmission, if the markets are efficient in their performance.

Empirical studies on market integration of poultry products typically use bilateral price relationships as an indicator of market integration which falls under the law of one price (LOOP). The law states that the identical commodity should sell for the same price in each region of a country which can be measured empirically (Moodley et al., 2000). Despite extensive studies carried out in India on market integration with respect to foodgrains, fish and horticultural commodities, only a few studies dealt with the price integration of poultry products. Among them, none of the studies have empirically tested the LOOP among regional egg markets which are highly geographically concentrated. Given the importance and the structural transformation of the poultry sector in general and egg markets in particular after the establishment of the NECC, the present study has been carried out with the objective of finding out the extent of market integration and price transmission within India and whether the LOOP holds true in the case of egg markets.

MATERIALS AND METHODS

Spatial market integration is a situation in which prices of a commodity in spatially separated markets move together due to arbitrage and the price signals and information are transmitted smoothly across the markets. With the free flow of information in a competitive market, the difference in prices of a product in the two markets would be equal to or less than the transportation cost between them (Vasciaveo et al., 2013). Hence, spatial market performance may be evaluated in terms of the relationship between the prices in spatially separated markets. Estimation of bivariate correlation coefficients between price changes in different markets has been employed as the most common methodology (Cummings, 1967; Lele, 1967; 1971) for testing market integration. But, this gives the integration of markets only in the short-run. Hence, cointegration analysis is suggested to know the long-run integration.

Data source

The study is purely based on the secondary data. Time series data on wholesale daily prices of eggs were collected from the NECC

portal from January 2011 to November 2012 for major seven markets across different states viz., Hyderabad, Mumbai, Delhi, Chennai, Namakkal, Kolkata and Bangalore and used for the present analysis.

Instability in prices

Instability index has been used to examine the extent of variation and risk involved in prices. It is measured by Cuddy-Della Valle Index (Cuddy and Della Valle, 1978). This method is superior to others as it de-trends the time series while computing the instability in the selected variable. The index is computed as,

Cuddy-Della Valle Instability Index (%) $= CV \times \sqrt{(1 - \overline{R^2})}$

Where, CV is the coefficient of variation in per cent, and $\overline{R}^2$ is the coefficient of determination estimated from a time trend regression adjusted to its degrees of freedom.

Market integration and price transmission (Cointegration test)

Cointegration test has been the most popular and widely adopted methodology among economists to study the integration between commodity markets. A number of mathematical improvements has been done including the methodology suggested by Hendry and Anderson (1977), Engle and Granger (1987), Johansen (1988, 1991, 1994, 1995) and Goodwin and Schroeder (1991). Among the available methodologies, Johansen's technique was considered to be the superior technique (Kumar and Sharma, 2003) since it permits the testing of cointegration as a system of equation in one step without any prior assumption of endogenous or exogenous variables. In addition, it does not impose any restrictions beforehand: test and estimation of the number of cointegration relationships can be carried out simultaneously.

Johansen's maximum likelihood method of cointegration

Before testing for cointegration, the time series has been checked for its stationarity. The stationarity properties and the exhibition of unit roots in the time series are substantiated by performing the Augmented Dickey-Fuller (ADF) test. This test was conducted on the variables in level (original price series) and first differences. The variables that are integrated of the same order may be cointegrated, while the unit root test finds out which variables are integrated of order one, or I(1) (Vasciaveo et al., 2013). The following ADF regression equation was tested for stationarity:

$$\Delta Y_t = \beta_1 + \beta_2 t + \delta Y_{t-1} + \alpha_i \sum_{i=1}^{m} \Delta Y_{t-i} + u_t$$

where, Y_t is a vector to be tested for cointegration, t is the time, $\Delta Y_t = Y_t - Y_{t-1}$ and u_t is a pure white noise error term. The null hypothesis that $\delta = 0$ signifies unit root which means the time series is non-stationary, while the alternative hypothesis, $\delta < 0$ signifies that the time series is stationary, thereby rejecting the null hypothesis.

In a cointegrated equation system, $$\Delta Y_t = \sum_{i=1}^{k-1} \Gamma_i \Delta Y_{t-i} + \alpha \beta' Y_{t-k} + \varepsilon_t$$, where Y_t is the price time series, Δ is the first difference operator $(Y_t - Y_{t-1})$ and matrix $\Pi = \alpha \beta$ is (n x n) with rank r $(0 \leq r \leq n)$, which is the number of linear independent cointegration relations in the vector space of matrix. The Johansen's method of cointegrated system is a restricted maximum likelihood method with rank restriction on matrix $\Pi = \alpha\beta'$. The rank of Π can be determined by λ_{trace} test statistics, and is estimated by

$$\lambda_{trace} = -T \sum_{i=r+1}^{n} \ln(1 - \hat{\lambda}_i), \text{ for r = 0, 1,..., n-1}$$

Where, $\hat{\lambda}_i$'s are the Eigen values representing the strength of the correlation between the first difference part and the error-correction part. Following hypotheses were tested, H_0: rank of Π = r (null hypothesis), and H_1: rank of Π > r (alternate hypothesis), where 'r' is the number of cointegration equations. The above test was carried out with the assumption of linear deterministic trend in original data and only intercept in the cointegrating equation. The cointegrating equation has only the intercept (no trend) because of difference in the price series while checking for its stationarity, whereas; the original price series follows a trend since the mean and variance is non-constant over a period of time (non-stationary property). Integration between two markets can be checked in a similar fashion through bi-variate Johansen's test.

Determination of lag lengths

Johansen's cointegration test is very sensitive to price lag length. Hence, the choice of lag length (k) is determined using the multivariate forms of Akaike Information Criterion (AIC). The length of lag distribution is decided by choosing the specification minimizing the AIC. The model is given as:

$$AIC = T \ln |\hat{\Sigma}| + 2m \quad \text{and} \quad AIC = T \ln |\hat{\Sigma}| + m.\ln T$$

where, T is the length of the time series and m is the number of parameters. The lag length (k) is represented by

$$\hat{\Sigma} = \sum_{t-k}^{T} \frac{\hat{e}_t \hat{e}'_t}{T}, \text{ and } \hat{e}_t \text{ is (n x 1) residual vector.}$$

Speed of convergence (Error correction mechanism)

After testing for cointegration, the residuals show the deviation from equilibrium and this can be captured by the vector error correction model (Brosig et al., 2011). In this case, a linear deterministic trend model is run only in the cointegrated markets specifying the number of cointegration equations between them. The model is represented as:

$$\Delta A_t = \alpha_0 + \alpha_1 \Delta B_t + \alpha_2 u_{t-1} + \varepsilon_t$$

where A_t is the price of market A, B_t is the price of market B and u_t is the cointegration vector. The coefficient (α_2) of the error correction term (u_{t-1}) indicates the speed at which the series returns to equilibrium. If it is less than zero, the series converge to long-run equilibrium and if it is positive and zero, the series diverge from equilibrium. If the estimated coefficient of market B is negative (positive), it indicates that decrease (increase) in the previous period's equilibrium error leads to a decrease (increase) in the

current period price, and *vice versa*.

LOOP analysis

Price integration between spatially separated markets does not indicate the efficiency but it can be considered as one of the indicators of overall market performance. This is ideally reflected by the LOOP. Johansen and Juselius (1990) indicated that LOOP can be proved by testing the hypothesis on the cointegration coefficients of both α and β using the likelihood ratio. For this, restrictions can be placed and parameters can be tested from the resultant β matrix of the cointegration equation following the Johansen's approach. Hence, in the case of testing integration between two markets, the rank of π = αβ' will be equal to 1 and α and β matrices will be of order 2 x 1. Now, the LOOP for two markets is tested by imposing the restriction β' = (1, -1)', and this test can be considered a valid test for LOOP in the long-run as the β matrix contains the long-run parameters within the cointegrated system. Alternatively, finding n-1 cointegrating vectors indicate that all prices contain the same stochastic trend and hence pairwise cointegrated validating the LOOP (Gandhi and Koshy, 2006; Awokuse and Bernard, 2007).

RESULTS AND DISCUSSION

Price behavior

It is imperative to know the price trend in order to know the behavior of the variable in different markets across the country (Figure 1). The major egg markets have been chosen purposefully covering different regions of the country (Table 1). The figure shows the symmetric pattern in the movement of prices in all the markets of the country with Kolkata having the highest price and Hyderabad with the status of major producer in the country, has the lowest price (Table 2). The rest of the markets exhibited a similar price movement in a band. The egg markets in India are highly geographically concentrated which allows for a varying price relationship between high and low production centres vis-à-vis consumption centres. The average price during the study period was higher in Kolkata (INR 289.39/100 eggs) which registered the maximum price too. It is also due to the huge demand from consumers', a highly populous region with a meagre share in the country's production. It is explicitly evident from the table that standard deviation and variance was higher in the case of Delhi market. Owing to these statistics, the estimated instability was highest in Delhi (14%). It is also noted that the Delhi market prices are more volatile during the study period and Chennai market prices are less volatile. The variance statistic ranged from 745 in Mumbai to as high as 2141 in Delhi indicating the wide spread of the price data in the respective market. Egg prices also showed the presence of skewness and kurtosis in the selected market. It confirms the scientific fact related to fat tails and scattered extreme observations which is a common feature of a high frequency data. All the markets exhibited a positive skew distribution indicating most of the observations concentrate on the left of the mean extreme prices is less, and they are wider spread around mean.

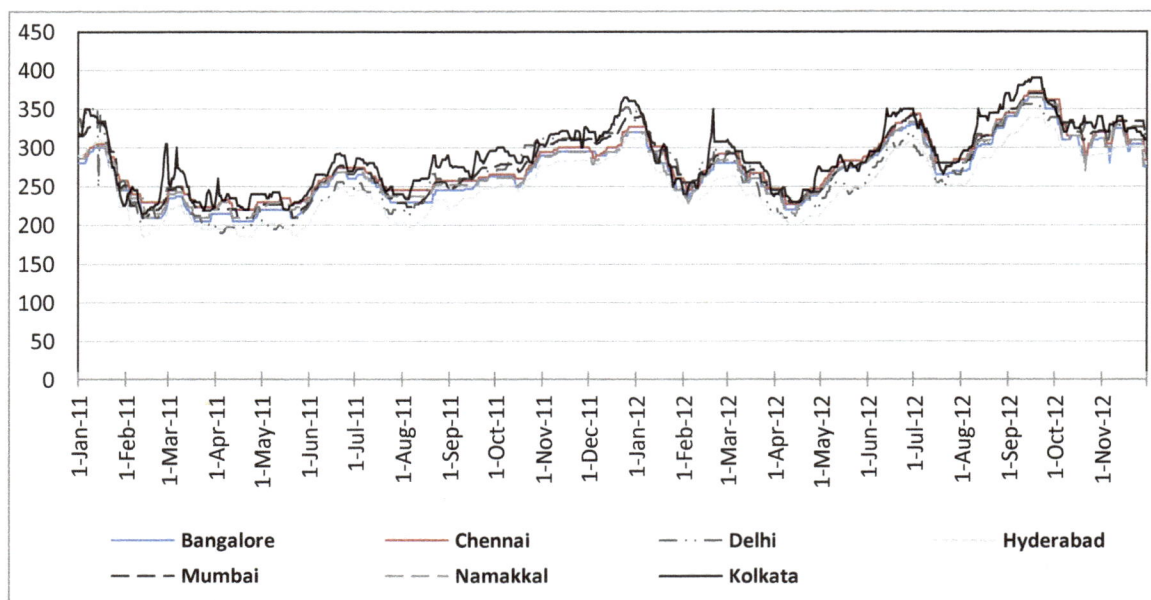

Figure 1. Price trend (INR/ 100 eggs) in major egg markets of India.

Figure 1. Price trend (INR/ 100 eggs) in major egg markets of India

Table 1. Selected egg markets from different states/union territory.

SN	Selected market	State / Union Territory	Basis for selection*
1	Bangalore	Karnataka	Consumption
2	Chennai	Tamil Nadu	Consumption
3	Delhi	Delhi	Consumption
4	Hyderabad	Andhra Pradesh	Production
5	Mumbai	Maharashtra	Consumption
6	Namakkal	Tamil Nadu	Production
7	Kolkata	West Bengal	Consumption

*Indicate the selection from the National Egg Coordination Committee (NECC) portal.

Table 2. Summary statistics of egg market prices.

Particulars	Bangalore	Chennai	Delhi	Hyderabad	Kolkata	Mumbai	Namakkal
Observations (days)	700	700	700	700	700	700	700
Maximum price (INR/100 eggs)	365	372	356	338	390	370	365
Minimum price (INR/100 eggs)	205	220	190	185	215	210	210
Range (INR/100 eggs)	160	152	166	153	175	160	155
Mean price (INR/100 eggs)	268.69	278.25	271.66	251.56	289.39	279.66	271.40
Standard deviation	38.78	37.18	46.27	40.28	41.89	41.77	37.79
Variance	1503.87	1382.47	2141.04	1622.88	1754.55	745.11	1428.39
Skewness	0.31	0.44	0.08	0.17	0.24	0.17	0.39
Kurtosis	-0.58	-0.49	-1.16	-1.04	-0.77	-1.00	-0.52
Instability index (%)	10.51	9.76	14.00	12.67	11.67	11.41	10.27

value, with extreme observations to the right. Kurtosis statistic turned negative for all the markets specifying the fat or short tailed (platykurtic) pattern of probability distribution in egg prices. This indicates that the prices

Table 3. Price correlation between major egg markets in India.

Market(n = 700)	Bangalore	Chennai	Delhi	Hyderabad	Mumbai	Namakkal	Kolkata
Bangalore	1	0.99*	0.92*	0.98*	0.98*	0.99*	0.95*
Chennai		1	0.91*	0.97*	0.97*	0.99*	0.94*
Delhi			1	0.96*	0.95*	0.91*	0.93*
Hyderabad				1	1.00*	0.97*	0.96*
Mumbai					1	0.97*	0.96*
Namakkal						1	0.95*
Kolkata							1

*Indicates the significance of Karl Pearson's correlation coefficient at one per cent level of probability.

Table 4. Estimated ADF statistic for unit root test.

Market	Level series		1st differenced series	
	ADF statistic	AIC lag length	ADF statistic	AIC lag length
Bangalore	-2.87	2	-8.70*	4
Chennai	-2.98	3	-10.83*	2
Delhi	-3.09	2	-17.76*	1
Hyderabad	-3.46	4	-8.87*	3
Mumbai	-3.47	4	-11.75*	1
Namakkal	-3.08	2	-15.18*	1
Kolkata	-3.74	4	-14.36*	3

*Indicates the significance at one per cent level of MacKinnon (1996) one-sided probability value.

show a flatter distribution than a normal distribution with a wider peak. Further, the probability of extreme egg prices is less, and they are widely spread around the mean price.

Market integration

Market integration is the co-movement or long-run relationship between spatial prices. For the present study, major egg markets in India were purposively selected on the basis of production and consumption criteria (Table 1), and tested for cointegration analysis using Johansen's (1988) approach. Before cointegration, correlation between different egg markets was carried out to know the short-run integration. Correlation analysis revealed a positive co-movement between the egg price series, a priori (Table 3). The results indicated a high degree of significant positive correlation between all the major egg markets that are spatially separated.

Before testing for cointegration relationship between different egg market prices, it becomes mandatory to check the order of integration of the level variables. Therefore, unit root tests of each variable at their levels as well as first differences of non-stationary level variables were conducted for each market (Table 4). The results indicated the presence of a unit root at their levels, that is, non-stationarity of each market price time series. However, all the non-stationary variables are found to be stationary at their first differences, and therefore, are integrated of order one, I (1) if statistically tested, supporting the findings of Saran and Gangwar (2008). The conformation that each level series is I (1) helped to carry out the Johansen's cointegration test.

The cointegration test results furnished in Table 5 revealed the Eigen value and the trace statistic for the selected markets. The test rejected the null hypothesis of no cointegration relationship to at most three relationships between the egg markets (r = 0 to r ≤ 3) at 5% level of probability indicating the presence of four possible cointegration relationships among the selected seven markets. The extent of price transmission and integration could be influenced by the market structure. Increased geographical concentration and vertical coordination of poultry markets may be the plausible reasons for strong market integration (Awokuse and Bernard, 2007). The purpose of this analysis is to know whether the egg markets are integrated in the long-run and thereby price transmission holds true. The flow of market information across markets helps to realise the law of one price in Indian eggs barring the transportation cost in each market. However, the speed of information flow can be estimated through the error correction model.

Johansen's test revealed the long-run equilibrium

Table 5. Estimates of the Johansen's multivariate cointegration test.

Data period	01/01/2011 to 30/11/2012
Included observations	696 after adjustments
Trend assumption	Linear deterministic trend
Lag length	3
Selected markets: Hyderabad, Mumbai, Delhi, Chennai, Namakkal, Kolkata, and Bangalore	

Null hypothesis	Eigen value	Trace statistic	Critical value at 5%	Probability**
r = 0*	0.1250	303.8397	125.6154	0.0000
r ≤ 1*	0.1194	210.8885	95.7537	0.0000
r ≤ 2*	0.0768	122.3820	69.8189	0.0000
r ≤ 3*	0.0534	66.7931	47.8561	0.0003
r ≤ 4	0.0243	28.6006	29.7971	0.0682
r ≤ 5	0.0138	11.4699	15.4947	0.1842
r ≤ 6	0.0026	1.7881	3.8415	0.1812

Trace test indicates four cointegrating equations at five per cent level of probability,* denotes rejection of the null hypothesis at five per cent probability and ** shows the MacKinnon-Haug-Michelis (1999) probability values.

between all selected egg markets, justifying the use of vector error correction model (VECM) for showing the short-run dynamics. For this, bi-variate (2 markets) analysis was done and the integrated markets were tested for error correction mechanism. The results of the VECM indicated that barring Hyderabad and Mumbai, the rest of the market pairs registered expected coefficients (Table 6). In those pairs, Market 'A' exhibited negative coefficients whereas market B exhibited positive coefficients. This indicated that the price series with positive coefficients diverge from equilibrium in the short-run; and price series with negative coefficients converge to the long-run equilibrium. In other words, within a short time, positive changes tend to persist whereas; negative changes tend to move the price series towards equilibrium in the long-run. Further, it should be noted that the average time taken for correcting the errors due to positive news is much lesser in comparison to negative shocks which is evident from the magnitude of the error correction coefficients. However, price adjustments will occur only when there is a wide deviation from the equilibrium which is expected to be more than the transaction costs causing arbitrage activities to be profitable (Kaur et al., 2010).

The vector error correction coefficient was estimated at -0.0173 for market A (Bangalore) and 0.1209 for market B (Chennai). This indicated that how quickly the dependent variables such as Chennai and Bangalore prices absorb and adjust themselves for previous period disequilibrium errors. In other words, the coefficient measures the ability of the prices to incorporate shocks or price news available in the market. In this case, Chennai and Bangalore market absorb 12.09 and 1.73% respectively to bring about the equilibrium in prices. The information flow is more in Chennai market as evident by

the magnitude of the coefficient (0.1209). Hence Chennai is more efficient than the Bangalore market in terms of reaction to news on price. Similar kind of interpretation can be done for rest of the markets.

Law of one price (LOOP)

Trace tests showed four cointegrating vectors for all the selected egg markets, the number of common stochastic trends turned out to be three of these seven markets (Table 7). The number of common stochastic trends was determined by deducting the number of cointegrating vectors from the dimension of the impact matrix given by the number of variables (n) included in the cointegration test (Gandhi and Koshy, 2006). Despite price linkage, the findings of $n - 1$ cointegrating vectors implied that different stochastic trend existed and hence the LOOP does not hold true for all the major egg markets across India. The analysis of the LOOP has a significant role in determining the efficient functioning of the egg markets across India underlying the importance of transaction costs. If their role is ignored, the results of integration may be misleading. Hence, the NECC which is responsible for egg market information should take additional responsibility for collecting and disseminating various components of transaction costs *viz.*, loading and unloading costs, transport costs, insurance and financing.

CONCLUSIONS AND POLICY IMPLICATIONS

The degree of market integration assessed using the Johansen's cointegration test indicated that major egg markets across the country are cointegrated in the long-

Table 6. Estimates of the Johansen's bivariate cointegration test and error correction model.

Markets under test for cointegration	Null hypothesis	Lag length criteria		Cointegration test		Error correction estimates		Log likelihood
		AIC value	Order of lags	Eigen value	Trace statistic	Market A	Market B	
A. Bangalore	r = 0*	-12.19	3	0.0570	45.3155	-0.0173	0.1209	4246.92
B. Chennai	r ≤ 1*			0.0064	4.4542	(0.0337)	(0.0284)	
A. Bangalore	r = 0*	-10.34	4	0.0230	20.2133	-0.0320	0.0344	3610.61
B. Delhi	r ≤ 1*			0.0058	4.0483	(0.0104)	(0.0179)	
A. Bangalore	r = 0*	-11.95	8	0.0343	32.6833	-0.0044	0.0638	4163.07
B. Hyderabad	r ≤ 1*			0.0123	8.5411	(0.0189)	(0.0154)	
A. Bangalore	r = 0*	-12.14	8	0.0352	34.0860	-0.0225	0.0656	4228.16
B. Mumbai	r ≤ 1*			0.0134	9.3196	(0.0235)	(0.0175)	
A. Bangalore	r = 0*	-11.53	7	0.0377	32.9160	-0.0733	0.1478	4020.52
B. Namakkal	r ≤ 1*			0.0091	6.3124	(0.0312)	(0.0402)	
A. Bangalore	r = 0*	-10.54	7	0.0567	47.0708	-0.0349	0.1121	3674.03
B. Kolkata	r ≤ 1*			0.0096	6.6831	(0.0137)	(0.0205)	
A. Chennai	r = 0*	-10.58	4	0.0237	20.9956	-0.0313	0.0322	3691.56
B. Delhi	r ≤ 1*			0.0062	4.3314	(0.0093)	(0.0181)	
A. Chennai	r = 0*	-12.11	3	0.0262	18.4606	-0.0405	0.0309	4225.35
B. Hyderabad	r ≤ 1*			0.0071	4.9251	(0.0153)	(0.0151)	
A. Chennai	r = 0*	-12.30	8	0.0298	28.8425	-0.0137	0.0526	4282.77
B. Mumbai	r ≤ 1*			0.0114	7.9026	(0.0159)	(0.0140)	
A. Chennai	r = 0*	-12.26	3	0.0943	73.4250	-0.1895	0.1579	4272.31
B. Namakkal	r ≤ 1*			0.0064	4.4849	(0.0283)	(0.0555)	
A. Chennai	r = 0*	-10.79	7	0.0616	50.9119	-0.0447	0.1125	3763.05
B. Kolkata	r ≤ 1*			0.0099	6.8961	(0.0122)	(0.0208)	
A. Delhi	r = 0*	-10.80	8	0.0350	31.4770	-0.0009	0.0456	3768.74
B. Hyderabad	r ≤ 1*			0.0098	6.8290	(0.0206)	(0.0097)	
A. Delhi	r = 0*	-10.97	8	0.0271	26.2419	-0.0062	0.0328	3827.792
B. Mumbai	r ≤ 1*			0.0104	7.2322	(0.0188)	(0.0081)	
A. Delhi	r = 0*	-10.04	3	0.0318	26.0887	-0.0314	0.0352	3511.37
B. Namakkal	r ≤ 1			0.0052	3.6222	(0.0136)	(0.0091)	
A. Delhi	r = 0*	-9.50	8	0.0443	36.7426	-0.0274	0.0706	3316.88
B. Kolkata	r ≤ 1*			0.0079	5.4542	(0.0161)	(0.0146)	
A. Hyderabad	r = 0*	-15.54	8	0.0250	20.0332	-0.1358	-0.1121	5406.43
B. Mumbai	r ≤ 1			0.0037	2.5585	(0.0354)	(0.0322)	
A. Hyderabad	r = 0*	-11.56	8	0.0334	32.0735	-0.0506	0.0320	4033.53
B. Namakkal	r ≤ 1*			0.0124	8.6257	(0.0132)	(0.0179)	
A. Hyderabad	r = 0*	-10.97	8	0.0675	58.4773	-0.0237	0.1486	3820.47
B. Kolkata	r ≤ 1*			0.0146	10.1967	(0.0135)	(0.0242)	
A. Mumbai	r = 0*	-11.76	8	0.0316	31.0341	-0.0458	0.0489	4099.61
B. Namakkal	r ≤ 1*			0.0127	8.8496	(0.0139)	(0.0206)	
A. Mumbai	r = 0*	-11.14	8	0.0636	54.0138	-0.0133	0.1443	3879.43
B. Kolkata	r ≤ 1*			0.0124	8.5910	(0.0119)	(0.0232)	
A. Namakkal	r = 0*	-10.30	7	0.0556	46.9449	-0.0484	0.1055	3592.23
B. Kolkata	r ≤ 1*			0.0106	7.3703	(0.0159)	(0.0212)	

Figures within parentheses indicate the standard error of the estimate. * The critical value for rejecting the null hypothesis, H_0: r=0 is 15.49 and r ≤ 1is 3.81at five per cent level of probability.

run with the possibility of four cointegration equations which is evident by the trace statistic. Price transmission occurred from one market to another market due to the flow of market information owing to the development in

Table 7. Confirmation of the LOOP for egg markets.

Selected markets	Number of cointegrated vectors	Number of stochastic trends	Confirmation of LOOP
Hyderabad, Mumbai, Delhi, Chennai, Namakkal, Kolkata and Bangalore (7 markets)	4	7 – 4 = 3	No

information tools and technology as well as the establishment of NECC. The speed of convergence of egg prices between markets depends more on the government policies and investment on infrastructure facilities. Further, market integration relies heavily on the efficient functioning of markets. In fact, markets that are spatially integrated themselves serve an indicator of market efficiency. However, the law of one price does not occur in the egg markets indicating the significance of transaction costs.

The present study suggests some policies. This kind of analysis on market integration and price transmission enlighten the significance of commodity based research. Such kind of studies is equally important as they provide better information on which decisions can be taken for scarce resource allocation. Government and private investors can invest and allocate more resources for efficient markets that are highly integrated. On the contrary, additional investment on infrastructures for less integrated markets will bring down the transaction costs and further improve the degree of integration between markets. Obviously, efficient markets provide less or no market distortions, thereby making the resource allocation more efficient and increase the welfare of producers and consumers.

ACKNOWLEDGEMENT

The authors are thankful to the referee for their constructive and valuable comments for improving the presentation of this paper.

REFERENCES

Awokuse OT, Bernard JC (2007). Spatial price dynamics in U.S. regional broiler markets. J Agric. Appl. Econ. 39(3):447-456.

Brosig S, Glauben T, Gotz L, Weitzel E, Bayaner A (2011). The Turkish wheat market: Spatial price transmission and the impact of transaction costs. Agribusiness 27(2):147-161.

Cuddy JDA, Della Valle PA (1978). Measuring the instability of time series data. Oxf. Bul. Econ. Stat. 40(1):79-85.

Cummings RW (1967). Pricing Efficiency in the Indian Wheat Market. Impex India, New Delhi.

Engle RF, Granger CWJ (1987). Cointegration and error-correction: Representation, estimation and testing. Econometrica 55:251-276.

Fackler P (1996). Spatial Price Analysis: A Methodological Review. Mimeo, North Carolina State University.

FAO Stat (2011).Food and Agriculture Organization of the United Nations. http://faostat.fao.org/site/342/default.aspx.

Gandhi VP, Koshy A (2006). Wheat marketing and its efficiency in India.

Working Paper No.: 2006-09-03, Indian Institute of Management, Ahemadabad, India.

GoI (Government of India) (2012). Economic Survey 2012-13. Ministry of Finance, Economic Division, New Delhi.

Goodwin BK, Schroeder TC (1991). Cointegration tests and spatial price linkages in regional cattle markets. Am. J. Agric. Econ. 73:452-64.

Granger C (1981). Some properties of time series data and their use in econometric model specification. J. Economet. 16:121-130.

Hendry D, Anderson G (1977). Testing dynamic specification in small simultaneous models: An application to a model of building society behaviour in the United Kingdom. Front. Quant. Eco. Pp.361-383.http://www.apeda.gov.in/apedawebsite/SubHead_Products/Poultry_Products.htm

Johansen S (1988). Statistical analysis of cointegration vectors. J. Econ. Dyn. Control 12:231-254.

Johansen S (1991). Estimation and hypothesis testing of cointegration vectors in Gaussian vectors auto regression models. Econometrica 59:51-80.

Johansen S (1994). The role of the constant and linear terms in cointegration analysis of nonstationary variables. Economet. Rev. 13:205-229.

Johansen S (1995). Likelihood-based Interference in Cointegrated Vector Autoregressive Models. Oxford: Oxford University Press.

Johansen S, Juselius K (1990). Maximum likelihood estimation and inference on cointegration with applications to the demand for money. Oxf. Bul. Econ. Stat. 52 (2):169-210.

Karthikeyan R, Nedunchezhian VR (2013). Vertical integration paving way to organised retailing in Indian poultry industry. Int. J. Bus. Manage. Invention 2(1):39-46.

Kaur B, Arshad FM, Tan H (2010). Spatial integration in the Broiler market in Peninsular Malaysia. J. Int. Food Agribus. Market. 22:94-107.

Kumar P, Sharma RK (2003). Spatial price integration and price efficiency at farm level: A study of paddy in Haryana. Ind. J. Agric. Econ. 58(2):201-217.

Lele UJ (1967). Market integration: A study of sorghum prices in western India. J. Farm Econ. 49:147-159.

Lele UJ (1971). Food Grain Marketing in India: Private Performance and Public Policy. Cornell University Press, Ithaca, New York.

Moodley D, Kerr WA, Gordon DV (2000). Has the Canada - US trade agreement fostered price integration? Rev. World. Econ. 136:334-354.

Saran S, Gangwar LS (2008). Analysis of spatial cointegration amongst major wholesale egg markets in India. Agric. Econ. Res. Rev. 21:259-263.

Vasciaveo M, Rosa F, Weaver R (2013). Agricultural market integration:price transmission and policy intervention. Paper presented at the 2nd AIEAA Conference on "Between Crisis and Development: which Role for the Bio-Economy", 6-7 June, 2013 at Parma, Italy.

23

Assessment of rural and experimental dairy products under dryland farming in Sudan

F. M. El-Hag[1], M. M. M. Ahamed[2], K. E. Hag Mahmoud[3], M. A. M. Khair[4], O. E. Elbushra[1] and T. K. Ahamed[5]

[1]Agricultural Research Corporation (ARC), Dryland Research Center (DLRC), Soba, Khartoum.
[2]Institute of Environmental Studies, University of Khartoum, Khartoum, Sudan.
[3]State Ministry of Agriculture, Animal Resources and Irrigation, Kordofan State, El-Obeid, Sudan.
[4]Agricultural Research Corporation (ARC), Wad Medani, Sudan.
[5]Food Research Centre, Shambat, Sudan.

Rural dairy processing situation in western Sudan (North Kordofan) was first assessed through a structured questionnaire. Some rural dairy products were sampled and assayed for bacteriological and chemical composition. The objectives were to investigate traditional dairy products and evaluate their nutrients composition and hygienic situation. Laboratory cheese making trials were then conducted to study the effects of milk type (goat vs. cow) and cheese type (white soft vs. braided) on cheese characteristics. Descriptive statistics were used for the statistical analysis of the survey data, a randomized complete block design for the cheese samples data and a 2×2 factorial experiment in a randomized complete block design for the laboratory trials data. There were seasonal fluctuations in quantities of milk processed. Most of the producers (62.7%) used mixed cow, sheep and goat milks for cheese processing. Braided cheese had a high cost of production compared with white soft cheese. Major production constraints stated were marketing, fluctuations in milk supply and shortage of water. Milk sources reported were from nomadic and transhumant herds, and to a lesser extent from villages and only very few of the producers had their own dairy animals. Cheese samples contained variable chemical constituents (total solids, fat and protein) that varied from location to another.

Key words: Rural dairy products, cows, goats.

INTRODUCTION

Milk composition is found to be affected by production systems by mechanisms likely to be linked to the stage and length of the grazing period, and diet composition, which will influence subsequent processing, and sensory and potential nutritional qualities of the milk (Chen et al., 2004; Lock et al., 2004; Gillian et al., 2005; Dewhurst et al., 2007). Food processing is an important measure for the preservation of food constituents as sources of nutrients and cash for many people in the world. Milk as a food is an ideal medium for the growth of bacteria and if kept at above 16°C the bacteria present will multiply rapidly thereby causing deterioration in quality (O'Connor, 1993). Therefore, surplus milk needs to be processed to preserve its valuable constituents for a long time. The fatty acid (FA) and fat-soluble antioxidant composition in milk fat is known to affect processing and sensory quality of dairy products (Jones et al., 2005; Kristensen et al., 2004) and may also affect their nutritional value (Thorsdottir et al., 2004; Havemose et al., 2006). The degree of saturation in milk fat has a bearing on the hardness, texture and taste of manufactured dairy products, particularly butter and cheese (Chen et al., 2004).

In Sudan, milk is traditionally preserved into different products e.g. ghee, butter and cheese to mention but three. The traditional cheese making is concentrated in the White Nile and Kordofan States. However, in some other areas where there is surplus fresh milk, no suitable method of preservation is available (Macquat and Bujanble, 1960; Ibrahim, 1970). In these areas where refrigeration or chilling means are not available, the importance of fermentation is obvious as it is the only method of preserving most of the milk nutrients. In Kordofan region, traditional cheese making is a seasonal activity. During the rainy season when plenty of milk is available, few people are actively engaged in making cheese. This, perhaps, is the most commonly practiced method of preservation of milk in the Sudan (Ibrahim, 1970).

Rural development projects in the area have many mobile laboratories or units for cheese making, scattered along the migration routes and camping areas of transhumant tribes in North Kordofan State (Ahmed, 1985). However, no thorough investigation of traditional processing methods and quality of dairy products in the area has been undertaken. This work was carried out with an ultimate objective of assessing rural dairy products situation and quality in North Kordofan, and comparing cheeses made from goat and cow milk.

MATERIALS AND METHODS

Study area

This study was carried out in western Sudan (North-Kordofan State) under semi-arid conditions; latitude 11° 15' N, longitude 27° 32' E, altitude 560 above sea level (asl). Average temperature varied between 30 to 35°C with peaks above 40°C. Summer rainy season extends from July to October during which many animal herders are engaged in cheese making from surplus milk.

Assessment of cheese quality products of local producers

A total of 51 local dairy producers were interviewed using purposive sampling technique. The interview focused on milk type (animal source), handling, processing; marketing and production constraints. 24 samples of dairy products included 6 braided cheeses and their whey, 6 white cheeses and their whey. Samples were collected from three different geographical locations and analyzed for bacterial contamination and biochemical composition.

Trials within the station

At the research station, two types of white cheese (soft and braided) were processed from cows and goats using rennet and table salt (NaCl). The trials were replicated six times for six consecutive days. Milk from both sources were first heated to 72°C cooled and divided into two equal portions, one possessed for braided and the other for soft white cheese.

The white cheese was made by heating the milk to 40°C table salt (8%) and rennet (one tablet/50 kg milk) added. This was stirred with milk for 5 min and left to coagulate. The curd was cut into cubes, placed in a mesh and left to drain the whey for 20 to 30 min. The whey was kept for further use, whereas the cubes were transferred into clean wooden moulds lined with cheese cloth and pressed overnight. The cubes were finally stored in plastic containers containing the previously preserved whey heated to 72°C for 1 min and cooled.

The braided cheese was prepared as the soft cheese but the cubes produced were scolded until the required acidity (0.46 to 0.60) when kneading was reached. Ripening was assessed by testing the ability of the curds to be kneaded into ropes. After draining off the whey, the ripened cubes were placed in a wooden plate and cut into slices to which Nigella sativa seeds were added. The slices were then hand-kneaded and pulled into braided ropes washed and stored with the whey in plastic containers.

Chemical analyses

Chemical analyses of milk, cheese and whey were carried out. Fat was determined by the Gerber method, crude protein (CP) according to Kejldahl method, total solids (TS) and water by the draft oven method (Marshall, 1993). Lactose was determined as outlined by Taylor (1970), while pH and ash were estimated according to AOAC (1990).

Bacteriological analyses

Media preparation and chemical tests for bacteria were carried out as shown by Cowan and Steel (1981). Viable bacterial counts were done according to Schalm et al. (1971).

Panel test

Ten untrained panelists were chosen to judge the quality of cheese (color, texture, flavor and taste) using hedonic scale of 1 to 4 (Watts et al., 1989).

Statistical analyses

Descriptive statistics were used for the questionnaire. For chemical composition data, random complete block design was utilized. The data of the laboratory cheese making trials and panel test used a 2 × 2 factorial randomized block design. Duncan multiple range test was used to test mean separation (Steel and Torrie, 1980).

RESULTS

Questionnaire of rural dairy products

Fifty-one rural cheese processing producers were interviewed through a structured questionnaire. The majority of the respondents (78.4%) reported that they produced cheese during rainy season, a time of high milk availability (Table 1). There was a complete consensus among the interviewed farmers that there were fluctuations

Table 1. Production season, milk supply situation and processing of rural dairy products as reported by the respondent farmers (%).

Evaluation parameters	Frequency	Percent
Production season		
One season	40	78.4
All around the season	11	21.6
Fluctuation in milk quantity		
Yes	51	100.0
No	0	00.0
Time between milk receipt and processing		
Direct manufacture	51	100.0
Storage	0	00.0
Treatment of milk before processing		
Yes	0	00.0
No	51	100.0
Type of cheese produced:		
White	13	25.5
Braided	15	29.4
White + Braided	9	17.6
White + Rome	1	2.0
Braided + Rome	2	3.9
White + Braided + Rome	11	21.6
Any other type of dairy products		
Yes	28	54.9
No	23	45.1
Type of other dairy products		
Ghee	2	7.14
Hard fermented milk (HFM)	1	3.57
Butter	1	3.57
Soft fermented milk (SFM)+ ghee	8	28.58
Ghee + HFM	9	32.14
HFM + SFM + ghee	6	21.43
HFM + SFM + ghee + butter	1	3.57

in milk quantity and handling during the period of production. Also there was no treatment made to the milk from collection to the time of processing. Braided cheese recorded the highest production rate (29.4%) followed by white cheese (25.5%), while the combined production of both shuddered cheese and white cheeses recorded the lowest percentage (2.0%). One-third of persons interviewed (32.14%) reported that they also produced ghee and fermented hard milk, 28.58% produced soft fermented milk and ghee while farmers producing soft fermented milk, ghee, hard fermented milk and butter recorded the lowest percentage (3.57%) (Table 1).

The bulk of milk for these rural dairy processing units was reported to be purchased from nomads (54.9%), and to a lesser extent from villages (19.6%). Only very few (2.0%) of the producers had their own dairy animals. The majority of the respondents (58.8%) reported that they mixed cow, sheep and goat milks for cheese processing, whereas the use of goat + cow milk recorded the lowest percentage (3.9%). Quantities of milk processed into cheese in these units ranged from 50 to 200 L/day in the majority (39.2%) of units surveyed, while 7.8% of the interviewed producers stated that they processed quantities of milk in the range of 500 to 650 L/day. Half

Table 2. Milk sources, quantities used and equipment used in Cheese processing as reported by respondent farmers (%).

Evaluation parameters	Frequency	Percent
Milk source		
Village	10	19.6
Nomads	28	54.9
Own dairy animals	1	2.0
Village + nomads	9	17.0
Village + own dairy animals	1	2.0
Nomads + own dairy animals	2	3.9
Type of milk used in processing		
Cow	19	37.3
Cow + goat	2	3.9
Cow + goat + sheep	30	58.8
A mount of milk used per day		
50 – 200 liter	20	39.2
201- 350 liter	9	17.6
351 – 500 liter	12	23.5
501 – 650 liter	4	7.8
More than 650 liter	6	11.8
Milk containers		
Plastic barrel	26	51.0
Metal barrel	15	29.4
Plastic + metal barrel	10	19.6
Type of cheese cloth used		
Cotton cloth	41	80.4
Synthetic	7	13.7
None	3	5.9
Cheese storage containers		
Plastic barrel	29	56.9
Metal barrel	17	33.3
Plastic + metal barrel	5	9.8

(51.0%) of the respondents reported that they used plastic barrels as milk containers and for cheese storage, whereas the use of both plastic and metal barrels recorded the lowest percentage (10%). The majority (80.4%) of the producers reported that they used cheese cloth to separate the whey from coagulated cheese (Table 2).

More than two-thirds (68.6%) of the interviewed producers reported that braided cheese had a higher cost of production than white cheese. Most of the respondents (60.78%) used to sell their dairy products at El-Obeid market due to availability of marketing facilities especially the purchasing power. The rest of interviewees (5.88%) who lack the means of transportation used local markets through which marketing of dairy products was done. Although poisoning of milk and milk products could easily happen due to the vulnerability to contamination by germs and bacteria, 94.1% of the interviewed producers reported that there were no poisoning cases in their local processing units. The most important problems mentioned by respondents were marketing problems, shortage of milk and shortage of water which recorded 20.00, 18.57 and 8.56% of the interviewees' answers, respectively. Other minor constraints reported were difficulty in coagulation, rise in milk prices and unavailability of transportation means to marketing centers and storage especially during the rainy season (Table 3).

Table 3. Relative production cost of different cheese types, markets and production constraints as reported by the interviewed producers (%)

Evaluation parameters	Frequency	Percent
High production cost		
White	16	31.4
Braided	35	68.6
Markets		
Local	3	5.9
El-Obeid town	31	60. 8
Out of State	4	7.8
Local + town	10	19.6
Town + out of State	3	5.9
Cheese poisoning		
Yes	3	5.9
No	48	94.1
Constraints to cheese production		
Yes	21	41.2
No	30	58.8
What are the constraints?		
Marketing	14	20.0
Shortage of milk	13	18.6
Shortage of water	6	8. 6
Coagulation difficulties	5	7.1
Prices reduction	3	4.3
Rise in milk prices	5	7.1
Failure of production	2	2.9
Lack of capital	2	2.9
shortage of canning equipment	5	7.1
Veterinary facilities and follow up not enough	1	1.4
Shortage of store in production area	1	1.4
Transportation to stores during high rainfall	5	7.1
Shortage of rennet tablet	3	4.3
Milk adulteration	3	4.3
Mobility of milk producer	2	2.9

Chemical composition and bacteriological profile of dairy products

Processing dairy units in the area of study

Samples of dairy products taken from different locations in the area of study were analyzed for their nutrient contents, chemical composition and bacterial constituents. It was observed that, cheese obtained from location 2 had the highest (P <0.05) fat content, whereas the braided cheese showed higher (P <0.05) fat and protein contents than the white cheese (Table 4). On the other hand, the whey of the braided cheese had higher (P <0.05) fat content than that of the white cheese (Table 5). There was no significant difference between both white and braided cheese in bacteriological profile (Table 6).

Station dairy products trials

Fresh milk samples from cows and goats, analyzed for their nutrient contents and chemical composition showed that, cow milk had higher (P <0.05) total solids (TS), fat and protein contents, whereas goat milk had higher (P

Table 4. Chemical composition of sampled white and braided cheese produced in the area of study.

Factor	pH	TS[1] (%)	Ash (%)	Fat (%)	Protein (%)	Lactose (%)
Location						
1	4.9	49.7	4.1	16.8	12.3	16.4
2	4.3	54.4	5.8	24.5	13.4	9.9
3	5.1	48.5	6.6	20.5	13.2	8.3
± SE	0.28^{NS}	4.3^{NS}	0.33^{NS}	0.80*	0.44^{NS}	3.4^{NS}
Cheese type						
White	4.4	45.3	5.6	19.1	8.5	11. 9
Braided	5.2	56. 5	5.4	22.1	17.4	11.1
±SE	0.13*	2.9^{NS}	0.77^{NS}	0.51*	0.61*	2.1^{NS}
Location x Cheese type						
±SE (Interaction)	0.22^{NS}	5.2^{NS}	1.3^{NS}	0.88^{NS}	1.0569^{NS}	3.7^{NS}

NS = not significant (P > 0. 05), * Significant at P < 0.05. [1] total solids.

Table 5. Chemical composition of white and braid cheese whey samples produced in the area of study.

Factor	pH	TS (%)	Ash (%)	Fat (%)	Protein (%)	Lactose (%)
Location						
1	4.9	17.9	10.6	0.33	0.58	6.5
2	5.1	16.7	8.8	0.36	0.65	6.9
3	5. 5	14.8	8.2	0.35	0.50	5.7
± SE	0.50^{NS}	1.1^{NS}	0.29^{NS}	0.30^{NS}	0.03^{NS}	0.90^{NS}
Cheese type						
White	5.8	17.1	9.3	0.62	0.52	6.8
Braided	4.6	15.8	9.1	0.08	0.63	5.9
± SE	0.35^{NS}	1.08^{NS}	0.53^{NS}	0.42*	0.049^{NS}	1.5^{NS}
Location x Cheese type						
±SE Interaction	0.60^{NS}	1.8^{NS}	0.92^{NS}	0.73^{NS}	0.08^{NS}	2.55^{NS}

NS = not significant (P > 0. 05), * Significant at P < 0.05.

Table 6. Chemical composition of milk.

Constituent	Cow milk	Goat milk	SE±
pH value	6.6	6.7	0.04*
Total solids (%)	13.2	12.2	0.28*
Ash (%)	0.7	0.9	0.03*
Fat (%)	3.9	3.5	0.10*
Protein (%)	3.7	3.4	0.09*
Lactose (%)	4. 8	4.5	0.22^{NS}

NS = not significant (P > 0.05), * Significant at P < 0.05

<0.05) pH and ash values (Table 7). Effects of milk and cheese types on coagulation time and yield, showed that, goats' milk had longer (P <0.05) coagulation time but lesser yield (P <0.001) than cows' milk, whereas braided cheese was higher yield (P <0.001) than the white cheese. Interactions of type of milk X type of cheese

Table 7. Effects of milk type and cheese type on coagulation time (hr) and yield of cheese.

Factor	Coagulation Time (h)	Cheese Yield (kg/100 kg milk)
Type of milk		
Cow	6.5	13.7
Goat	10.6	11.9
± SE (Type of milk)	1.06*	0.12 ***
Type of cheese		
White	9.1	15. 9
Braided	7.9	9. 8
± SE (Type of cheese)	1.06[NS]	0.12***
Type of milk × Type of cheese		
Cow–White (CW)	5.1[b]	17[a]
Goat–White (GW)	13.2[a]	14.8[b]
Cow–braided (CB)	7.8[b]	10.8[c]
Goat–braided (GB)	8.1[b]	9.2[d]
± SE (interaction)	1.5*	0.17*

[abcd]Values within the same column bearing different superscripts vary significantly at (P<0.05). NS = not significant (P>0.05). *Significant at P<0.05, *** highly significant at P<0.001.

Table 8. Effects of milk type and cheese type on cheese chemical composition.

Factor	pH	TS[1] (%)	Ash (%)	Fat (%)	Protein (%)	Lactose (%)
Type of milk						
Cow	5.3	58.3	7.9	24.6	20.8	4.9
Goat	5.4	55.2	9.1	21.8	19.6	4.8
±SE	0.07*	0.62*	0.17**	0.42**	0.39*	0.38[NS]
Type of cheese						
White	5.1	54.0	8.5	23.5	16.6	5.4
Braided	5. 7	59.5	8.5	22.8	23.8	4.4
±SE	0.07***	0.62***	0.17[NS]	0.42[NS]	0.39***	0.38[NS]
Type of milk × type of cheese						
Cow-white (CW)	4.9	56.3	8.37[bc]	24.9	17.3	5.8
Goat-white (GW)	5.1	51.7	8.8[ab]	22.2	15.9	4.9
Cow-Braided (CB)	5.7	60.3	7.64[c]	24.2	24.3	4.0
Goat-braided (GB)	5.7	58.7	9.4[a]	21.3	23.3	4.7
±SE	0.10[NS]	0. 9[NS]	0.25**	0.59[NS]	0.55[NS]	0.5[NS]

[abc]Values within the same column bearing different superscripts vary significantly at (P<0.05). NS = not significant (P>0.05). *Significant at P<0.05, ** highly significant at P<0.001 *** very highly significant at P<0.001. [1] total solids.

showed significant (P <0.05) longer time for coagulation of white cheese from goats but at the same time higher yields for the white cheese from goat milk. The lowest yield was obtained for braided cheese of goat milk (Table 8).

Except for lactose, significant differences due to milk type could be detected. TS, and ash (P <0.01) as well as fat and protein (P <0.05) were higher in cheese produced from cow than those from goats, whereas the pH was higher (P <0.05) in cheese produced from goats. Braided cheese showed higher (P <0.001) pH, TS and protein values than white cheese. Interaction due to type of milk X type of cheese was only significant (P <0.05) for ash where braided cheese from goats was higher than that of both white and braided cheeses from cows (Table 9).

Efficiencies of milk fat and protein recovered were

Table 9. Effects of type of milk used and type of cheese produced on nutrient recovery in different cheeses.

Factor	Efficiency of fat recovery (%)	Efficiency of protein recovery (%)	Efficiency of lactose recovery (%)
Type of milk			
Cow	85.9	78.7	14.3
Goat	78.7	74.7	14.8
± SE (type of milk)	2.22*	1.75*	1.23[NS]
Type of cheese			
White	91.7	72.0	17.4
Braided	72.8	81.4	11.7
± SE (type of cheese)	2.22**	1.75**	1.23**
Type of milk × type of cheese			
±SE (interaction)	3.15[NS]	2.48[NS]	1.73[NS]

NS = not significant (P > 0.05) *Significant at P<0.05, ** highly significant at P<0.001.

Table 10. Effect of milk type and cheese type on first whey (the whey before ripening) chemical composition.

Factor	pH	TS[1]%	Ash%	Fat%	Protein%	Lactose%
Type of milk						
Cow	5.8	10.3	4.0	0.08	1.06	5.1
Goat	5.7	11.0	4.6	0.09	1.07	5.3
±SE	0.14[NS]	0.40[NS]	0.41[NS]	0.27[NS]	0.066[NS]	0.28[NS]
Type of milk cheese						
White	5.9	14.5	8.0	0.14	0.99	5.3
Braided	5.6	6.9	0.55	0.02	1.13	5.1
±SE	0.14[NS]	0.40***	0.41***	0.03**	0.07[NS]	0.28[NS]
Type of milk x type of cheese						
Cow-white (CW)	6.0	14.0	7.5	0.13	1.1[ab]	5.3
Goat-white (GW)	5.8	14.9	8.5	0.15	0.88[b]	5.4
Cow-braided (CB)	5.7	6.6	0.50	0.02	1.03[ab]	5.1
Goat-braided (GB)	5.5	7.1	0.59	0.02	1.3[a]	5.2
±SE	0.19[NS]	0.57[NS]	0.58[NS]	0.03[NS]	0.09*	0.40[NS]

[ab]Values within the same column bearing different superscripts vary significantly at (P<0.05).   NS = not significant (P>0.05). *Significant at P<0.05, ** highly significant at P<0.001 *** very highly significant at P<0.001. [1] total solids.

higher (P <0.05) in milk from cows than those recovered from goats, whereas braided cheese showed higher (P <0.01) protein recovery but less (P <0.01) fat recovery (Table 10).

First whey characteristics were shown to be significantly affected by both cheese type and interaction of milk type X cheese type. Fat (P <0.01), TS and ash (P <0.001) were higher in braided than white cheese. Interaction was only significant (P <0.05) for the protein content where braided cheese first whey from goat milk had higher protein content than that of the white cheese (Table 11). Second whey showed higher fat and protein contents (P <0.001) as well as lactose values (P <0.05) in the white cheese whey compared to braided cheese (Table 12). The microbial profile for the dairy products' whey was not affected by either type of milk or type of cheese or their interaction (Table 13).

Organoleptic scoring of white and braided cheese

Cheese of cow milk had higher scores for taste and texture (P <0.05) as well as flavor (P <0.001) than goat milk. However, type of milk did not show significant effect

Table 11. Effect of milk type and cheese type on chemical composition of second whey (the whey after cheese ripening)

Factor	pH	TS[1]%	Ash%	Fat%	Protein%	Lactose%
Type of milk						
Cow	4.8	13.1	8.4	0.15	1.0	3.5
Goat	4.9	13. 6	9.2	0.24	1.1	3.0
±SE	0.12[NS]	0.52[NS]	0.44[NS]	0.03[NS]	0.08[NS]	0.29[NS]
Type of milk cheese						
White	4.4	14.1	8.1	0.37	1.04	3.9
Braided	5.3	12.6	9.4	0.03	0.45	2.7
±SE	0.12**	0.52[NS]	0.44[NS]	0.03***	0.08***	0.29*
Type of milk x type of cheese						
±SE	0.17[NS]	0.74[NS]	0.63[NS]	0.04[NS]	0.11[NS]	0.42[NS]

NS = not significant (P>0.05). *Significant at P<0.05, ** highly significant at P<0.001 *** very highly significant at P<0.001. [1] total solids.

Table 12. The microbiological profile of milk types used and cheese processed in the laboratory.

Factor	*Staphylococcus*	*Bacillus*	*Coliform*	Total Bacteria Count
Milk type				
Cow	1.2×10^4	0	0	12×10^4
Goat	2.2×10^4	1.3×10^4	0	3.5×10^4
Cheese type × Milk type				
Cow-White (CB)	0	4.9×10^2	0	4.9×10^2
Goat-White (GW)	0	6.1×10^2	0	6.1×10^2
Cow-Braided (CB)	0	3.1×10^2	0	3.1×10^2
Goat-Braided (GB)	0	3.3×10^2	0	3.3×10^2

Table 13. Organoleptic scoring of white and braided cheese made from milk of cows and goats.

Factor	Color	Texture	Flavor	Taste
Type of milk				
Cow	2.81	3.53	3.42	3.06
Goat	2.94	3.03	2.50	2.33
±SE	0.214[NS]	0.153*	0.134***	0.192*
Type of cheese				
White	3.11	3.36	2.97	2.86
Braided	2.64	3.19	2.94	2.53
±SE	0.214[NS]	0.153[NS]	0.134[NS]	0.192[NS]
Type of milk x Type of cheese				
Cow-White (CW)	2.78	3.44	3.22[a]	3.00
Goat-White (GW)	3.44	3.28	2.72[ab]	2.72
Cow-Braided (CB)	2.83	3.61	3.61[a]	3.11
Goat-Braided (GB)	2.44	2.78	2.29[b]	1.94
±SE	0.303[NS]	0.217[NS]	0.189[NS]	0.271[NS]

[ab]Values within the same column bearing different superscripts vary significantly at (P<0.05). NS = not significant (P>0.05). *Significant at P<0.05, ** highly significant at P<0.001 *** very highly significant at P<0.001.

on organoleptic scoring. Interactions showed that white cheese of cow milk had higher (P <0.05) flavor score than goat braided cheese (Table 13).

DISCUSSION

Dairy rural products

Most of the rural dairy producers surveyed in this study agreed that there were fluctuations in milk quantities during the rainy season depending on rain amounts and availability of pasture, a situation which led to mixing of milk from different animal species (cows, goats, and sheep). The high cost of braided cheese as reported by the respondents was due to the extra cost needed for heating before milk processing. Other constraints were shortage of water and marketing.

Variations in the nutrient contents and chemical composition observed in dairy products obtained from different locations could be contributed to factors such as percentage of differences in type of milk mixed and manufacturing conditions.

Dairy products processed at station

Nutrient contents of fresh milk and milk products processed at station showed higher fat, protein and total solid (TS) for cow's milk and milk products than goats milk. This could be attributed to species differences where goats' milk is characterized by small fat globules and low protein content and hence lower TS. Similar observations were obtained in the milk composition of Nigerian cattle, sheep and goats indicating significant variations in all constituents except protein percentage. Caprine milk contained the highest percentages of fat (5.80%), total solids (15.37%) and ash (0.77%), and bovine milk contained the least percentages of fat (0.68%) and lactose (1.84%) (Aduli et al., 2002). Also, it has been pointed out that as fat content increases, moisture content decreases (El Erian, 1976). Also goat's milk showed longer time to coagulate, the poor cheese making ability with goats milk could be due to the specific properties of casein micelles such as composition, hydration and size compared to cow's milk (Abdel-Razig, 1966). The paste rennet, usually produced by shepherds themselves from the abomasum of the lambs or kids of their flock, or the vegetable rennet from the cardoon (*Cynara cardunculus*) flowers locally harvested, are technological factors that strongly influence the quality of the local dairy products (Scintu and Piredda, 2007). It has also been found that renneting time for goat milk is shorter than for cow milk, and the weak consistency of the gel is beneficial for human digestion but decreases its cheese yield (Park et al., 2007). On the other hand, the high yield of cheese from cow's milk could be attributed

to the high TS content (Moneib et al., 19881; Ahmed and Khalifa, 1989).

Bacterial profile

Bacillus was found in all samples of cheese either made in laboratory or from rural processing units. This was attributed to the wide range of pH for the growth of *Bacillus*: 4.9 to 9.3 at salt content of 7.5 to 10% (Buchanon and Gibbons, 1974). Other studies on isolates of lactic acid bacteria, from local (Jordan) white cheese made from sheep raw milk indicated that the presence of heterofermentative *Lactobacillus* with variable levels of contamination and frequency (Haddadin, 2005). Also samples collected from different farms in Italy was found to affect bacterial profile of Caprine cheese, isolated species were: coliforms, lactococci, lactobacilli and halotolerant (Foschino, et al., 2002). While those obtained in traditional Egyptian soft Domiati cheese were: *Leuconostoc mesenteroides*, *Lactococcus garvieae*, *Aerococcus viridans*, *Lactobacillus versmoldensis*, *Pediococcus inopinatus*, and *Lactococcus lactis* (El-Baradei et al., 2007).

Staphylococcus aureus was found in samples of white cheese obtained from the traditional processing units but not in pasteurized milk. The major factors that contribute to the presence of *S. aureus* in cheese were the use of un-pasteurized milk with high infected starter culture (Santos and Genigeorg, 1981). Santos et al. (1980) reported that the use of raw milk for manufacture of cheese contains high count of *S. aureus*. And if the milk was collected by non-refrigerated truck and brought to the plant after 3 to 5 h, *Staphylococci* and other organisms will multiply rapidly. Lack of satisfactory sanitary practices in the dairies may contribute to heavy contamination of the milk. Preformed enter toxin can survive milk pasteurization. Therefore, pasteurization cannot be substituted for sanitary milk production. Microbes enter milk and milk products via air, handling, equipments, and high environmental temperatures (> 16°C) (O'Mahony, 1988). It was found that the source of microbial milk contamination was its handling in the time from leaving the udder (65,000 bacteria/ml) until reaching the refrigeration farm tank (in the case of machine-milking, with 362,000 bacteria/ml) or the bulk tank of the cooperative (in the case of hand-milking, with 262,000 bacteria/ml). Farms with fewer animals (<100 animals) that practiced hand-milking had a better hygiene-sanitary quality (Delgado-Pertiñez et al., 2003).

The average total bacteria count for cow and goat milk used for cheese manufacturing at the station were 1.2×10^4 and 3.5×10^2 CFU/ml. These were low counts compared with those reported by Zeng and Escobar (1996) who obtained maximum bacterial counts of 6.4×10^5 CFU/ml. Also O'Ocnnor (1993) stated that plate count of bacteria should not exceed 50000 bacteri a per milliliter. The average of *Staphylococcus* counts found in cow and goat milk were 1.2×10^4 and 2.2×10^4 per ml, respectively. This count was in the range of the recommended number of *Staphylococcus* $< 10^3$ to 10^6 CFU/ml, depending on origin of milk (Zeng and Escobar, 1996). Similarly, samples cheese collected from different goat breeds showed that the counts of coliform were within the normal range, although there were significant differences between farms in different regions (Muostafa et al., 2009)

Organoleptic scores of white and braided cheese made from cow and goat milk

Cow milk cheese recorded the highest score points for texture flavor, and taste. Generally, white cheese recorded relatively the highest score points in color, texture, flavor and taste compared to braided cheese. White cheese made from cow milk had the highest scores score points in color, flavor and taste. These results were in accord with the findings of Abdalla and Abdel-Razig (1997) who reported that white cheese made from cow milk significantly scored the best texture and flavor, while the color, saltiness and sourness were not significantly affected by type of milk. It is worth noting here that the cheese obtained in this study was of high standard quality, good color, attractive and glossy with smooth but firm body and texture, with better consistency, richness, much clean and had a good flavor, and without gas holes.

REFERENCES

Abdel-Razig KA (1996). The Production of White Soft Cheese from Different Milk Sources. M.Sc. Thesis, University of Khartoum, Sudan.

Abdalla MO, Abdel-Razig K.A (1997). Effect of type of milk on the quality of white soft cheese. University of Khartoum J. Agric. Sci. 5:47.

Aduli E, Othniel M-A and Yakubu R A (2002).Comparative study of milk compositions of cattle, sheep and goats in Nigeria. Anim. Sci. J. 73:541–544.

Ahmed AM (1985). Bacteriological and Chemical Characteristics of Sudanese White Cheese Produced and Stored under Different Conditions, Ph.D. Theses. University of Khartoum, Sudan.

Ahmed TK, Khalifa NA (1989). The Manufacture of white soft cheese from recombined milk. Sudan J. Anim. Prod. 2:63-69.

AOAC (1990). Official Methods of Analysis (15th Edn.). Association of Official Analytical Chemists (AOAC). Washington, D.C., USA

Cowan ST, Steel RL (1981). Cowan and Steel's Manual for the Identification of Medical Bacteria. 2nd edition.

Chilliard Y, Rouel J, Leroux C (2006) .Goats alpha-s1 casein genotype influence its milk fatty acid composition and delta-9 desaturation ratios. Animal Feed Sci. Technol. 131:474-487.

Delgado-Pertiñez M, Alcalde MJ, Guzmán-Guerrero JL, Castel JM, Mena YF, Caravaca F (2003). Effect of hygiene-sanitary management on goat milk quality in semi-extensive systems in Spain. Small Rum. Res. 47:51–61.

Dewhurst RJ, Shingfield KJ, Lee MRF and Scollan ND (2006). Increasing the concentrations of beneficial polyunsaturated fatty acids in milk produced by dairy cows in high-forage systems. Anim Feed Sci Technol. 131:168–206.

El-Barade G, Delacroix-Buchet A, Ogier JC (2007). Biodiversity of Bacterial Ecosystems in Traditional Egyptian Domiati Cheese. Appl Environ Microbiol. 73(4):1248–1255.

El-Erian AFM, Abdel T, GadElrab IE (1976). The Effect of Brine Salting Method on the Quality and Chemical Composition of Domiati Cheese from Unsalted Milk. Agric. Res. Rev. 54(7):173-179.

Baradei F, Soryal K, Zeng S, Van Hekken D, Bah B, Villaquiran Sanz Sampelayo MR, Chilliard , Schmidely PH, Boza J (2007). Influence of type of diet on the fat constituents of goat and sheep milk. Small Rum. Res. 68:42-63.

Foschino R , invernizzi A, barucco R and stradiotto K (2002). Microbial composition, including the incidence of pathogens, of goat milk from the Bergamo region of Italy during a lactation year. J. Dairy Res. 69:2:213-225.

Macquat G, Bujanbe SM (1960). Technology of Ewes Milk and Goat's Milk Products. Dairy Sci, Abstr. 22:1.

Moustafa EL, El-Demerdash ME, Hashem ME (2009). Survey study on goat milk from different breeds raised under Egyptian conditions. Egypt. J. Sheep. Goat Sci. 4(1):89-98.

O'Connor CB (1993). Traditional Cheese Making Manual. ILCA (International Livestock Centre for Africa), Addis Ababa, Ethiopia.

O'Mahony F (1988). Rural Dairy Processing: Experiences in Ethiopia. ILCA (International Livestock Centre for Africa) Manual No. 4. ILCA, Addis Ababa, Ethiopia. P. 64.

Santos EG, Genigeorgis CC (1981). Survival and Growth of *Staphylococcus aureus* in Commercially Manufactured Brazition Minas Cheese. J. Food Prod. 44,177-184.

Schalm WO, Carrol EJ, Jain NC (1971). Bovine Mastitis, a book. School of Veterinary Medicine, University of California Press, Davis, USA.

Scintu MF, Piredda G (2007). Typicity and biodiversity of goat and sheep milk products. Small Ruminant Res. 68(1):221-231.

Steel RGD, Torrie JK (1980). Principle and Procedures of Statistics A biometrical Approach (2nd Edn.). McGraw-Hill Books CO. NY. USA.

Talpur FN, Bhanger MI, Memon NN, (2009). Milk fatty acid composition of indigenous goat and ewe breeds from Sindh, Pakistan. J. Food Comp. Anal. 22:59-64.

Taylor GC (1970). A rapid enzymatic method for the determination of lactose in cheese. Austr. J. Dairy Technol. 25:7.

Thorsdottir I, Hill J, Ramel A, Short communication: seasonal variation in *cis*-9, *trans*-11 conjugated linoleic acid content in milk fat from Nordic countries. J. Dairy Sci. 87:2800-2802 (2004).

Zeng SS, Escobar EN (1996). Effect of Breed and Milking Method on Somatic Cell Count, Standard Plate Count and Composition of Goat Milk. Small Rum. Res. 19:169-175.

24

Least cost diet plan of cows for small dairy farmers of Central India

S. N. Goswami, A. Chaturvedi, S. Chatterji, N. G. Patil, T. K. Sen, T. N. Hajare and R. S. Gawande

National Bureau of Soil Survey and Land Use Planning, Amravati Road, Nagpur, Maharashtra 440 010, India.

Small dairy farmers of Central India (Madhya Pradesh, Vidarbha region of Maharashtra and Chhattisgarh) "are constrained by inadequate supply of protein source during the dry season" is extracted from Chakeredza et al. (2007). Poor economic conditions of the dairy farmers do not allow them to purchase commercial protein concentrates. Locally available non-conventional protein sources can be used as alternative cheaper protein sources. Formulation of a cost efficient diet balanced for key nutrients has emerged as the biggest problem to the farmers. The present study presents a least cost balanced diet formulation plan for the small dairy farmers. The Linear Programming Technique was applied to formulate the least cost ration plan for daily feeding for the cross bred and local dairy cows separately. The least cost ration plan formulated for daily feeding for cross bred dairy cows yielding 5 to 10 L of milk per day included 3.50 kg paddy straw, 10.60 kg Napier grass, 1.35 kg soybean cake, 2.08 kg wheat bran, and 0.06 kg mineral mixture, costing 19% less in comparison to the routine feeding plan followed by the farmers. Similarly, the least cost daily feeding ration plan formulated for the local lactating cows yielding 3 L of milk per day included 3.06 kg paddy straw, 7.60 kg Napier grass, 0.86 kg soybean cake, 1.20 kg wheat bran, and 0.05 kg mineral mixture, reducing the feed cost by 22% as compared to the existing ration plan followed by the farmers . The least cost ration plan formulated through this study is recommended for use by the small dairy farmers of Central India to reduce the feed cost of dairy animals.

Key words: Least cost, fodder, dairy feed, small scale, linear programming.

INTRODUCTION

Cost of feeding is the single most important factor affecting the profitability of a dairy enterprise. The economical feeding of cows is a major component of a dairy farmer's decision making. Feed typically accounts for 60-80 per cent of variable cost of milk production (Webster, 1993; Patil, 2010). Without good feeding programmes, the benefits of good breeding and management programmes cannot be realized (Chakeredza et al., 2008).

The economical feeding of dairy cows is a relatively complex problem as it has to cater for the nutritional requirement to maintain the health of the cow and milk production. The nutrients to be supplied in a feeding programme include energy, protein, minerals and vitamins (Pond et al., 1995). Carbohydrates and fats are the major source of energy. In Central India (Madhya Pradesh, Vidarbha region of Maharashtra and Chhattisgarh) sorghum, maize, paddy straw, and wheat bran are the major sources of energy. Minerals and vitamins are incorporated in the diet as pre-mixes. Protein is included in the dairy meal concentrate and mainly supplied through soybean cake.

Composition of the ration currently fed to the cows is decided based on dry matter intake required for getting a required level of milk production per day. Due to high cost and non-availability of concentrates and protein sources, locally available supplements are required to optimally feed dairy cows. These should be supplied in higher quantities as a replacement to concentrates to reduce the feed cost. However, there is a problem to formulate diets that are balanced with respect to protein, energy, vitamins, and minerals and at the same time being low cost.

Review of existing literature offers numerous examples of utilizing operation research technique for solving nutrition management problem. The most common is the least cost ration optimization based on linear programming technique. It has been widely used in modeling the least cost ration problem (O'Coner et al., 1989; Munford, 1996; Alexander et al., 2006; Chakeredza et al., 2007). No study has until now been reported from Central India that emphasizes cost minimization for feeding dairy cows under constraints faced by the dairy farmers. Therefore, an attempt is made in the present study to formulate a least cost ration plan for the dairy farmers of Central India.

MATERIALS AND METHODS

A total of 200 small dairy farmers, 100 each having cross bred and local lactating cows, were randomly selected through guidance from animal nutrition experts from different parts of Madhya Pradesh, the Vidarbha region of Maharashtra, and Chhattisgarh in such a way so that they could represent a homogeneity regarding type of animal rearing, feeding, and management situation of dairy farmers of Central India. Data were collected during 2009 to 10 through pretested questionnaire by interview method.

For each breed/type maintained in the farm, an individual lactating animal was selected, whose per day milk yield was the nearest to the average milk yield per day 5 L for cross bred and 3 L for local during lactation period. Major sources of dietary energy in the study area include wheat bran, broken wheat, maize, soybean cake and roughages (dry fodder). The source of nutrients provided by the majority of dairy farmers was retained in the existing plan to ensure minimum switch over or changes and utilization of locally available material. One method that can be used to derive least-cost rationing is linear programming (Torez, 2000). The linear programming technique was used to work out the least-cost combination of feeds and fodders under the specified nutrient restrictions, which were estimated from the actual feeding of the lactating animal. The following situations were identified for programming least cost combination of feeds and fodder:

Situation I: Cross-bred cows (jersey) yielding 5 to 10 L of milk per day. Situation II: Local cows (mix of breeds) yielding below 3 L of milk per day.

The cost minimizing model used in this study was of the following form:

$$\text{Minimize} \quad Z = \sum_{j=1}^{n} C_j X_j$$

Subjected to linear constraints such as:

(i) $\sum_{j=1}^{n} a_{1j} x_j \leq B_1$

(ii) $\sum_{j=1}^{n} a_{2j} x_j \leq B_2$

(iii) $\sum_{j=1}^{n} a_{3j} x_j \geq B_3$

(iv) $\sum_{j=1}^{n} a_{4j} x_j \geq B_4$

(v) $\sum_{j=1}^{n} a_{5j} x_j \geq B_5$

(vi) $\sum_{j=1}^{n} a_{6j} x_j \geq B_6$

And non-negativity constraints $x_j \geq 0$ where, z = total cost of feed mix in rupees; x_j = quantity of jth feed material in the feed mix in kg; c_j = unit cost of feed material x_j in rupees per kg; a_{ij} = amount of ith nutrient available in one kg of x_j feed material; i = 1, 2, 3, 4, 5, and 6.

B_1, B_2, B_3, B_4, B_5, and B_6 are required levels of nutrients such as dry matter supplied by roughages, dry matter supplied by concentrates, total digestible nutrients (TDN), digestible crude protein (DCP), calcium (Ca) and phosphorus (P) for specific level of milk yield per day per lactating animal. Recommended level of nutrients suggested by the animal nutrition experts for maintaining optimum daily milk yield were estimated from the collected data based on average body weight of animal. INRA (France) feeding standard was used.

Objective function

The objective function was to minimize the cost of feed and fodders fed to the milch cows for the specified two programming situations.

Technological matrix

The initial technological matrices for cost minimization problems under the above mentioned two situations were formulated by using the simplex method of linear programming (Anderson et al., 2000). Initial matrices were formulated with nutrient restrictions, purchase price of feeds and fodders, alternative feeds and fodders (or) real activities, disposal activities and artificial activities. These components of initial matrix are discussed below:

Nutrient restrictions

In the present study, two types of nutrient restrictions' were used namely, maximum restrictions and minimum restrictions.

Maximum restrictions

Maximum restrictions were applied to dry matter as the belly capacity of the animal is fixed under an assumption that, the animals are being fed full belly capacity and thus, the dry matter intake was calculated based on actual amount of roughages and concentrates fed to the lactating cows at the specified level of milk production. The dry matter restrictions were imposed separately for roughages and concentrate fed to the lactating animals at the specified level of milk production. Thus, it was ensured that the optimal plan does not violate the requirements of existing nutritional norms.

Minimum restrictions

Minimum restrictions were applied to:

(i) Total digestible nutrients (TDN) which is the system of measuring available energy of feed and energy requirement of animals involving a complete formula of measured nutrients and it is estimated as digestible energy/0.044,
(ii) Digestible crude protein (DCP) which is the amount of crude protein actually absorbed by the animal (NDDB, 2012),
(iii) Calcium (Ca) and,
(iv) Phosphorus (P) that was estimated on the basis of actual feeding for different situations specified and imposed as requirements to satisfy optimal conditions.

Real activities (alternate nutrient sources)

Real activities are alternative nutrient sources available in the locality (Chandler and Walkir, 1972). While selecting the alternative nutrient sources (real activities), the following points were taken into account:

(i) It should be sufficiently available in the locality throughout the year,
(ii) It must be palatable to the dairy animals,
(iii) The information about the digestibility coefficient and chemical composition of feeds and fodders should be available. Based on the above criteria, feeds and fodders considered as real activities included paddy straw, sorghum straw, sorghum fodder, Napier grass, berseem fodder and other fodder, soybean cake, wheat bran, maize, tapioca chips, wheat protein, molasses and mineral mixtures.

Disposal activities

Disposal activities are also termed as non-use activities or slack activities (Babbar, 1956). These are activities to deal with the inequalities of linear programming. In the present study, the inequalities problem was solved by introducing six disposal activities such as dry matter supplied by roughages, dry matter supplied by concentrates, TDN, DCP, Ca and P.

Artificial activities

Artificial activities are used to get a better feasible plan (Anderson et al., 2000) and are used for each restriction which has the disposal activity with minus coefficient. Artificial activity has a positive coefficient for the restriction, denoting that it uses the restrictions. In the present study, four artificial activities were introduced for the four restrictions viz. TDN, DCP, Ca and P.

C_J values

C_j values are the purchase value (cost) per unit of real activity. For farm produced fodders, the cost of production in rupees per unit was used as C_j values. In the case of fodders supplied by other agencies, the market price was considered as the C_j value. For the concentrate activities, the prevailing market prices in the study area in rupees per unit were considered as C_j values as it was observed that, the prices of concentrates did not vary because of its supply from government agencies.

Input coefficients

The specified nutrients viz. dry matter, TDN, DCP, Ca, and P available in different feeds and fodders were obtained from the animal nutrition experts of the study area. The optimal plans for feeding the dairy cows were developed for the two specified programming situations.

RESULTS

Situation 1

The least cost feeding plan reduced the ration cost for cross bred cows from 111.12 to 90 Rupees (Table 1). The least cost ration only used 5 ingredients compared with 13 in the original feed plan, and 10 kg less was needed to meet calculated requirements than what was originally fed to the cows. The principle component of the least cost ration was Napier grass, compared with sorghum fodder in the original diet.

Paddy straw and Napier grass were the principal components of energy, while soybean cake supplied protein. Mineral mixture is essential for lactating cows as it contains Ca, P, Mg, Fe, and Zn. Thus, the feed cost was reduced while keeping the essential source of nutrients for supplying energy, protein and minerals.

Marginal cost of nutrients

Marginal cost of nutrients or shadow price under the given set of conditions indicates the potentiality of nutrients (Table 2). It could be observed that, the activities in the solution at non-zero values have zero shadow prices; those reported at zero level have a negative shadow price indicating reduction in cost by that amount when one unit of the particular nutrient is decreased.

The level of slack activity for phosphorus was 0.068 kg, which indicated that phosphorus restriction was ineffective. This means that the least cost combination of feeds and fodders which meets the dry matter, TDN, DCP and calcium requirements also exceeds the phosphorus requirement by 0.068 kg without any cost implication.

The shadow prices on the slack activities at zero level indicate by how much the cost of the ration would be reduced when the constraint is relaxed by one unit. A decrease in one unit (kg) of dry matter supplied by

Table 1. Optimal plan compared with existing plan for daily feeding 5 L-yielding cross bred cows.

S/N	Source of nutrients	Existing plan		Optimum plan	
		Quantity (Kg)	Cost (Rs)	Quantity (Kg)	Cost (Rs)
1	Paddy straw	4.00	8.00	3.50	7.00
2	Sorghum straw	1.00	2.00	-	-
3	Sorghum fodder	10.00	30.00	-	-
4	Napier grass	3.00	12.00	10.60	42.40
5	Berseem (*Trifolium alexandrinum* L.)	2.00	4.00	-	-
6	Other fodder	3.00	6.00	-	-
7	Soybean cake	1.20	13.03	1.35	21.10
8	Wheat bran	1.25	7.70	2.08	13.35
9	Maize	0.70	8.09	-	-
10	Tapioca chip	0.50	3.33	-	-
11	Wheat broken	0.75	4.82	-	-
12	Molasses	0.25	2.15	-	-
13	Mineral mixture	0.10	10.00	0.06	6.00
		-	111.12 (100.00)	-	90.00 (81.00)

Figures in bracket show the percentage of the total feed cost of the existing plan.

Table 2. Level of slack activity and shadow price for cross-bred cow diet nutrients.

S/N	Particulars	Level of slack activity (kg)	Shadow price (Rs/d)
1	Dry matter from roughages	0	-2.58
2	Dry matter from concentrates	0	-126.81
3	Total digestible nutrients (TDN)	0	-9.35
4	Digestible crude protein (DCP)	0	-270.40
5	Calcium	0	-950.78
6	Phosphorus	0.068	0

roughages constraints reduced the cost of optimal plan by Rs 2.58. For the dry matter supplied by concentrate constraints, one unit decrease will result in a reduction of cost by Rs 126.81 in the optimum plan. The marginal cost of TDN was Rs. 9.35. That is, for every decrease in one unit of restrictions, cost will decrease by Rs 9.35 and vice versa. The marginal cost of DCP and calcium were Rs 270.40 and Rs. 950.38. This means that for every decrease in one unit of these restrictions, the cost will reduce by Rs 270.40 for DCP and Rs 950.38 for calcium

Situation II

The least cost feeding plan reduced the ration cost for local cows from 75.57 to 58.56 rupees (Table 3). The least cost ration only used 5 ingredients compared with 13 in the original feed plan, and 7 kg less was needed to meet calculated requirements than what was originally

fed to the cows. The principle component of the least cost ration was Napier grass, compared with sorghum fodder in the original diet.

Marginal cost of nutrients

It can be observed in Table 4 that, supply of P is surplus, amounting to 0.061 kg with zero shadow price. Marginal cost of dry matter supplied from roughages is Rs 2.36 and it is 120.10 for dry matter supplied by concentrates. For both of these restrictions, shadow prices imply that every decrease in one unit of the activity will result in reduced cost of their respective shadow prices. In the case of TDN, DCP, and Ca, the marginal costs estimated are Rs 10.25, Rs 251.20, and Rs.845.32, respectively. Every decrease in one unit of these three constraints will cause a decline in cost of their respective shadow prices.

Table 3. Optimal plan compared with existing plan for daily feeding 3 liter-yielding local cows.

S/N	Source of nutrients	Existing plan		Optimum plan	
		Quantity (Kg)	Cost (Rs)	Quantity (Kg)	Cost (Rs)
1	Paddy straw	4.00	8.00	3.06	6.12
2	Sorghum straw	1.00	2.00	-	-
3	Sorghum fodder	8.00	24.00	-	-
4	Napier grass	3.00	12.00	7.60	30.40
5	Berseem (*Trifolium alexandrinum* L.)	1.50	3.00	-	-
6	Other fodder	0.00	-	-	-
7	Soybean cake	0.70	7.60	0.86	9.34
8	Wheat bran	0.75	4.81	1.20	7.70
9	Maize	0.40	4.62	-	-
10	Tapioca chip	0.30	2.00	-	-
11	Wheat broken	0.35	2.25	-	-
12	Molasses	0.15	1.29	-	-
13	Mineral mixture	0.04	4.00	0.05	5.00
		-	75.57 (100.00)	-	58.56 (77.49)

Figures in bracket show the percentage to the total feed cost of the existing plan.

Table 4. Level of slack activity and shadow price for local cow diet nutrients.

S/N	Particulars	Level of slack activity (kg)	Shadow price (Rs/d)
1	Dry matter from roughages	0	-2.36
2	Dry matter from concentrates	0	-120.10
3	Total digestible nutrients (TDN)	0	-10.25
4	Digestible crude protein (DCP)	0	-251.20
5	Calcium	0	-845.32
6	Phosphorus	0.061	0

DISCUSSION

The existing feeding plan followed by the dairy farmers contained large number of ingredients due to which the feed cost was observed to be high. The animal nutritionists are of the opinion that there is scope to reduce the feed cost by formulating an optimum feeding plan by minimizing the number of ingredients and suitably reallocating their quantities. Therefore, there is a need to formulate an optimum least cost ration plan to reduce the feed cost. Least cost feed formulation is combining many feed ingredients in a certain proportion to provide the target animal (both crossbred and local lactating cows) with a balanced nutritional feed at the least possible cost.

This paper suggested an optimum ration plan through incorporation of locally available feed resources at recommended level for minimization of cost for cross bred and local lactating cows. Through the adoption of the plans it is possible to reduce the food cost while maintaining a balanced diet for the crossbred and local lactating cows. A number of workers, for example Munford (1996), Torez (2000), Djumaera et al. (2009) and Griffith (2010), have advanced the use of linear programming in formulation of least cost diet plans for dairy animals. The optimum plans for crossbred and local lactating cows presented here are user friendly as the user is quite familiar with the ingredients incorporated into the plan. In the optimum plans high cost concentrates are replaced by Napier grass. The higher quantity of Napier grass entered into the final plan is mainly because of its higher production, thereby reducing the per unit cost of production and at the same time providing the essential nutrients required for increased milk production.

The preference to Napier grass in both the plans is due to high dry matter production, reasonably good fodder quality, drought tolerance and its persistence under frequent harvesting. Incorporation of soybean cake into both the optimal feeding plans increased substantially thereby supplying more protein. Soybean cake has a high crude protein content of 44 to 50% and a balanced amino acid composition for feed formulation (FAO, 2004). A high level of inclusion (30 to 40%) is used in high performance monogastric diets (FAO, 2004).

Paddy is the dominant crop of the region and experience has shown that, paddy straw can be used in rations to provide dry matter. Paddy straw can provide neutral detergent fiber, acid detergent fiber and crude protein. Inclusion of wheat bran concentrates in both the optimum ration plans is required, in spite of having a higher price per kg than per kg price of wheat broken and molasses, because it is high in total digestible nutrients. Increase in the cost due to increase in the quantity of wheat bran in the two optimal plans is compensated by the entry of low cost locally available feed resources like Napier grass. Incorporation of mineral mixture in the optimal plan is essential to provide minerals like Ca, P, Na, Cl, K, S and Mg which are essential for increasing milk production. The dairy animal is more likely to suffer from lack of both Ca and P than from a lack of any other mineral, with the possible exception of salt (www.agriculture.kzntl). Paddy straw and Napier grass supplied mainly the energy; soybean cake provided the protein and mineral mixtures supplied minerals like Ca, P, Mg, Zn, Fe, Cu, Iodine etc.

The shadow price of nutrient constraints included in both the models implied that, every increase in one unit of the nutrient will result in reduction of their respective shadow prices. In this study all the nutrient constraints except P have positive shadow prices when the minimum constraint has been reached. The least cost will be increased by the amount of the shadow price if the minimum constraint is forced to be one unit higher. The zero shadow price of P reveals that, the least cost combination of feeds and fodder after meeting all the requirements also exceed the P requirement by 0.068 kg without any cost implication. Similar feed ingredients are incorporated into both the optimum ration plans of cross bred and local dairy cows.

Cost reduction to the extent of 19 and 23% is noticed in the optimum plans for cross bred and local cows as compared to the existing plans. The least-cost feeding plan once formulated through the model will continue for a reasonably long time as there are no frequent changes in price of the ration items as a result of which dairy farmers are continuing the same feeding plan for a sufficient long time. The model does a reasonably good job as the feed ingredients included in the least cost plan are not new to the farmers as these are already used by them in their present feeding plan. The extension workers should try to implement the optimum plans suggested through this study for reduction of feed cost, and feedback from farmers should be obtained for further improvement. However, long term studies on the effect of using higher quantities of Napier in production rations for reducing feed cost need to be carried out.

Conclusion

The optimal plan showed how the locally available cheap ingredients can be combined to formulate a least cost feed plan. The results suggested that, in both situations, there was considerable reduction in total feed cost in the optimal plan while supplying all the nutritional requirements

to the animals. This indicates that, there is considerable scope for minimizing the cost, under the given situations and restrictions. Confronted with the situation of growing resource scarcities at farm level, formulation and adoption of optimal plans should form an integral part of farm planning for these farms. In both situations p is available in excess quantities. The marginal quantities of feed items observed in the optimal plans act as a guide for efficient use of existing resources.

The results of this study can be of significant value to dairy farmers of the region. The amount of savings in dairy feed cost could have a large positive impact on reducing animal maintenance cost and thus, the profitability of dairy cooperation.

ACKNOWLEDGEMENTS

The authors acknowledge the assistance rendered by the World Bank through National Agriculture Innovation Project (NAIP) of Indian Council of Agricultural Research (ICAR). The work reported here was conducted as a part of sub-project entitled" Efficient land use based integrated farming systems for rural livelihood security". The authors also gratefully acknowledge the help and guidance received from the Director, National Bureau of Soil Survey and Land Use Planning (ICAR), Nagpur, Maharashtra (India), without which it would not have been possible to complete the study.

REFERENCES

Alexander DLJ, Morel PCH, Wood GR (2006). Feeding strategies for maximizing gross margin in pig production. In: Global Optimization: Scientific and Engineering Case Studies, Springer, USA pp. 33-43.

Anderson DR, Sweeney DJ, Williams TA (2000). An Introduction to Management Science, 9th. Ed., West, St. Paul, MN, Chaps pp. 2-4.

Babbar MM (1956). A note on aspects of linear programming technique. J. Farm. Econ. 38(2):607.

Chakeredza S, Hove L, Akinnifegi FK, Franzel S,Ajayi OC, Sileshi G (2007). Managing fodder trees as a solution to human-livestock food conflict and their contribution to income generation of smallholder farmers in Southern Africa. Nat. Resour. Forum 31:286-296.

Chakeredza S, Akinnifegi FK, Ajayi OC, Sileshi G, Mngoba S, Gondwe MT (2008). A simple method for formulating least cost diet for small holder dairy production in Sub- Saharan Africa. Afr. J. Biotechnol. 7(16):2925-2933.

Chandler TP, Walkir HW (1972). Generation of nutrient specifications for dairy cattle for computerized least cost ration formulation. J. Dairy Sci. 55(12):1741-1749.

Djumaera D, Djanibekov N, Vlek, PLG, Martius C, Lamers JPA (2009). Option of optimizing dairy feed ration with foliage of trees grown in the irrigated dry land of Central Asia. Res. J. Agric. Biol. Sci. 5(5):698-708.

FAO (2004). Protein sources for the animal feed industry.ID 166962.Book.(http://www.fao.org/docrep/007//y5019e/5019e00.htm).

Griffith M (2010). Minimum cost feeding of dairy cows in Northern Victoria. Center for Policy Studies, Monash University, Australia. General Paper No. G-195. http//.www.monash.edu.au/policy/.

Munford AG (1996). The use of iterative linear programming in practical applications of animal diet formulation. Math. Comput. Simul. 42:255-261.

National Dairy Development Board (NDDB) (2012). Nutritive value of commonly available feed and fodder in India. Compendium, National

Dairy Development Board, Anand-388001.

O'Coner J, Sniffen CJ, Fox DG, Miligan RA (1989). Least cost dairy cattle ration formulation model based on the degradable protein system. J. Dairy Sci. 72:2733-2745.

Patil V (2010). Cost of milk production: A case study of Maharashtra. www. afcindia. org. in /aug/2010/ 33-25.pdf.

Pond WG, Church DC, Pond KR (1995). Basic Animal Nutrition and Feeding. 4th Edition. John Wiley and Sons, Inc, P. 615.

Torez PZ (2000). Least cost ration formulation for Holstein dairy heifers by using linear and stochastic programming. J. Dairy Sci. 83:443-451.

Webster AJF (1993). Understanding the Dairy Cow. 2nd Edition. Blackwell Scientific Publications, London Edinberg, Boston. 2nd Ed. pp. 288-333.

25

Benefits of donkeys in rural and urban areas in northwest Nigeria

Hassan M. R.[1], Steenstra F. A.[2] and Udo H. M. J.[2]

[1]Department of Animal Science, P. M. B 1044, Ahmadu Bello University; Zaria, Nigeria.
[2]Animal Production Systems Group, Wageningen University; P. O. Box 338, 6700 AH Wageningen, The Netherlands.

The objective of this study is to explore the benefits of donkeys for rural and urban smallholder farmers in northwest Nigeria. We visited 112 smallholder donkey farmers located in rural and urban areas from four states in northwest Nigeriathrough four focus group meetings, interviews with individual farmers and in depth interview with 12 key informants. In addition, 80 citizens were interviewed about their perception on donkeys. Donkeys were used more intensively in urban than in rural areas. The number of donkeys was higher (p<0.001) in urban (4.1) than in rural areas (1.9). The number of days per week working with donkeys was also higher (p<0.05) in urban (6.4) than in rural (2.9) areas. However, farm sizes werelower (p<0.001) in urban (0.5 ha) than in rural (1.0 ha) areas. Farmers in urban areas received 16% higherannual income from their donkeys than those in rural areas. Donkeys were mainly appreciated by farmers for their low purchasing price, low-cost equipment, ease of management, and role in ceremonies. The main constraints facing the farmers were lack of information on donkey keeping, lack of access to clean water and proper feed, and lack of money to expand the business. About 50% of citizens associated donkeys with poverty.It was concluded that donkeys play important socio-economic roles in the farming systems and should therefore be included in future livestock policy planning in Nigeria.

Key words: Constraint, donkey, northwest Nigeria, perception, socio-economic role.

INTRODUCTION

Donkeys are one of the ancient domesticated livestock. In developing countries donkeys are valued in particular for their ability to survive under harsh conditions (Blench et al., 1990; Swai and Bwanga, 2008), yet they are often regarded as animals of low social status and neglected by research and development organizations (Starkey, 1995). There are 41.5 million donkeys worldwide (Desalegne et al., 2011). Nigeria is one of the countries witha relatively large (800,000) donkey population (Mabayoje and Ademiluyu, 2004). Cross-border movements by the pastoral Fulani from Niger, Chad, Burkina Faso, Mali and Cameroon, have increased the number of donkeys in Nigeria (Blench, 2004). Yet, most donkey breeding is practiced in the neighbouring countries. In Nigeria, donkeys are concentrated mainly in the northern states because of the savannah type of vegetation and fewer disease vectors such as tsetse flies (RIM, 1992). In southeastern Nigeria, donkeys are used as meat animals and about 16,000 donkeys are transported annually from the northern states for this purpose (ATNESA, 1997; Blench, 2004).

Donkey development in Nigeria was started with the introduction of different donkey breeds through trans-Sahara caravan trade across the Nile via the Sudan and

Chad (Fielding and Starkey, 2004). However, in 1970s donkey population was drastically reduced due to oil exploration and trade (Blench et al., 1990). At that time, it was relatively cheaper to use other means of transport such as trucks and motorbikes due to excellent road networks. In the late 1980s, there was economic hardship due to changes in government policy which led to rising costs in transport fares as a result of unstable fuel prices and degraded road networks (Blench, 2004). Presently, donkeys have started gaining popularity again among the smallholder farmers for employment opportunities and as reliable option for poverty reduction (Fielding and Starkey, 2004).

In recent years, studies about donkeys have been of considerable interest among researchers. For example, donkeys are being used for income generation activities through local transportation of goods (Pearson et al., 2000). Donkeys cansurvive in new environments under poor management (Jones, 2009) and help to facilitate marketing of goods in some African countries through the use of cart drawn implements (Pritchard, 2010). In Europe, donkey milk is, based on its composition in lipids and proteins, considered as a valid alternative for human milk (Vincenzetti et al., 2008). Recently, emphasis on education and training of donkey owners in management strategies have been reported (Stringer et al., 2011). In Nigeria, donkeys help to transport people, carry water from deep wells and rivers, and serve Fulani herdsmen during seasonal migration throughout Nigeria. In urban areas, donkeys provide small-scale services, such as transportation of building materials and grains, particularly in the northern part of the country. However, donkeys are not promoted by any governmental agency. Consequently, donkeys are perceived by policy makers in particular and society in general as less valuable than other livestock. There is no reliable information on their roles and benefits for rural and urban households. The northwest Nigerian states vary greatly in terms of economic development, population density and infrastructural facilities. It is hypothesized that the use of donkeys varies between the states and between rural and urban areas depending on their economies. The main objective of this study was to explore the benefits of donkeys to sustainable livelihoods among rural and urban smallholder farmers and to identifycitizens' perceptions on donkeys as well asconstraints and opportunities in donkey keeping in four states of northwest Nigeria.

MATERIALS AND METHODS

Area of the study

Geographical description

The field research was conducted between January and March 2008 in the semi-arid zone in northwest Nigeria.The mean annual temperature in this area is about 27°C. There is a single rainy season from May to October with mean annual rainfall of 508-1016

mm. The length of growing period is 100 to 150 days. The vegetation pattern ranges from open woodland and scattered trees to dense vegetation. Human population is over 35 million people (NPC, 2006).

Livelihoods

The major inhabitants of northwest Nigeria are mixed crop-livestock farmers and livestock herders. More than 80% of Nigerian livestock population (cattle, sheep, goats, poultry, rabbits, guinea pigs, pigs, horses, donkeys and camels) is concentrated in this region because it is free from tsetse flies and the rainfall pattern is unimodal (RIM, 1992).The research was carried in the states of Jigawa, Kaduna, Kano, and Katsina. Katsina and Jigawa border Niger Republic; Kaduna is the most southern state among these four states. Characteristics of the study area are given in Table 1. In all four states agriculture is important. The economic activities and labour demands are higher in Kano and Kaduna compared to Katsina and Jigawa. In Kano, the distance to the state capital is the smallest compared to the other three states.

Smallholder donkey owners and their farming system in northern Nigeria

We identified smallholder donkey owners as owners that owned five or lessdonkeys in their herds. We recorded information on their farming system such as types of crops grown, types of livestock kept, number of donkeys owned and farm size.We collected data about how household incomes were obtained. Cash received by the family is used to purchase farm inputs such as new donkeys, materials such as ropes, rakes, spade, saddle, sacks, vaccines, fertilizers, hired labour and firewood.

Research approach and methodology

A rapid rural appraisal (RRA) was carried out in the four states duringthree months. RRA is a research methodology which enables researchers to meet, associate and collect information from stakeholders affected by a particular problem in the most cost effective way (Chambers, 1981; Mohammed et al., 2012). Aspects about farm size, education level of farmers, time spent in off-farm economic activities, prices of inputs used in donkey management and utilization, and the income generation potential from the use of donkeys were investigated. Data collected included: age of respondent, family size, years of working experience with donkeys, and working days per week with donkeys. The research was carried out in five stages which included: familiarization visit to the study area, direct participatory observation, individual interviews, focus group discussions with farmers and in depthinterviews with key informants for confirmation and additional information concerning the socio-economic roles of donkeys.

After selection of the four states namely; Jigawa, Kaduna, Kano and Katsina, a two stage sampling technique was used using the method described by Berhanu et al. (2012). In the first stage, two Local Government Areas and locations (rural and urban) were selected based primarily on distribution and population of donkeys. In the second stage, respondent households were randomly selected from the locations using systematic sampling procedures. Donkey's contribution to farmers' livelihoods was measured using livelihood indicators such as contribution of donkeys to household income, number of children attending school, number of livestock owned, purchase of luxury items (motorcycle, mobile phones, radio, television etc) and type of roofing materials in the house (iron sheet, thatched, mud etc) following the method of Smith (2004).

In total, 112 households representing the smallholder rural and

Table 1. Details of selected characteristic features of four states in northwest Nigeria.

Items	Jigawa	Kaduna	Kano	Katsina
Population density	Low	High	High	Medium
Commercial activities	Low	Medium	High	Low
Informal sector (e.g. industries)	Low	Medium	High	Low
Infrastructure (e.g. roads)	Medium	High	Low	Low
Ethnic diversity	Low	High	Low	Low
Formal sector (e.g. schools)	Medium	High	Medium	Medium
Agricultural activities	High	High	High	High

urban donkey farmers were interviewed using a structured questionnaire, with open- and closed-ended questions; 14 in rural areas and 14 in urban areas in each of the four states surveyed. Sample selection during the study was based on the number of donkey owning households, road accessibility, gender and geographical position within the states, particularly considering the scale of north-south axis for better representation of the information. Only male farmers were interviewed during the study due to cultural barriers between males and females in these states. Donkey households were identified with the help of key informants: farmers' leaders, local community leaders, agricultural extension agents, university researchers, marketing agents, government authorities, veterinary doctors, school teachers, youth and the community elders.

Four focus group discussions with farmers and key informants were organized at community-level in which Strengths, Weaknesses, Opportunities and Threats (SWOT) issues raised by farmers were thoroughly discussed. The SWOT issues raised by the farmers were further discussed during the farmers' group discussions, involving 28 farmers in each state. The SWOT group discussions were led by the researcher and lasted for two hours and the discussions were fairly informal. In order to create a conductive atmosphere for the discussion of particular issues of concern among the farmers, eight separate focus group interviews were conducted using preference ranking and scoring procedures (Watson and Cullis, 1994; Starkey, 2000). Also, 20 citizens (non-donkey keepers) were randomly selected from each state and interviewed about their perceptions on donkeys in the society. Citizens were selected based on their willingness to participate in the research because many of them were sceptical about the purpose of the survey on donkeys.

Data analyses

Total annual income from donkeys was estimated as the gross outputs minus the variable costs. Total family income was estimated from annual income from donkeys, sales of manure and offspring,and off-farm income. Univariate General Linear Model procedure (nested ANOVA) was used in SPSS 15.0 (2003) to analyze total family income and annual income from donkeys as dependent variables and state and location (rural or urban) as fixed factors, including interactions between state and location. Means with significant F-values were separated by least significant difference (LSD) test.

Chi-square test for a two-way contingency table was used to test the hypothesis that there was no difference between the states in perception of citizens about donkeys. The outcome was considered significant when fewer than 20% of the cells in the table have an expected cell count of less than 5 and none of them has an expected cell count of less than 1.

RESULTS

Farming systems with donkeys

Cattle, sheep, goats, donkeys, and poultry were the most commonly found livestock in these states. However, in some parts of Kaduna, local pigs were reared in the backyard. Camels were mainly restricted to Jigawa and Katsina. The most common crops grown included: sorghum, maize, groundnut, cowpea (beans), rice, vegetables, and sesame. In Kaduna, the farmers cultivated sugar cane, onions and tuber crops because of differences in the amount of rainfall and soil conditions. Millet and cotton were mainly cultivated in Jigawa and Katsina, respectively, while paddy rice was becoming popular in Kano both under irrigation and rain-fed conditions. Crops were grown in mixtures.In the four states, crops were cultivated by hand using traditional hoes. The household obtains income from farm outputs such as manure, draught, offspring, farm produce, hiring out of donkeys, donkey sales, gifts, exchange and off-farm labour.

The focus group discussions indicated that donkeys were the primary pillars in the farming system of smallholder donkey farmers. They provided manure to crops in both rural and urban areas. Manure serves as an alternative to chemical fertilizers, thereby lowering the cost of crop production. Also, donkeys helped to carry out hard labour, such as conveying farm produce from farms to homes or markets, to fetch water from deep wells, and they were also used intensively for commercial pack transport. Donkeys could be hired out for some hours or days. Further, donkeys served as the major source of income to the farmers in cash and in kind from the sale of manure, offspring and herd replacement.The cash received (from donkey services/sales)was used to purchase farm inputs such as new donkeys, materials such as ropes, rakes, spade, saddle, sacks, vaccines, and fertilizers, hired labour and firewood.

Farmers said that the reasons for preferring donkeysrather than cattlewere thatdonkeys cost less to purchase and maintaincompared to cattle and were more easily managed even by women and children in rural areas. The minimum price of a donkey was 6,000-7,000

Table 2. Least squares means and standard deviations (SD) of household (hh) and farm characteristics for smallholder rural and urban donkey farmers from four states in northwest Nigeria.

Factors		Farm size (ha)		Time spent off-farm (h)		Age of head (y)		Family size (n)	
		Mean	SD	Mean	SD	Mean	SD	Mean	SD
Overall		0.7	0.5	7.4	1.5	43.5	7.7	12.4	4.5
State									
Jigawa		0.8[a]	0.5	6.5[b]	0.9	43.9	8.5	12.9	3.9
Kaduna		0.8[a]	0.3	8.4[a]	1.9	43.1	7.9	12.5	6.1
Kano		0.4[b]	0.5	8.0[a]	1.2	41.6	7.8	11.9	4.3
Katsina		0.9[a]	0.5	6.7[b]	0.8	45.6	6.0	12.3	3.3
Location									
Rural		1.0[a]	0.4	7.3	1.1	44.7	7.6	12.6	3.9
Urban		0.5[b]	0.3	7.5	1.2	42.4	7.2	12.2	4.7
Interaction p-value		ns[1]		ns		Ns		ns	

		Livestock number (TLU)		Livestock density (TLU/ha)					
		Mean	SD	Mean	SD				
Overall		14.5	15.0	21.8	30.7				
State									
Jigawa	Rural	9.6[c]	8.7	9.7[b]	6.2				
	Urban	7.7[c]	4.2	15.8[b]	10.7				
Kaduna	Rural	27.8[b]	12.9	29.5[b]	12.4				
	Urban	21.5[b]	17.7	38.8[a]	26.0				
Kano	Rural	28.6[a]	24.7	56.0[a]	66.0				
	Urban	4.0[a]	2.6	0.0[c]	0.0				
Katsina	Rural	7.4[c]	2.6	7.3[b]	4.6				
	Urban	9.7[c]	2.4	17.3[b]	9.1				
Interaction p-value		0.000			0.000				

[a, b, c] Different superscripts denote significant differences between means within columns (p<0.001), [1] ns: non significant (p>0.05).

Naira (1 US$= 117 Naira (2008) while the minimum price of a young bull was about 40,000 Naira. Also, donkeys could be trained easily, especially at their young age, whereas cattle couldonly be trained by the first owner when they reached the age of maturity. Another reason was that before using cattle, the farmer must purchase a cart, which was more expensive than the saddle or pannier used for donkeys. Also, the use of cattle for activities in urban areas would be dangerous because of the horns and the hostile nature of cattle, especially when they were frightened. The theft of donkeys was much less common than cattle, because donkeys were able to recognize theirowner even at night. Another reason the farmers presented was that they wanted to preserve the culture of their forefathers rather than using cattle, which were practically unaffordable. However, farmers said that they could not afford to purchase inputsfor their donkeys such as donkey carts, drugs and medications, manufactured feeds and supplements.

Farming characteristics of the respondents

All farmers interviewed were from the Hausa and Fulani tribes which are the dominant tribes of the area studied. Table 2 presents the farming characteristics of rural and urban donkey farmers in the four states. In the rural areas, average farm size was twicecompared to the urban areas (p<0.001): 1.0 to 0.5 ha. The time spent in off-farm activities per week was significantly different between the states with Kano and Kaduna having spent about one and a half hour per person more (p<0.001) compared to Jigawa and Katsina. Interactions in the livestock number and density between the state and location were statistically significant (p<0.001), indicating that the differences in livestock number and density between the rural and urban locations were different in the different states.Overall, Kaduna and Kano had a higher (p<0.001) livestock density (28-38 TLU/ha) compared to the other two states (12.5 TLU/ha). Especially, farms in Kano (0.4 ha) were

Table 3. Least squares mean and standard deviation (SD) of use of donkeys by smallholder rural and urban donkey farmers from four states in northwest Nigeria.

Factors		Donkeys (n)		Working days (days/week)	
		Mean	SD	Mean	SD
Overall		3.0	1.9	4.7	2.0
State					
Jigawa		3.3	2.1	4.9[a]	2.1
Kaduna		2.4	1.8	4.1[b]	2.0
Kano		3.1	2.1	5.4[a]	2.3
Katsina		3.1	1.6	4.2[b]	2.4
Location					
Rural		1.9[b]	0.7	2.9[b]	1.7
Urban		4.1[a]	2.1	6.4[a]	0.2
Interaction p-value		ns[1]		ns	

		Working with donkeys (h/day)		Experience with donkeys (y)	
		Mean	SD	Mean	SD
Overall		5.2	2.0	11.0	3.6
State					
Jigawa	Rural	3.6[b]	1.0	10.1	3.8
	Urban	5.9[a]	0.5	12.1	3.3
Kaduna	Rural	3.1[b]	0.6	11.9	3.8
	Urban	7.7[a]	0.7	9.1	3.8
Kano	Rural	4.9[b]	2.1	10.6	3.7
	Urban	7.4[a]	1.0	10.4	3.4
Katsina	Rural	3.2[b]	1.5	12.9	2.5
	Urban	6.3[a]	0.5	10.5	3.3
Interaction p-value		0.001		0.046	

[a, b, c] Different superscripts denote significant differences between means within columns ($p < 0.001$),[1] non significant ($p > 0.05$).

significantly ($p<0.05$) smaller than in the other states (0.8 ha).

Table 3 shows that rural farmers (1.9 donkeys) had significantly ($p<0.001$) fewer donkeys than urban farmers (4.1 donkeys). The mean working days with donkeys per week was higher ($p<0.001$) in Kano and Jigawa (5.2 days) than in Kaduna and Katsina (4.2 days). Respondents from urban areas worked more days per week (6.4 days) with donkeys than those from rural areas (2.9 days)($p<0.001$). In particular, the difference in working hours with donkeys between the rural and urban respondents in Kaduna was higher than in the other states ($p<0.001$). This may be related to the number of hours spent on transit in conveying firewood and other materials to urban areas.

Economic impact of donkeys for smallholder donkey keepers in Nigeria

Mean annual income from donkeys (Table 4) was six times higher for urban donkey farmers than for rural donkey farmers, with highest income in Kano (558,000 Naira). The working hours with donkeys and the number of donkeys had a positive effect ($p<0.001$) on income from donkeys. The regression coefficient for the number of donkeys indicated that increasing the number of donkeys by one donkey increased annual income from utilizing donkeys by 90% in rural areas and 21% in urban areas. Farm size had a significant ($p<0.01$) negative effect on income from donkeys in both rural and urban areas. This implies that farmers with larger farms use fewer donkeys for economic activities. On average, total family income was 32,903 Naira in rural areas and 211,771 Naira in urban areas. After donkeys, main income sources were cropping in rural areas, and off-farm income in urban areas. The type of off-farm activities depended on the location. In Jigawa and Kaduna, many rural farmers were engaged in small scale food processing industries, while in the urban areas of Kaduna and Kano, farmers worked mainly in large scale commercial farms around cities.

Table 4. Least squares means and regression coefficient for farmers' annual income from donkeys resulting from farm resources, for smallholder rural and urban farmers in northwest Nigeria (1000 Naira).

Factors	Rural		Urban	
	l.s. mean	S.D.[1]	l.s. mean	S.D.[1]
Overall average	32.903	55.431	211.771	254.424
States				
Jigawa	16.571[b]	9.108	60.914[c]	19.276
Kaduna	10.354[b]	3.401	182.251[b]	164.205
Kano	99.794[a]	80.010	558.320[a]	248.376
Katsina	4.891[b]	2.254	45.600[c]	17.924
	Regression	s.e[3]	Regression	s.e[3]
	Rural		**Urban**	
Family members (n)	821	1.630	11.686*	5.587
Farm size (ha)	-43.076**	12.610	-194.007**	58.325
Livestock number (TLU)	171	286	-1.356	2.259
Off-farm activities (h)	2.196	9.177	61.315	32.314
Working with donkeys (h/wk)	1.772***	433	15.141***	3.179
Years of experience (y)	1.240	1.863	95	7.594
Number of donkeys (n)	29.748***	7.347	44.654***	11.841
Cash input per donkey (× N1000)[4]	2	1	4	2
R^2 full model (%)[2]		68		65

[1]: standard deviation, [2]: coefficient of determination, [3]: standard error of mean, [4]: 1US$ = 117 Naira, l.s. Means with different superscripts within columns are significantly different *: $p<0.05$, **: $p<0.01$, ***: $p<0.001$, NB: a, b, c means indicate state differences while * shows differences within rows.

However, the availability of such jobs is uncertain and the income they receive is small compared to the income from their donkeys, provided there is work available. So anothersource of income waspetty trading, mostly done by the farmers' wives and children after school hours (e.g. selling bread, fried groundnuts or kerosene). The financial gains from such activities (≤15% of the total family income) were used to purchase daily family needs.

Citizens' perception about donkeys

Table 5 shows the perception of 80 citizenson donkeys. The results indicated that citizens had clear different perceptions ($p<0.05$) about donkeys regardless of the state. However, citizens in all the states had similar opinion ($p>0.05$) about association between donkeys and poverty, impression of donkeys to tourists, donkeys were wicked animals, productivity of Nigerian donkeys and donkeys should be promoted (Table 5). In general, citizens had a positive ($p<0.05$) opinion on the use and characteristics of donkeys, and they agreed with increasing the numberand use of donkeys in Nigeria. However, they were overall negative ($p<0.05$) about the ease of handling donkeys, and agreed that donkeys caused overgrazing and road accidents. The citizens interviewed were on average 32 years, ranging from 16 to 60 years. Sixty-onepercent of the respondents agreed ($p<0.001$) with the proposition that donkeys have social and economic values in the society, and 45% did not think that donkey owners were the poorest of the poor. Fifty-one percent of the interviewed citizens agreed that donkeys were friendly.

Potentials and constraints limiting donkey productivity in northwest Nigeria

The SWOT discussions (Table 6a) showed that farmers wereconstrained by four major factors: technical constraints (e.g. feed shortage), financial constraints, socio-cultural constraints (e.g. low status) and policy constraints. The smallholder donkey farmers were constrained by the livestock development policy of the federal government of Nigeria because donkeys were not valued compared to other livestock species. The farmers lacked donkey drawnequipment (e.g. ridger, cart or wagon) to ease their work with donkeys.This could be related to inadequate knowledge about the opportunities using donkey drawn implements. Some profitable transport options using donkey carts such aswater transportand firewood businesses, rural transportation of people and goods and donkey wagons for tourists were among the best options identified by the farmers during the discussion meetings. Another option was the use of donkey carts for easy transportation of water pumps in irrigation fields during the dry season. The farmers had some internal weaknessesthat were directly under the control of the farmers and therefore could be avoided at their own will. The first weakness identified was thatsome farmers lacked motivation to improve the management of their donkeys. Donkeys were reared extensively in all the states. Another weakness identified was that farmers failed to organize themselves into donkey farmers associations, especially in rural areas. In areas where such associations existed, there were mostly few members. Another weakness was that farmers in rural

Table 5. The perception of citizens (n = 80) about donkeys (D) in four states in northwest Nigeria (n).

Statements	Strongly agree	Agree	Undecided	Disagree	Strongly disagree	p-value[2]
D[1] have socio-econ. value	21	40	6	1	12	0.000
D are associated with poverty	18	18	8	14	22	0.136
D look ugly	26	18	11	12	13	0.047
D look bad to tourists	14	17	11	15	23	0.287
D are friendly	20	31	12	12	5	0.000
D are hard working	41	32	4	2	1	0.000
D are wicked animals	12	19	14	15	20	0.579
D are stubborn	15	37	17	5	6	0.000
Nigerian D are unproductive	16	15	8	17	24	0.087
D provide cheap transport	40	26	2	7	5	0.000
D cause overgrazing	11	15	10	16	28	0.012
D cause road accidents	19	27	12	13	9	0.013
Use of D is necessary	23	22	10	15	10	0.043
D should be promoted	17	13	16	18	16	0.928
D population should increase	21	29	13	11	6	0.000
D should be fully utilized	15	35	15	9	6	0.000
D are difficult to control	11	41	10	11	7	0.000
D cope with distractions	15	33	9	11	12	0.000
D cost less than cattle and easy to manage	25	31	12	8	4	0.000

[1]D: Donkeys, [2]p-values: Level of significance across the contingency table, p-values are chi square probabilities.

areas were tethering their donkeys near the main roads, thereby causing road accidents. Also, in both rural and urban areas, farmers did not bury their dead donkeys in a pit; they disposed them outside the town, thereby causing air pollution to citizens.

Farmers mentioned the main threats facing them at the time of the survey (Table 6b). The first threat in rural areas was competition between the donkey owners and commercial motorcyclist and wheelbarrow pushers in conveying goods to various destinations. Another threat was increasing market prices for donkeys and their equipment. The farmers said the market price of donkeys washigh due to the slaughtering of donkeys for meat in the southeastern part of the country. The farmers also complained aboutthe new policies and programmes of the state ministries of environment and agriculture and natural resources.In these programmes, the farmers were prevented from undertaking their normal business with donkeys due to the imposed environmental sanitation in the last Saturday of every month and/or due to heavy tax paid to the government as part of revenue generation.

DISCUSSION

Socio-cultural and economic benefits of donkeys in northwest Nigeria

In the past, donkeys were owned by farmers for personal uses such as transportation, drawing water from deep wells, conveying manure to the farmsand transporting farm produce to home or local markets. This study indicates that donkeys still play significant roles in the life of smallholder farmers in northwest Nigeria. They are used to generate income, as gifts and as entertainment during ceremonies. However, farmers differed in their opportunity to own and utilize donkeys due to differences in resources, wealth, economic activities, and labour demands between the rural and urban locations. For example, respondents from rural areas had larger farm size (Table 2) because they are less affected by urbanization, compared to those in urban areas where there is virtually less land (ECA, 2004). However, competition for family labour between farm activities and working with donkeys could potentially reduce annual income from utilizing donkeys. Huang et al. (2009) observed a negative relationship between total household income and working off farm in China.

During the group discussions, farmers emphasized the use of donkeysfor income generation rather than other draught animals, like the popular White Fulani bulls (Bawa and Bolorundoru, 2008). The rising demand for services of donkey transport in urban areas is the main driving force for the large number of donkeys in urban areas (Table 3). The income generating activities using donkeys was impressive in both rural and urban areas of northwest Nigeria. The time spent in working with donkeys, 25 h per week, shows the intensity to which

Table 6. Strengths and weaknesses of Nigerian agricultural system with donkeys, (b) Opportunities and threats of Nigerian agricultural system with donkeys.

Internal	Strengths(+)	Weaknesses(-)
	(a)	
Socio-cultural	(i) Donkeys are used during annual celebrations e.g. Durbar* (ii) Donkey owners work in groups (iii) Farmers form cooperative societies (iv) Theft of donkeys is minimized. associated with donkeys in North-West Nigeria (v) Citizens prefer to use donkeys to convey certain goods and farm produce because they are safer and cheaper (vi) Donkeys are given as gifts to friends, families and less privileged individuals in the society (e.g. lepers). (vii) Donkeys serve as starting capital to farmers which shift to other professions later in life. (viii) Donkeys are source of off-farm income to farmers (ix) All materials required by donkey users are locally produced and therefore cheap	(i) Donkeys have low status in the society (ii) Poor management of female donkeys (iii) High abortion rate (iv) Loss of foal for replacement (v) Decrease in donkey population (vi) Over-utilization of donkeys (vii) Lack of improvement in the management of donkeys (viii) Bad welfare condition and mal-treatment of donkeys (ix) Lack of proper information about their activities and possible constraints (x) Inability to organize themselves into successful farmers organizations especially in rural areas (xi) Perception that donkey owners are associated with poverty especially in rural areas (xii) Farmers have to replace their donkeys occasionally (xii) High rate of disease incidence e.g. colic diseases.
Economic	(i) Donkeys with good body condition cost more than the emaciated ones (ii) Donkey trading offers employment to some people e.g. donkey traders (iii) Donkeys are source of food (meat) for some people (iv) Farmers enjoy monopoly of market using donkeys especially in urban areas (v) Utilization of donkey manure helps to recycle products in mixed crop-livestock system	(i) Use of extensive management system (ii) Farmers use donkeys for long distance journeys (iii) Donkeys have no access to clean drinking water and good quality feed. (iv) Farmers lack motivation to use donkeys as draught animals (v) High costs of medical care (vi) Low investment in donkeys especially in rural areas
Ecological	(i) Donkeys cause less environmental degradation compared to other livestock (ii) Farmers have better option to use their manure due to lack of artificial fertilizers (iii) Donkeys are adapted to their environments	(i) Land sizes are fragmented and small (ii) High rate of deforestation due to firewood business (iii) Donkeys cause so many road accidents (iv) Dead donkeys cause environmental pollution
	(b)	
Socio-cultural	(i) Considerable interest in donkey utilization among youths in north-west Nigeria (ii) Increased opportunities in donkey utilization in urban centers due to rapid urbanization (iii) State ADPs in northern states are considering the use of donkeys as sources of farm power	(i) High taxes paid to the government for revenue generation (ii) Citizens do not yet realize the advantages of donkeys (iii) Extension work is done mainly on other livestock. Donkeys are neglected.

Table 6. Contd.

	(iv) Image of donkey farmers is improving gradually in the society	(iv) Donkey population is decreasing due to slaughtering for meat in the south-eastern states
		(v) Escalating prices for donkeys and their equipment in the market
		(vi) Unpredictable returns to investments in donkeys especially in rural areas
Economic	(i) Farmers form cooperative organizations	(i) High costs of replacement of donkeys
	(ii) Less conflicts between donkey owners and other people (e.g. traffickers)	(ii) High competition between donkey owners and commercial motor cyclists and wheelbarrow owners
	(iii) Socio-economic role of donkeys is becoming recognized again in the society	(iii) High costs of some implements such as shovel digger saddle and axe
	(iv) There could be high demands for donkey milk in the future due to its nutritious and medicinal values	(iv) High incidence of court trial cases in urban centers due to conflicts between farmers and citizens
	(v) The desire of some local governments to give priority to donkey farmers may give opportunities	
Ecological	(i) Increased demand of donkey manure in rural areas	(i) New policies and programmes of Ministry of environment at federal, state and local government levels (e.g. control of livestock movements in urban centers).
	(ii) Improvement of farming systems research for better integration of mixed crop-livestock system	

Durbar* = Colourful event with horses to celebrate certain occasion or welcome an important visitor.

donkeys are utilized (Table 3). In a rural area in Ethiopia, Crossley (1991) found that donkeys were used for only 8 h per week. This result suggests that donkeys in Nigeria have the potential to provide their owners with a steady income, provided they are well managed (Berhanu and Yoseph, 2011). In Kano, farmers worked on average 36 h/w, because of high demands for services of donkey farmers, resulting also in attractive prices. Kano is the most densely populated state and has the smallest distance to the state capital compared to the other three states. The latter makes movement of people and goods easier, thereby boosting commercial activities.

Constraints to donkey utilization in northwest Nigeria

Despite the significant role of donkeys in income generating activities in northwest Nigeria, the productivity of donkeys in terms of income generated by the farmers per annum still remains low due to some technical constraints. Farmers mentioned the problem of feed shortage which is linked to other problems such as general management, diseases and high costs for veterinary care and materials. Therefore farmers, animal scientists, health providers and policy makers need to work together to address the problem of feed scarcity and low veterinary care (Pritchard, 2010). Another problem outlined was shortage of funds and lack of credit facilities. Hence collaborated efforts are needed from both

the government and non-governmental organizations (NGOs) to assist farmers with credit facilities so as to raise the efficiency of the system and improve farmers' livelihoods (Shomo et al., 2010). In Ethiopia, Pearson et al. (2000) reported that about 20% of the respondents perceived lack of finance and feed shortage as two most important constraints for donkey management. Another major limitation mentioned by the farmers was the issue of low status in the society. Donkeys are not being used as meat animals in northwest Nigeria and therefore some citizens consider them as animals with very low social values. A similar observation was made in Zambia (Mofya, 2004). The situation could be improved by the donkey farmers' associations through activities that promote the image of donkeys in the society (Starkey, 2001). Despite the contribution of donkeys towards food security, improved livelihoods and nation's building, there are no government policies directed toward protecting, promoting and utilizing donkeys in Nigeria. Therefore policy makers in Nigeria need to appreciate the contribution of donkeys to the nation's building by formulating policy instruments in line with farmers' needs which could bring about the required social change (Berhanu and Yoseph, 2011).

Status of donkeys in northwest Nigeria

Citizens' perception about donkeys may have

considerable effects on the smallholder farming system with donkeys, and should be addressed to achieve sustainable promotion of donkeys in Nigeria. The perception of citizens showed that donkeys still had low status. This might be connected to the local traditional beliefs of the people in both urban and rural areas. There are a lot of misperceptions about donkeys in northwest Nigeria which are passed down from generation to generation. For example, citizens believed that donkeys are generally dull animals that possessed little or no talent at all. One misperception about donkeys is that donkeys are owned by the poor compared to horses which are owned by the rich, elites and traditional rulers. In the past, donkeys were not used during festive occasions such as traditional durbar (a colourful event with horses to celebrate certain occasions or welcome an important visitor), but now they are being used.

The current change in perception about donkeys by some citizens might be related to the present economic circumstances in Nigeria as a result of new economic policies (Bryceson, 1999, 2000), which led the citizens, especially the youths, to invest in the use of donkeys for different activities in both rural and urban areas.This was shownin the diverse response of citizens about the relationship of donkeys with poverty (Table 5). However, some citizens still hadnegative perceptions about donkeys and their socio-economic roles in the study areas.For example, 25% of the citizens perceived the use of donkeys in the Nigerian farming system as needless. This result was expected because most citizens give little or no attention to donkeys, anddonkeys are not promoted.In general, citizens had divergent opinion about the use of donkeys. In future, the welfare of donkeys in northwest Nigeria may become better as 65% of the citizens in this study recognized the social status of donkeys.These types of negative perceptions have also been reported in South Africa (Fielding and Starkey, 2004).

The image of smallholder donkey farmers is being gradually improved in the study area as a result of changes in peoples' lifestyle through education, wealth, travellingand political activities. This is similar to the situation reported in Ethiopia where the citizens realized the advantage of donkeys and their owners in carrying out daily activities in the society, although government officials and planners had different perceptions about donkeys (Sisay and Tilahun, 1997). Since farmers hadassociations in both rural and urban areas, it would be good if they could be advertising their activities to the citizens in both government and commercial radio and television stations in all the states. Previous studies have shown that local farmers associations succeeded in promoting donkey marketing without the intervention of the government (Starkey, 2001). Proper market information about donkeys should be provided to the farmers through local radio extension programmes and mobile phones in the future.

Prospects for utilizing donkeys in northwest Nigeria

Donkeys helped to provide employment opportunities to unemployed youths in northwest Nigeria. The level and intensity of utilizing donkeys for income generating activities serve as a means of employment opportunities for the unemployed youths. Income generated from utilizing donkeys is spent on other aspects of household needs. The daily income generated from donkeys in rural areas was only 250 Naira, but this wasabove the poverty line of US$1 for Nigeria. In the urban areas, the daily income was three times higher than in rural areas. A prospect for the smallholder donkey farmers was the increase in urbanization in all parts of northwest Nigeria, which requires the use of donkeys in transporting building materials. Donkeys are likely to remain as the main source for transporting building materials in all parts of northwest Nigeria in the years to come because of the recent increase in fuel prices. Therefore, since the use of vehicles is limited by high fuel prices, donkeys will fill the gap. The lack of proper management, technological backwardness, financial constraints and unfavourable government policies may impose some limitations to this prospect.

Conclusion

It can be concluded from this study that donkeys play significant socio-economic roles in terms of income generation, employment opportunities and improvement of livelihoodsofmany smallholder farmers and their families in northwest Nigeria. However, farmers find it difficult to effectively utilize their donkeys in both rural and urban areas. Both the government and private individuals should help to invest resources on donkeys in support of their income generating opportunities and poverty reduction among the youths. Productivity of Nigerian donkeys may not be improved without credit facilities, more knowledge on proper management and promotion of services. Also, welfare of Nigerian donkeys needs attention.

ACKNOWLEDGEMENTS

The study was jointly financed by the Netherlands Fellowship Programme (NFP) in collaboration with Wageningen University, The Netherlands and Ahmadu Bello University, Zaria, Nigeria. We would like to thank the farmers' organizations and local experts who participated in this research.

REFERENCES

ATNESA (1997). Improving donkey utilization and management. Report of the International ATNESA workshop held 5-9th May, 1997, DebreZeit, Ethiopia.

Bawa GS, Bolorunduro PI (2008). Draught animal power utilization in small holder farms – A case study of Ringim Local Government Area of Jigawa State, Nigeria. J. Food Agric. Environ. 6(2):299-3 0 2.

Berhanu A, Yoseph S (2011).Donkeys, horses and mules - their contributionto people's livelihoods in Ethiopia.The Brooke, Addis Ababa, Ethiopia, P. 72.

Berhanu T, Thiengtham J, Tudsri S, Abebe G, Tera A, Prasanpanich S (2012). Purposes of keeping goats, breed preferences and selection criteria in pastoral and agro-pastoral districts of South Omo Zone. Livestock Res. Rural Develop. 24: 213. Retrieved, from http://www.lrrd.org/lrrd24/12/berh24213.htm

Blench R, de Jode A, Gherzi E (1990). Donkeys in Nigeria: history, distribution and productivity. In: Starkey, P. and Fielding, D. (Eds), Donkeys, People and Development. A resource book of the Animal Traction Network for Eastern and Southern Africa (ATNESA) ACP-EU Technical Center for Agricultural and Rural Cooperation (CTA), Wageningen, The Netherlands 244:24-32.

Blench R (2004). Natural resource conflicts in northwest Nigeria. A handbook and case studies. MallamDendo Ltd Cambridge, United Kingdom. P. 106.

Bryceson DF (1999). African Rural Labour, Income Diversification and Livelihood Approaches: A Long-term Development Perspective. Rev. Afr. Pol. Econo. 80:171-189.

Bryceson DF (2000). Rural Africa at the crossroads: Livelihood practices and policies. Overseas Development Institute (ODI) U.K. P.14.

Chambers R (1981). Rapid rural appraisal: Rationale and repertoire. Pub. Admin. Develop. 1(2):95–106

Crossley P (1991). Transport for rural development in Ethiopia. In: Donkeys, mules and horses in tropical agricultural development (eds D. Fielding and R.A. Pearson), Centre for Tropical Veterinary Medicine and School of Agriculture, University of Edinburgh. pp. 48-61.

Desalegne A, Bojia E, Ayele G (2011). Status of parasitism in donkeys of project and control areas in central region of Ethiopia: a comparative study. Ethiopian Veterinary J. 15:2

Economic Commission for Africa (ECA) (2004). Land Tenure Systems and their Impacts on Food Security and Sustainable Development in Africa. ECA/SDD/05/09, P.140.

Fielding D, Starkey P (2004). Donkeys, people and development. A resourcebook of the Animal Traction Network for Eastern and SouthernAfrica (ATNESA). Technical Centre forAgricultural and Rural Cooperation (CTA), Wageningen, The Netherlands. P.211 ISBN 92-9081-219-2.

Huang J, Wu Y, Rozelle S (2009). Moving off the farm and intensifying agricultural production in Shandong: a case study of rural labor market linkages in China. Agric. Econo. 40(2):203-218.

Jones, P.A. (2009). Adaptation in donkeys. *Draught Animal News*, *47*:12-26.

Mabayoje AL, Ademiluyi YS (2004). A note on animal power and donkey utilization in Nigeria. Animal Traction Network for Eastern and Southern Africa (ATNESA) P. 2.

Mofya R (2004). Social consequences of introducing donkeys into Zambia. In: Starkey, P. and Fielding, D. (eds), Donkeys, people develop. P. 140.

Mohammed YHA, Noor RHA, Junaenah S, Abdullah HA, Ong PL (2012). Participatory Rural Appraisal (PRA): An Analysis of Experience in Darmareja Village, Sukabumi District, West Java, Indonesia. *Akademika* 82(1):15-19.

National Population Commission (NPC) (2006). The 2006 Nigeria census figures. http://nigeriaworld.com /articles/2007/jan/112.html (Accessed on 4th July, 2008).

Pearson RA, Alemayehu M, Tesfaye A, Allan EF, Smith DG, Asfaw M (2000). Use and Management of Donkeys inPeri-Urban areas of Ethiopia. Report of Phase One of the CTVM/EARO Collaborative Project April 1999-June 2000.

Pritchard JC (2010). Animal traction and transport in the 21st century: Getting the priorities right. The veterinary J. 186:271-274.

RIM (1992). Nigerian National Livestock Resource Survey.(VI vols). Report by Resource Inventory andManagement Limited (RIM) to Federal Department of Livestock and Pest Control Services (FDL&PCS).Abuja, Nigeria.

Shomo F, Ahmed M, Shideed K, Aw-Hassan A, Erkan O (2010). Sources of technical efficiency of sheep production systems in dry areas in Syria. Small Ruminant Res. 91:160–169.

Sisay Z, Tilahun F (1997). The role of donkey pack-transport in the major grain market (YehilBerenda) of Addis Ababa. Paper given at the Animal Traction Network of Eastern and Southern Africa (ATNESA) Workshop "Improving donkeyutilisation and management" 5-9 May 1997, DebreZeit, Ethiopia.

Smith D (2004). Use and management of donkeys by poor societies in Peri-Urban Areas of Ethiopia. Final Technical Report R7350, p.34.

SPSS (2003). Statistical Package for Social Sciences. PC version 15.0, Michigan Avenue, Chicago IIIinois, USA.

Starkey P (1995). The donkey in South Africa: myths and misconceptions: In: Animal traction in South Africa: empowering rural communities. A DBSA-SANAT publication. ISBN 1–874878-67-6. Halfway House, South Africa. 160p ISBN 1-874878-67-6.

Starkey PH (2000). Rapid appraisal methodologies for animal traction. In: Kaumbutho P.G; Pearson R.A; and Simalonga T.E (eds) 2000. Empowering Farmers with Animal Traction. Proceedings of the ATNESA workshop held 20-24th September, 1999, Mpumalanga, South Africa P. 344.

Starkey P (2001). Local transport solutions: people, paradoxes and progress-Lessons arising from the spread ofintermediate means of transport. Report for Sub-Saharan Africa Transport Policy Program (SSATP) and Rural Travel and Transport Program (RTTP). The World Bank. p.71.

Stringer AP, Bell CE, Christley RM, Gebreab F, Tefera G, Reed K, Trawford A, Pinchbeck GL (2011). A cluster-randomised controlled trial to compare the effectiveness of different knowledge-transfer interventions for rural working equid users in Ethiopia. Prev Vet Med. 100(2):90–99.

Swai ES, Bwanga SJR (2008). Donkey keeping in northern Tanzania: socio-economic roles and reported husbandry and health constraints. Livestock Res. Rural Develop. 20(5).

Vincenzetti S, Polidori P, Mariani P, Cammertoni N, Fantuz F, Vita A (2008). Donkey's milk protein fractions characterization. Food Chem. 106(2008):640–649.

Watson C, Cullis A (1994). Participatory Rural Appraisal and livestock development: some challenges. In: Bebington, T; Guijt, I; Pretty, J; and Thompson, J; eds, RRA Notes. Special issues on Livestock, No. 20, IIED, London, pp. 5-7.

26

Life history and predatory potential of eleven spotted beetle (*Coccinella undecimpunctata* Linnaeus) on cotton mealybug (*Phenacoccus solenopsis* Tinsley)

Asifa Hameed[1], Muhammad Saleem[2], Haider Karar[3], Saghir Ahmad[1], Mussarat Hussain[1], Wajid Nazir[1], Muhammad Akram[1], Hammad Hussain[1] and Shuaib Freed[4]

[1]Cotton Research Station Multan, Pakistan.
[2]Entomological Research Institute Faisalabad, Pakistan.
[3]Entomological Research Substation Multan, Pakistan.
[4]Bhauddin Zakariya University Multan, Pakistan.

Cotton mealybug (*Phenacoccus solenopsis* Tinsley) proved a menace to subcontinent south East Asia economy since 2005. After introduction of this notorious Caribbean pest it was necessary to identify biological control agents in country which exist in prevailing environment and successfully suppress the pest. In this study eleven spotted ladybird beetle (*Coccinella undecimpunctata* Linnaeus) proved the best predator whose population structure, biological parameters and predatory potential were determined using no choice feeding trials. It was concluded that 1st instar larvae of eleven spotted beetle 1st instar is an effective bio-control agent which consumed on an average 91.99 1st instar cotton mealybug wheras 2nd, 3rd instar and adult consumed 45.00, 44.00, 5.44 cotton mealybug respectively. *C. undecimpunctata* L. 2nd instar larvae devoured 97 1st instar, 35.66 2nd instar and 45.00, 3rd instar cotton mealybug and 7.11 adult stage cotton mealybug respectively, whereas 3rd instar *C. undecimpunctata* took in 121.66 1st instar, 51.66 2nd instar and 54.33 3rd instar cotton mealybug and 8.21 adult stage cotton mealybug respectively. The larvae of 4th instar *C. undecimpunctata* preyed 93.00 1st instar, 35.00 2nd instar and 33 3rd instar cotton mealybug respectively and 7.33 adult stage cotton mealybug respectively. Adult female of this beetle consumed higher number of mealybugs than adult male during its whole life. Regarding biological parameters it was proved from the results that *C. undecimpunctata* is an effective bio control agent of cotton mealybug which can be used in integrated pest management program successfully.

Key words: *Coccinella undecimpunctata* L, cotton mealybug instars, predatory efficiency, life cycle.

INTRODUCTION

Cotton is known as "Silver Fibre" crop of Pakistan. It is attacked by a number of insect pests, which not only reduce the cotton yield but also deteriorate the lint quality. Among these 150 delimiting pests of cotton crop, cotton mealybug proved a menace to Pakistan economy since 2005 (Centre for AgroInformatics Research, 2007). In 2005, *Phenacoccus solenopsis* Tinsely (Sternorrhyncha: Coccoidea: Pseudococcidae) was found causing serious damage to cotton crop in Punjab and Sindh Provinces, Pakistan (Abbas et al., 2010). Pakistan is the third largest exporter of cotton in world. The outbreak of this major pest is of economic importance. The infestation was recorded from 11 out of the 18 cotton-growing areas covering 45,000 sq.km. This outbreak of mealybug was observed on both Bt cotton and non-Bt cotton and the growers response has been to use large amounts of pesticides (US$ 121.4 million worth in the Punjab in two months in 2007. Such amount of pesticides increasing management costs, development of insecticide resistance, rising environmental consciousness. Biological control with *Coccinellids* has contributed greatly and suppressed the pests below economic damage level (Hoy and Nguyen, 2000). Efforts had been emphasized on evaluation of predators of such noxious pest, biology, control potential and other important parameters of predators and parasitoids (Mahmood et al., 2011). Twenty three species of predators have been reported from cotton field including *Coccinellids*, *Chrysopids*, *Lagaeids* and *formicids* (Cheema et al., 1980) in Pakistan, but *Coccinella undecimpunctata* L., has novel importance in cotton pests management (Nielsen, 1997; Marshall, 2005). These beetles are of extremely diverse habits found from ornamental (Wheeler et al., 1981), orchards to cash and fiber (cotton) crops, which can seriously manipulate crop economy, help in maintaining natural balance in ecosystem (Soares and Serpa, 2007) and enrichment of biodiversity in ecosystem (Orbycki and King, 1998).

Ecological studies on *Coccinellid* in subcontinent Ecosystem revealed that *C. undecimpunctata* laid eggs near prey (Khan and Suhail, 2001), increased in numbers when prey density increased, and became quiescent when the prey species declined (Kenneth and Hagen, 1970). Species increased its population size in a fairly short time under suitable weather conditions (Hameed and Hussnain, 1984). *Coccinellid* fecundity increased with cool temperature and increase in prey density. Various authors across the subcontinent particularly entomologists emphasized that eleven spotted beetle population should be enhanced in cash crops as of cotton and wheat to get an effective control over the sucking pests complex for better crop yields (Fayyaz, 1998; Bellows, 2001). However introduction of invasive species cotton mealybug delimitate need for enhancement of cotton production (Government of Pakistan, 2008), identification of biological control agents (Tanwar et al., 2011), evaluation of their predatory potential (Ghafoor et al., 2011), identification and maintenance of conservation resources of predators and parasitoids, life history studies and ecological studies of such beneficial organisms. Quantitative assessment of efficacy of *Coccinellid* for pest species in agricultural system, relation of food source to Biological parameters, adult's longevity, fecundity, oviposition, food and environmental relations to *C. undecimpunctata* L. life parameters is the urge of time in ecosystem.

Essence of life history and predatory potential of predators and parasitoids can be estimated from the fact that thousands of dollars have been spent in subcontinent of South East Asia to control invasive Caribbean pest through Biological Control agents (CGS projects, 2008).

Keeping in view needs of Country's economy the present studies were conducted to determine predatory efficiency and life history of indigenous predator *C. undecimpunctata* L for the management of cotton mealybug to reduce overreliance on insecticides and to provide base line population data for further experimentation under field.

MATERIALS AND METHODS

The experiment was conducted in Cotton Mealybug laboratory in Entomological Research Institute, Faisalabad. For the purpose of life history and predation studies temperature and humidity were maintained at 25±2°C and 65±5% R.H through the use of air conditioner and humidifier (Honeywell Quicksteam 3-Gallon Warm Moist Humidifier connected with thermo-hygrometer) at 4000 Lux maintained through tube lights connected with lux meter (Testo 540 Lux meter, JMW Limited, Calibration lab Warwick house England). The experiment was laid out in completely randomized split design consisting of 20 treatments, and each treatment comprised of 4 replicates. Predating efficiency was calculated at each instar stage on mealybug 1st, 2nd, 3rd and adult stage. Life table parameters were studied in plastic vials, 15/16*100 mm fitted with plastic lid. Predating efficiency was evaluated. Photographs of each instar was taken through EM-310M digital microscope eyepiece camera with USB 2.0 output 3.2 M / Resolution 2048×1536, 110 mm(H) x 55 mm (D) and 23 mm adapter having Sensor: 1/2" and enhanced color CMOS mounted on Labomed Model digizoom digital zoom stereo microscope.

Rearing of *P. solenopsis* (Tinsely)

Cotton mealybug was reared on bottle gourd Lagenaria siceraria in cages measuring 45 × 30 × 12 cm. The culture was used for experimentation.

Collection of adult beetles and rearing

Adult beetles were collected from cotton fields as well as from other crops during 2nd week of February, 2010 through sweep net technique. The specimens were brought to laboratory and placed in cages measuring 45 × 30 × 12 cm and were fed on mealybugs. The experiment was kept under observation and sexual balance was maintained.

Eggs

Eggs were collected on towel tissue paper and were placed in 9 cm diameter Petri-dishes which were kept on moist tissue papers. The data regarding color, duration and size were recorded.

1st instar

1st instar of *C. undecimpunctata* on emergence, were placed in plastic vials, 15/16*100 mm fitted with plastic lid. Cotton mealybug

Table 1. Average consumption of cotton mealy bug by predator *C. undecimpunctata.*

Instars of predator	Average cotton mealy bugs instars consumed by C. *undecimpunctata*			
	1st instar	**2nd instar**	**3rd instar**	**Adult**
1st instar	91.999[e]	45.333[d]	44.00[d]	5.44[c]
2nd instar	97.00[c]	35.667[e]	45.00[d]	7.11[c]
3rd instar	121.666[d]	51.666[c]	54.33[c]	8.21[c]
4th instar	93.001[e]	35.00[e]	33.00[e]	7.33[c]
Pupae	0.000[f]	0.000[f]	0.000[f]	0.000[d]
Adult male	388.11[b]	123.02[b]	87.01[b]	46.08[b]
Adult female	451.21[a]	141.21[a]	93.05[a]	58.11[a]
LSD value	3.398	2.799	3.167	3.203

each instar 1st, 2nd 3rd and adult were released in each cage 30, 20, 15 and 10 each day respectively to evaluate predating efficiency of beetle. First instar larvae size, color, duration was also recorded.

2nd instar

After moulting 1st instar larvae of *C. undecimpunctata*, 2nd instar appeared. Its diameter, size, color, duration were recorded and were offered cotton mealybug 40, 30, 20 and 10 for each replicate. Predating efficiency was calculated.

3rd instar

The 3rd instar larvae of *C. undecimpunctata* color, size, and duration were measured and were offered Cotton mealybug 1st, 2nd, 3rd and adult instar at 40, 15, 20 and 10 respectively. Predating efficiency were determined daily intervals.

4th instar

The 4th instar larvae of *C. undecimpunctata* parameters were studied as per description of earlier instars. Predating efficiency was determined day after intervals.

Pupae

Pupal size, color and weight of *C. undecimpunctata* were recorded. Daily changes in pupae skin texture were observed.

Adult

Adult size, color and weight of *C. undecimpunctata* were recorded. Differences between male and female were observed and pairing was made. Adult Pre ovi-positional period, post ovi-positional period, natality, fertility, fecundity, mortality and adults' survival rate were also determined.

Data collection and statistical analysis

Data were collected after 24 h interval for predating efficiency while for biological parameter it was conducted after 12 h intervals. Data were statistically analyzed using MSTAT-C program (Anonymous, 1989) and means were separated at significance level 0.05 using DMRT (Duncan Multiple Range Test Method).

RESULTS AND DISCUSSION

Predating efficiency

Total consumption by *C. undecimpunctata* L. on different instars of mealy bugs presented in Table 1 indicated that 3rd instar larvae of the predator consumed significantly higher numbers of 1st, 2nd and 3rd instars mealy bugs as compared to 1st,2nd and 4th instar of the predator. Results of present studies were similar to Sattar et al. (2007) who reported that 3rd instar larvae of the *Chrysoperla carnea* consumed significantly higher numbers of 1st, 2nd and 3rd instar of mealybugs as compared to 1st and 2nd instar. However in *C. carnea* 3rd instar had long duration as compared to other instars while in *C. undecimpunctata* 4th instar larvae is also present. Less feeding of fourth instar larvae might be due to short duration and pre-pupation period, in which insect feeding is ceased in most of cases.

Per day mean consumption of C. *undecimpunctata* larvae on 1st, 2nd and 3rd instars of the *P. solenopsis* were depicted in Tables 2, 3 and 4, respectively. The out comes of present studies revealed that there was a significant difference in per day consumption of *C. undecimpunctata* on different instars of cotton mealy bugs. The 1st instar larvae of the predator consumed 23.00, 11.33 and 11.00 mealy bugs of 1st, 2nd and 3rd instars, respectively. The results of present studies were in conformity with results of Noia et al. (2008) and Mari et al. (2005), who reported that *C. undecimpunctata* 1st instar consumed 55.10 mustard aphids (*Lipaphis erysimi*), 2nd instar consumed 32.333, 11.89 and 15.00 mealybugs of 1st, 2nd and 3rd instars, respectively. The results of present studies were similar to Noia et al. (2008), who reported intra-guild and extra-guild prey densities. Results were also similar to Mari et al. (2005) depiction that 2nd instar *C. undecimpunctata* consumed81.00 mustard aphids.

The results of present studies were in contradiction to Moura et al. (2006) and Cabaral et al. (2006) evaluations on predating efficiency of *C. undecimpunctata* on aphids. Low consumption of *P. solenopsis* than mustard aphid

Table 2. Average per day consumption of 1st instar of mealybug by C. *undecimpunctata.*

Instars of C. *undecimpunctata*	Average daily mealy bug consumed by 1st instar predator				Mean
	1st	**2nd**	**3rd**	**4th**	
1st	23.333[a]	23.000[a]	23.333[a]	22.333[a]	22.99
2nd	33.000[a]	31.000[a]	33.000[a]	0.0000[b]	24.25
3rd	30.333[ab]	31.333[a]	31.000[a]	29.000[b]	30.41
4th	24.000[a]	23.667[a]	22.667[a]	22.667[a]	23.25
Grand total					100.90

Table 3. Average per day consumption of 2nd instar of mealy bug by C. *undecimpunctata.*

Instars of C. *undecimpunctata*	Average daily mealy bug consumed by 2nd instar predator				Mean
	1st	**2nd**	**3rd**	**4th**	
1st	9.667[b]	12.333[a]	10.000[b]	13.333[a]	11.333
2nd	12.000[a]	13.000[a]	10.667[b]	0.00[c]	11.889
3rd	13.333[a]	12.000[a]	13.000[a]	13.333[a]	12.916
4th	9.333[ab]	7.667[c]	10.000[a]	8.000[bc]	8.750
Grand total					44.888

Table 4. Average per day consumption of 3rd instar of mealy bug by C. *undecimpunctata.*

Instars of C. *undecimpunctata*	Average daily mealy bug consumed by 3rd instar predator				Mean
	1st	**2nd**	**3rd**	**4th**	
1st	5.667[a]	4.000[a]	4.333[a]	2.000[a]	4.00
2nd	9.000[b]	6.667[a]	7.333[b]	0.00[c]	5.75
3rd	12.333[bc]	9.000[a]	9.333[c]	3.667[b]	8.58
4th	8.667[ab]	11.000[a]	8.333[ab]	4.000[a]	8.00
Grand total					26.33

Table 5. Per day mean consumption of adult stage of mealybug by C. *undecimpunctata.*

Instars of C. *undecimpunctata*	Average daily mealy bug consumed by adult of C. *undecimpunctata*				Mean
	1st	**2nd**	**3rd**	**4th**	
1st	0.50[c]	0.59[c]	0.93[a]	0.85[b]	0.7175
2nd	1.12[c]	1.33[a]	1.21[b]	1.01d	1.1675
3rd	1.11[b]	1.00[c]	1.00[c]	2.21[a]	1.33
4th	1.66[a]	1.00[c]	1.02b	1.00[c]	1.17
Grand total					17.54

might be due to fact that mealy bugs are covered with waxy layer, which makes the prey unpalatable for consumption by predators (Jonathan, 2005). The predator C. *septempunctata* consumed less *B. brassicae* than other species due to waxy coating on B. *brassicae* (Ashraf et al., 2010).

Per day mean consumption of C. *undecimpunctata* adult male and female on 1st, 2nd and 3rd instars of the mealybug is presented in Tables 5 and 6 respectively. Per day consumption of adult male C. *undecimpunctata* on 1st, 2nd and 3rd instars of cotton mealybug was 1.400, 1.47 and 1.47, respectively and that of adult female C.

Table 6. Total consumption of mealybug by adult (male) *C. undecimpunctata* and female *C. undecimpunctata*.

Instars of mealy bug	Adult male	Adult female
1st	388.11	451.21
2nd	123.02	141.21
3rd	87.01	93.05
Adult	46.08	58.11
LSD	19.98	19.97

Table 7. Studies on Biology of eleven spotted beetle (*C. undecimpunctata* L.).

Stages	Size (L mm × W mm)	Color	Duration (days)	Morphological characters
Egg	0.5 × 0.25	Yellowish orange.	2-3	Egg shape is oval Laid in clusters Each cluster containing 10-15 eggs.
1st instar	1.5 × 0.5	Black like small alligator.	3-4	Head and legs are black in color. Body is dark grey. Thorax has one white dot surrounded by two black dots. Four longitudinal rows of hair are present on abdomen.
2nd instar	2 × 0.75	Black like small alligator.	2-3	Body is elongate. Two black dots are present. Four longitudinal rows of hair are present.
3rd instar	2.5 × 1.0	Black like small alligator.	3-4	Body is larger than 2nd instar.
4th instar	5.0 × 2.5	Black like small alligator.	3-4	Body is larger in size.
Pupae	4.0 × 2.0	Dark brown.	4-5	Firstly pure yellow later on changed to oranges brown and reddish brown.
Adult (male)	0.3 × 0.2	Light orange in color containing eleven black spots on each elytron.	32-41	Smaller in size and eleven spots are present on elytra.
Adult (female)	0.5 × 0.25	Dark orange in colour containing eleven black spots on elytron.	37-55	Larger in size and eleven spots are present on elytra as compare to male.

undecimpunctata was 1.07, 1.13 and 1.27, respectively.

Life history

Female of eleven spotted ladybird beetle, *C. undecimpunctata* L. laid clusters of yellowish orange eggs that turned into dark yellow before hatching. Each cluster had an average of 10-15 eggs. Data in Table 7 unveiled that eggs incubation period was about 2-3 days and size of a single egg was 0.5 × 0.25 mm. Table 7 also indicated that average duration of 1st, 2nd, 3rd and 4th larval instars were 3-4, 2-3, 3-4 and 3-4 days respectively and they were black in color and small alligator like. 1st, 2nd, 3rd and 4th larval instars were 1.5 × 0.5 mm, 2 × 0.75 mm, 2.5 × 1.0 mm and 5.0 × 2.5 mm in size respectively. The pupa was dark brown in color and pupal period was 4-5 days. Size of pupa was 4.0 × 2.0 mm.

Results of present studies are very similar to Solangi et al. (2007) who reported that the mean incubation period

Table 8. Percent emergence, sex ratio, total life period in days and mortality of adults in *C. undecimpunctata*.

Emergence (%)		Sex ratio	Fecundity (eggs)			Total life period (days)		Mortality (%)
Male	Female	Male : female	Lowest	Highest	Average	Male	Female	Average
47	53	1 : 1.5	507	679	593	50-64 (means)	54-77 (means)	3

of ten eleven spotted lady-bird beetle in the laboratory was 3.7±0.94 days within the range of 2-5 days, while 1st, 2nd, 3rd and 4th instar larvae period was 3.1±1.19, 3.1±0.87, 3.5±1.26 and 3.3±0.94 days within the range of 2-5, 2-4, 2-6 and 2-5 days respectively and pupal period was 5.6±0.96 days within the range of 4 to 6 days. In another study egg production per female averaged 142.33, incubation period of eggs 2-9 days, 4 larval instars and last larval stage duration 7.0, 7.5, 12.0, 16.0 and 23.0 days, pupal development average 2.5 days at 30°C and 7.5 days at 14°C and egg to adult life cycle duration 12, 14, 21, 27.5 and 38.5 days at 30, 26, 22, 18 and 14°C, respectively was reported by Eraky and Nasser (1995).

Results in the Table 8 revealed that mean adult male and female emergence was 40 and 60%, respectively. Male to female sex ratio was averaged 2:3. The results indicated that highest, lowest and average fecundity recorded was 679, 507 and 593, respectively. Total life period of adult male and female was 50-64 and 54-77 days, respectively and average mortality was 3.0%.

Results of present studies were in agreement to Solangi et al. (2007) who reported that the emergence of adult male and female was 7.4±2.63 (38.50±13.12%) and 8.9±3.66 (43.38±8.24%) and total life period of adult male and female was 36.5±4.47 and 46.0±9.14, respectively. Solangi et al. (2007) also reported that sex ratio (male: female), average and highest fecundity and average mortality of adults was 1:1.25±1:0.45, 593.4±86.5 and 740 eggs and 17.57±14.51, respectively.

CONCLUSION AND RECOMMENDATIONS

It is recommended by analyzing results of present studies on biology and predatory efficiency of *C. undecimpunctata* (Linneaus) on cotton mealybug (*P. solenopsis* Tinsely) that *C. undecimpunctata* is an efficient bio-control agent for pest control and it must be included in IPM program for suppression of Cotton mealybug in Pakistan, because of its high fecundity, easily rearing ability and efficient suppression of invasive Carrabin pest.

ACKNOWLEDGEMENTS

The authors of this paper are extremely thankful to Ministry of Agriculture and Punjab Agricultural Research Board for providing funds for CGS project, "Control of Cotton Mealybug with Bio-control Agents" in theme number 1 for granting funds 120.637 million rupees. Author of paper are highly obliged for sincere comments by Dr. Haider Karar, Assistant Entomologist, Entomological Research Institute and Dr. Shuaib Assistant Professor, Entomology Department Bhauddin Zakariya University for proof reading of manuscript.

REFERENCES

Abbas G, Arif MJ, Ashraf M, Aslam M, Saeed S (2010). Host plants distribution of cotton mealybug (*Phenacoccus solenopsis* Tinsley; Hemiptera: Pseudococcidae). Int. J. Agric. Res. 12:421-425.
Anonymous (1989). MSTAT-C. Micro computer statistical programme. Michigan State University, Michigan Lansing, USA.
Ashraf M, Ishtiq M, Asif M, Adrees M, Ayub MN (2010). Studies on laboratory rearing of ladybird beetle (*Coccinella septempunctata* L.) to observe its fecundity and longevity on natural and artificial diet. Int. J. Biol. 2(1):165-173.
Bellows TS (2001). Restoring population balance through natural enemy introductions. Biol. Cont. 21:199-205.
Centre for AgroInformatics Research (2007). Mealybug: Cotton crop's worst catastrophe in District Multan during 2005-2006. Published by FAST Notational Univ. Comput. Emerg. Sci. P. 81.
Cheema MA, Muzaffar N, Ghani MA (1980). Investigation on phenology, distribution, host range and evaluation of predators of *Pectinophora gossypiella* in Pakistan. *Pakistan Cotton*, 24:140-176.
Eraky SA, Nasser MAK (1995). Effect of constant temperatures on development and predation prey efficiency of ladybird beetle, *Coccinella undecimpunctata* L. (Coleoptera: Coccinellidae). Assiut. J. Agric. Sci. 24:223-231.
Fayyaz A (1998). Predatory efficacy of three coccinellid species against wheat aphids in laboratory and under semi-natural conditions. *M.Sc. Thesis*, Dept. Agri. Entomology, Univ. Agri. Faisalabad.
Ghafoor A, Saba I, Khan MS, Farooq HA, Zubaida, Amjad I (2011). Predatory potential of *Cryptolaemus montrouzierii* for cotton mealybug under laboratory conditions. J. Anim. Plant Sci. 21(1):90-93.
Hoy MA, Nguyen R (2000). Classical biological control of brown citrus aphid: Release of *Lipolexis scutellaris*. Cit. Ind. 81:24-26.
Jonathan G., Lundgren RN (2005). Wiedenmann Tritrophic Interactions among Bt (Cry3Bb1) Corn, Aphid Prey, and the Predator *Coleomegilla maculata* (Coleoptera: Coccinellidae). Environ. Entomol. pp. 1621-1625.
Kenneth I, Hagen H (1970). Predatory efficacy of the Coccinellids against the aphids. J. App. Entomol. 12:34-41.
Khan H, Suhail A (2001). Feeding efficacy, circadian rythems and oviposition of ladybeetle (Coccinellidae: Coleoptera) under controlled conditions. Int. J. Agric. Biol. 4(3):384-386
Mahmood R, Aslam MN, Solangi GS, Samad A. (2011). Historical Perspectives and achievements in biological management of cotton mealybug *Phennacoccus solenopsis* Tinsely in Pakistan. 5th Asian meeting ICAC. www.icac.org/tis/regional-networks/asian_network/meeting_5/document/papers/mahmood,R.pdf.
Mari JM, Rizvi NH, Nizamani SM, Qureshi KH, Lohar MK. (2005). Predatory efficiency of *Menochilus sexmaculatus* Fab and *Coccinella undecimpunctata* Linneaus (Coccinellidae: Coleoptera) on alfa alfa aphid *Theriophiis trifolii* (Monell). Asian J. Plant Sci. 4(4):354-358.

Marshall S (2005) The London and Essex ladybird survey. London Natural History Society (LNHS) and the Essex Field Club, UK., pp.1-25.

Moura R, Garcia P, Cabral S, Soares AO (2006). Does pirimicarb affect the voracity of the euriphagous predator, *Coccinella undecimpunctata* L. (Coleoptera: Coccinellidae). Biol. Control 38:363-368.

Nielsen GR (1997). Lady beetles. Plant and soil science. University of Vermont Extension (UVEXT), pp: 1-5.

Noia M, Borges I, Soares A (2008). Intraguild predation between the aphidophagous ladybird beetles *Harmonia axyridis* and *Coccinella undecimpunctata* (Coleoptera: Coccinellidae): the role of intra and extraguild prey densities. Biol. Control 46:140-146.

Orbycki JJ, King TJ (1998). Predaceous Coccinellidae in Biological Control. Ann. Rev. Entomol. 43:295-321.

Sattar M Hamed M, Nadeem S (2007). Predatory potential of *Chrysoperla Carnea* (Stephens) (Neuroptera: Chrysopidae) against cotton mealybug. Pak. Entomol. 29(2)

Soares OA, Serpa A, (2007). Interference competition between ladybird beetle adults (Coleoptera: Coccinellidae): effects on growth and reproductive capacity. Pop. Ecol. 49:37-43.

Solangi BK Lanjar AG., Lohar MK. (2007). Biology of 11spotted beetle *Coccinella undecimpunctata* L. (Coccinellidae: Coleoptera) on mustard aphid *Lipaphis erysimi* Kalt. J. Appl. Sci. 7(20):3086-3090.

Tanwar RK, Jeyakumar P, Singh A, Jafri AA, Bombawale OM (2011). Survey of cotton mealybug *Phennacoccus solenopsis* (Tinsely) and its natural enemies. J. Environ. Biol. 32:381-384.

Wheeler A GJR, Hobeke ER (1981). A revised distribution of *Coccinella undecimpunctata* L., in Eastern and Western North America (Coleoptera: Coccinellidae). Coleop. Bullet. 35:213-216.

27

Physical, chemical and sensory factors of Mexican and New Zealand sheep meat commercialized in Central of Mexico

M. D. Mariezcurrena[1], A. Z. M. Salem[2], C. Tepichín[1], M. S. Rubio[3] and M. A. Mariezcurrena[2]

[1]Universidad Autónoma del Estado de México. Facultad de Ciencias Agrícolas. Código Postal 50200. Instituto Literario N0. 100 Col. Centro. Toluca, Estado de México.
[2]Universidad Autónoma del Estado de México. Facultad de Medicina Veterinaria y Zootecnia, Código Postal 50200. Instituto Literario N0. 100 Col. Centro. Toluca, Estado de México.
[3]Laboratorio de Ciencia de la Carne, Secretaria de Producción Animal, Facultad de Medicina Veterinaria y Zootecnia. Universidad Nacional Autónoma de México. Código Postal 04510. Av. Universidad 3000, Del. Coyoacán, Ciudad Universitaria, Distrito Federal, México.

The object of this study was to determine the physicochemical and sensory quality of national and imported from New Zealand mutton sold in the highest mutton-selling region in Mexico (Capulhuac, Estado de Mexico). Samples were obtained from 6 wholesale points. In each outlet a piece of *longissimus dorsi* muscle was bought. The factors evaluated were moisture, protein, fat, objective color (L*, a* y b*) subjective color and shear force. There were differences (P < 0.05) in variables L*, a*, b*, moisture and protein between national and imported meat. There were also differences in the percentage of protein (national 21.27%; imported 20.27%) of fat (national 2.07%; imported 3.37%), and in shear force (national 3.73 kg; imported 2.09 kg). The sensory evaluation was done by 49 consumer judges. Results showed that consumers prefer imported meat over national meat.

Key words: Quality, ovine, lamb, mutton, meat quality, meat imported, sensory evaluation.

INTRODUCTION

Sheep meat is affected by certain *ante-mortem* factors such as genetics, age, gender, diet and stress among others, and by *post-mortem* factors such as freezing, refrigeration, maturation and electrical stimulation among others. These factors affect the quality of the meat and its physical and chemical composition (Torrescano et al., 2009).

In the center of Mexico, in the states of Hidalgo, Chihuahua, Jalisco, Estado de México and San Luis Potosí, ovine herds are a cross of Suffolk or Hampshire breeds. Of the total number of sheep bred in this area,

20% are finished with balanced concentrates, 40% are finished with a combination of grassland use and energy and protein supplements (Barrios, 2005) and finally 40% are finished with grazing. The latter often using inadequate reproductive and/or sanitary handling, premises in bad conditions and unsuitable ingestion systems, which generally cause malnutrition and parasitism (Martínez et al., 2010; Hinojosa-Cuéllar et al., 2009).

The imported sheep meat comes from New Zealand, where the breeds are mainly Borderdale, B-Leicester,

Coopworth, Corridale, Dorset Down, Drysdale, E-Friesian, Hampshire, Lincoln, Merino, Texel, Suffolk (New Zealand Sheep-breeders Association, 2011; Scerra et al., 2007). Farmers use *brassica, raigrás perenne, italian raigrás*, turnips, red clover, two types of cabbage and bananas to improve ovine production, seeing as these influence the color, tenderness and pH (New Zealand Sheep-breeders Association, 2011; Scerra et al., 2007). In Mexico there is a growing demand of over 85,000 tons of sheep meat, 40,000 tons of those amount are imported. In 2010 they were sold on the national market for $33 /kg carcass and the frozen imported meat for $42 /kg (Martínez et al., 2011). At present there are no studies comparing the national meat quality vs. imported is therefore vitally important to provide such information to assist domestic producers to market their product best

Usually sheep meat in Mexico is consumed in traditional recipes such as *barbacoa* and *pastor* (Arteaga, 2006). *Barbacoa* is a traditional Mexican lamb dish which involves cooking meat in a pit in the ground along with a container of water, causing the heat that is roasting the meat to be very damp and making the meat moist. *Pastor* is lamb marinaded in chillies and spices and then roasted).

In 2008 there was a deficit of almost 35,000 tons, which had to be imported principally from New Zealand but also from Australia and Chile. That year national production made up 48.9% of total consumption and imported sheep meat made up the remaining 51.1% (Martínez et al., 2010). *Barbacoa* producers prefer Mexican sheep meat due to its taste, however they choose imported meat because it implies less cutting of the carcasses (Martínez et al., 2009). The demand of ovine products is determined by the requirements of human populations of meat, wool and leather. These requirements have been steadily growing by 1.07% for the past 10 years, according to the estimations of the Comisión *Nacional* de *Población* or National Population Commission.

Furthermore it has not been clearly stated if the sheep meat from different sources has varying levels of quality depending on their chemical composition and consumer acceptance. Not being aware of the differences between national and imported meat and their impact on the national market leads to economic loss (5.00/kg) along the production-consumption chain, particularly in the area this study is focusing on, the municipality of Capulhuac, Estado de Mexico. The aforementioned location is the main national collection point for sheep meat and the principal *barbacoa* trader. Imported sheep meat comes in its majority from New Zealand but also from Australia and Chile, which could suggest it is of better quality than the meat from the states of Zacatecas, Jalisco, San Luis Potosí, Chihuahua, Michoacán, Durango and Aguascalientes, even though this has not been proven to date. Therefore the objective of this study was to determine the physicochemical and sensory quality of national and imported sheep meat sold in Capulhuac, Estado de Mexico.

MATERIALS AND METHODS

The testing was carried out in the months of September to November, 2010 in the municipality of Capulhuac, Estado de México The location was chosen for being the greatest ovine livestock center, where circa 400,00 heads are sacrificed per year and approximately 190 sheep meat retailers operate, there are 350 producers and suppliers from 7 states in the Mexican Republic (Zacatecas, Jalisco, San Luis Potosí, Chihuahua, Michoacán, Durango and Aguascalientes, among others), 700 *barbacoa* producing clients and distributors of froze, imported sheep meat from New Zealand, Australia and Chile (Center for livestock introduction and production of Capulhuac, Estado de México, S.P.R. de R.L. de C.V., 2011). These animals are slaughtered and frozen at -18°C, then are shipped and shipped via sea to Mexico (40 days), arriving at the Port of Vera Cruz, immediately are transported in trailers with Termoking to Capulhuac at 6 retailers wholesalers, maintaining the same storage conditions, where samples were taken for the present and taken to the laboratory experiment, while maintaining the same conditions. The animals are slaughtered in Mexico weighing approximately 45 Kg, hybrid line (wool / meat) fed with concentrates mostly slaughtered in backyard, sold unripe or freeze

Sampling

Shop owners were interviewed to find the wholesale points which sold national and imported sheep meat. Six principal wholesale points were identified and a directed sampling method was proposed (Scheaffer et al., 2007). At each point of sale a piece of meat (15 cm) from the longissimus *dorsi* muscle was bought. It was stored in a hermetic plastic bag (Ziploc®) and kept in ice at approximately 3°C while being transported to the laboratory where it was frozen at -2°C, awaiting analysis.

Instrumental analysis

The samples were defrosted for 19 h in refrigeration until they reached approximately 16°C. The surrounding fat and adjacent muscles were removed. The chop from the 5th rib was used to measure subjective color according to the scale (1 = pale pink, 2 = cherry red, 3 = dark red and 4 = deep purple) proposed by Sierra (1974). Objective color was measured with a Minolta Chroma meter CR-400, with observer (2 degrees, standard color matching CIE 1931: (x2λ, y2λ, z2λ) and lighting (C, D65) as well as with a 20° angle of vision made in Tokyo, Japan). The measurements were taken 15 min after cutting the piece of meat transversally (Honikel, 1998). The chops were ground using a Moulinex (France) food processor and tested for moisture using the dry oven method (AOAC, 1990). To determine protein content the Kjeldahl (AOAC,) method was used, and fat was measured using a chloroform-methanol mix 2:1 and the Soxtec Foss Tecator 2055 gravimeter, adapted with the official method AOAC #991 (Mariezcurrena et al., 2010). To test shear force, the chops were cooked on a grill to a final temperature of 70°C measured with a thermocouple (Omega Inc., Stamford, E.U.) and a portable recording thermometer (Omega Inc., Stamford, E.U.) (American Meat Science Association, 1995).

Sensory evaluation

A simple paired-comparison and then a degree of satisfaction test were carried out with 49 non-trained judges (consumers) in semi pilot test ranging in ages from 20 to 50 years old. In preparation for the tasting, the samples were defrosted for 19 h in refrigeration and then allowed to reach room temperature, and then they were grilled

Table 1. Physical and chemical composition of national and imported meat.

Variables	National meat	Imported meat	SEM
Protein (%)	21.27[a]	20.27[b]	0.27
Fat (%)	2.07[b]	3.37[a]	0.05
Moisture (%)	77.33[a]	64.53[b]	0.85
L*	33.63[b]	35.95[a]	0.74
a*	14.37[b]	16.06[a]	0.40
b*	6.13[b]	7.41[a]	0.28
Subjective color (1-4)	3.45[a]	3.55[a]	0.12
Shear force (kg$_f$)	3.73[a]	2.09[a]	0.39

Means with different superscripts [a y b] in the same row are significantly different (p<0.05). L* varies from 0 (black) to 100 (white), a* positive (a*>0, red) or negative (a*<0, green), b* positive (b*>0, yellow) or negative (b*<0, blue), EEM: Standard error in means.

Table 2. Simple paired-comparison test for flavor intensity.

Intensity of flavor	National meat	Imported meat	P-Value	SEM
Lamb	6.0[a]	7.0 [a]	0.248	0.416
Mutton	6.0[a]	6.0[a]	0.600	0.456
Fat	3.0[b]	5.0[a]	0.008	0.421
Metallic	4.0[a]	2.0[a]	0.987	0.405

Means with different superscripts [a y b] in the same row are significantly different (p < 0.05), SEM: Standard error in means, Hedonic scale from 0 to 9; with 0 = absent; 9 = extremely intense.

until the geometric center reached 70°C. Then the borders were removed and the meat was cut into uniform cubes (of approximately 2 cm$^{3)}$, put in plastic bags and placed on bain-marie (23°C; AMSA, 1995). Each judge received two samples on plates labeled with random 3-digit numbers and accompanied by neutral wholemeal crackers, water and the questionnaire. These questionnaires were designed to evaluate a simple paired-comparison which consisted of grading the intensity of taste of the following: lamb (Newborn) mutton, fat and metallic. These variables were measured using this hedonic scale: 0 (absent) to 9 (extremely intense). The meat was also tested for general satisfaction, juiciness, tenderness and taste using the following hedonic scale: 1 (dislike extremely), 2 (dislike very much), 3 (dislike moderately), 4 (dislike slightly), 5 (neither like nor dislike), 6 (like slightly), 7 (like moderately), 8 (like very much), 9 (like extremely). The aforementioned qualities of juiciness, tenderness and taste were also graded from 1 (extremely dry/tough/insipid) to 9 (extremely juicy/tender/tasty). Judges were also asked to state buying preference from 1 (I would definitely not buy it) to 5 (I would definitely buy it).

Statistical analysis

Samples from both imported and national meat were tested, using 5 samples of each. 8 variables were analyzed: objective color (L*, a* y b*), subjective color, fat, protein, moisture and shear force, a variance test was carried out on both types to determine physicochemical characteristics. Significant variation to P < 0.05 was observed and the averages were measured with the Tukey test (SAS, 2004). The results of the sensory evaluation were analyzed statistically using the U of Mann-Whitney (P < 0.05) test which was carried out on both national and imported meat. 49 repetitions for each type (national and imported) and 12 answer variables (intensity of lamb, mutton, fat and metallic taste; like/dislike,

juiciness, tenderness and taste; level of juiciness, tenderness, taste and purchase preference). With this analysis the effects of the two types of sheep meat and the sensory characteristics were discovered. Furthermore an analysis of main components for sensory evaluation was done to reduce the dimensionality of the data and thus discover the causes of variability within it and order them by importance.

RESULTS

Physical and chemical characteristics results

The averages and the standard deviation of the variables of the physicochemical composition of the two types of sheep meat evaluated are shown in Table 1. National meat showed lower scores in the variables L, a*, b*, which shows it is darker meat. It also displayed higher protein and moisture content and less fat content. There were no significant differences in subjective color and shear force.

Sensory evaluation

The consumer judges did not detect significant differences between national and imported meat in terms of intensity of lamb flavor, of mutton flavor or of metallic taste, however they did notice that imported meat tasted more intensely of fat than did national meat (Table 2).

Table 3. Simple paired-comparison test for juiciness, tenderness and flavor.

Characteristics	National meat	Imported meat	P-Value	SEM
Juiciness	5.0[b]	7.0[a]	0.00001	0.252
Tenderness	5.0[a]	7.0[b]	0.00080	0.252
Flavor	7.0[a]	7.0[a]	0.07500	0.324

Means with different superscripts [a y b] in the same row are significantly different (p<0.05), SEM: Standard error in means, Hedonic scale from 1 to 9; with 1=extremely dry/tough/ insipid; 5= not dry or juicy/ not tough nor tender/ not insipid nor strong-tasting; 9=extremely juicy/tender/strong tasting.

Table 4. Simple paired-comparison test for satisfaction.

Degree of satisfaction	National meat	Imported meat	P-Value	SEM
Juiciness	5.0[a]	7.0[b]	0.0003	0.254
Tenderness	6.0[a]	7.0[b]	0.040	0.290
Flavor	5.0[a]	6.0[b]	0.009	0.254
General satisfaction	5.0[a]	7.0[b]	0.002	0.297

Means with different superscripts [a y b] in the same row are significantly different (p<0.05), SEM: Standard error in means, Hedonic scale from 1 to 9; with 1=dislike extremely; 5=neither like nor dislike 9=like extremely.

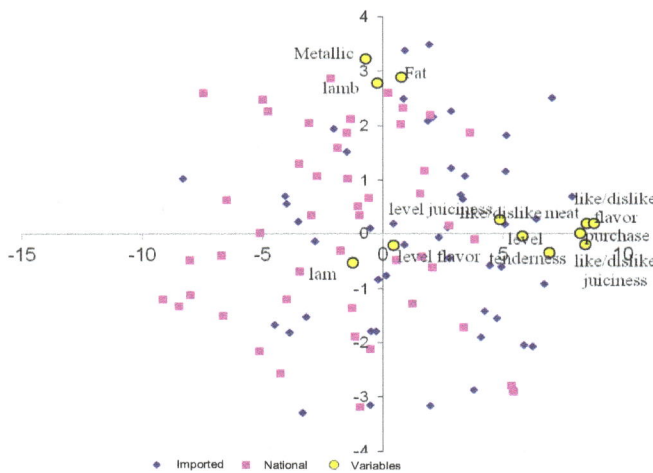

Figure 1. Main components of the sensory evaluation.

The consumer judges did not detect differences in intensity of flavor between national and imported meat, however they did rate imported meat higher on juiciness and tenderness (Table 3). The consumer judges preferred the imported meat over the national meat for juiciness, tenderness, flavor and general satisfaction. The consumer judges showed a preference (P<0.001) for buying imported meat over national meat (Table 4).

Main component analysis in sensory evaluation variables

The analysis of the main components (Table 1) showed

main component 1 accounts for 38.37% of the variance of the data obtained and main component 2 accounts for 19.48%, both adding up to 57.86% of variance. Main component 1 is largely determined by the degree of satisfaction test (juiciness, tenderness and general satisfaction regarding sheep meat) and the preference test. Main component 2 is determined by the taste test (lamb, mutton, fat and metallic). Figure 1 shows the decision to buy imported meat is influenced by the general degree of satisfaction in tenderness and taste variables. This allows us to see that the consumers involved in the testing would prefer to buy imported meat over national meat for its juiciness, tenderness, taste and general satisfaction characteristics. Consumers stated that national meat had a more intense taste of lamb, mutton and fat, and this would lead them to not buy it, a trend that is clearly shown in Figure 1.

DISCUSSION

This study showed that imported sheep meat has a larger percentage of intramuscular fat than national meat; this implies the moisture percentage of national meat is greater, as has been previously reported (Forrest et al., 1979). One of the factors affecting the quality of the meat is diet fed to sheep. Gutiérrez et al. (2005), Nuncio et al. (2001) and Hernández and Vidal (2001) reported that there are two production systems in Mexico: extensive (grazing and concentrates) and intensive when sheep fed on concentrates principally based on maize and sorghum. Costa et al. (2009) reported that sheep fed on concentrates, mainly maize and sorghum, show higher percentages of protein and moisture in the meat. There is virtually a constant between myofibrillar protein and water in meat (Sánchez, 1999) seeing as the majority of water lies trapped between the myofibrillar proteins (70%; Carballo et al., 2001). Farmers in New Zealand feed the sheep with leguminous forage plants (clover) which cause an increase in intramuscular fat, compared to sheep fed on concentrates of high amount of maize, (Research into Lamb Meat Quality, 2010; Schreurs et al., 2008; Scerra et al., 2007; Díaz et al., 2005).

These are the reasons imported sheep meat turned out to be tenderer than national meat. Realini et al. (2004) reported that sheep fed on grass yielded tenderer meat. Another influencing factor is that Mexican *barbacoa* producers prefer warm carcasses, thus there is no maturation of the meat. On the other hand, Costa (2002) reported that intramuscular fat is due to lipids deposited between and inside cells, and this is associated with increased flavor, tenderness and juiciness. This study shows that consumers found imported sheep meat to taste more of fat than national meat, this could be due to diet, seeing as Whittington et al. (2006) and Fisher et al. (2000) found that panelists detected more fat flavor in concentrate-fed sheep. This study found that consumers prefer imported sheep meat, as it is juicier and tenderer than national meat. According to Sañudo (2008), intramuscular fat content is directly related to juiciness and tenderness in sheep meat, this explains why the bromatological analysis showed more intramuscular fat in imported meat, thus preferred by consumers. Furthermore the content of intramuscular fat has bearing on tenderness, acting as a lubricant between teeth and mouth during chewing by diminishing friction.

Conclusions

This study showed that imported sheep meat, sold in the municipality of Capulhuac, Estado de Mexico contained more intramuscular fat than national sheep meat. This made it tenderer and thus preferred by consumers.

ACKNOWLEDGEMENT

We thank the National Council for Science and technology (CONACYT) for the financing of the project FORDECYT-CONACYT, registered under 00000000116234 and fund number 10014.

REFERENCES

American Meat Science Association (1995). Research guidelines for cookery, sensory evaluation and instrumental tenderness measurements of fresh meat. Chicago, IL: American Meat Science Association and National Livestock and Meat Board.

AOAC (1990). Helrick, K. (Ed.), Official methods of analysis of the association of official analytical chemists (15th ed.). USA: Arlington. P. 1230

Arteaga CJD (2006). Situación de la ovinocultura y sus perspectivas. Memorias de la primera semana nacional de ovinocultura; Hidalgo, México: AMTEO, pp. 60-73.

Barrios (2005). Guía practica de ovinocultura enfocada hacia la producción de carne. BacomITDA. Empresa del sector agropecuario y ambiental. Rancho de la oveja. Bogota P. 48.

Carballo B, López de Torre G, Madrid A (2001). Tecnología de la carne y de los productos cárnicos. AMV ediciones. Primera edición. ISBN 84-89922 – 52 –P. 7.

Centro de introducción y producción de ganado de Capulhuac, Estado de México, S.P.R. de R.L. de C.V. (2011). Introducción de ovinos en la zona centro del país: el caso Capulhuac, México. http://spo.uno.org.mx/wp-content/uploads/2011/03/jclj_introductores.pdf Citado 16/05/2012

Costa E (2002). Composição física da carcaça, qualidade da carne e conteúdo de colesterol no músculo Longissimus dorsi de novilhos Red Angus superprecoces, terminados em confinamento e abatidos com diferentes pesos. Revista Brasileira de Zootecnia. pp. 243-252

Costa RG, Sancha A, Madruga M.S, Gonzaga S, Cássia R, Araújo Filho J. T, Selaive A (2009). Physical and chemical characterization of lamb meat from different genotypes submitted to diet with different fibre contents. Small Rumin. Res. 81:29-34.

Díaz MT, Álvarez I, De la Fuente J, Sañudo C, Campo MM, Oliver MA, (2005). Fatty acid composition of meat from typical lamb production systems of Spain, United Kingdom, Germany and Uruguay. Meat Sci. 71(2):256-263.

Fisher AV, Enser M, Richardson RI, Wood JD, Nute GR, Kurt E, Sinclair LA, Wilkinson RG (2000). Fatty acid composition and eating quality of lamb types derived from four diverse breed times production systems. Meat Sci. 55:141-147.

Forrest J, Aberle E, Hedrick M, Judge R, Merkel RA (1979). Fundamentos de Ciencia de la carne. Editorial Acribia, Zaragoza. pp. 60-67, 97- 98.

Gutiérrez J, Rubio MS, Méndez RD (2005). Effects of crossbreeding Mexican Pelibuey sheep with Rambouillet and Suffolk on carcass traits. Meat Sci. 70:1–5.

Hernández JO, Vidal A (2001). Utilización de zeranol en borregos pelibuey en pastos y con concentrados energético Universidad y Ciencia. 17(34):57-64.

Hinojosa-Cuéllar JA, Regalado-Arrazola FM, Oliva-Hernández J (2009). Crecimiento prenatal y predestete en corderos Pelibuey, dorper, katahdin y sus cruces en el sureste de México. Revista Científica, FCV-LUZ 19(5):522-532.

Honikel KO (1998). Reference Methods for the assessment of physical characteristics of meat. J. Meat Sci. 49(4):447-457.

Mariezcurrena MA, Braña D, Partida JA, Ramírez E, Domínguez IA (2010). Estandarización de la metodología para la determinación de grasa en la carne de cerdo. Rev. Mex. Cienc. Pecu. 1(3):269-275.

Martínez GS, Aguirre OJ, Zepeda GJ, Ulloa CR, Figueroa MR, Macías CH, Moreno FLA (2009). La ovinocultura de Nayarit, México. En: Ganadería y seguridad alimentaria en tiempo de crisis. Universidad Autónoma Chapingo. México. pp. 305-309.

Martínez GS, Aguirre OJ, Jaramillo LE, Macías CH, Carrillo DF, Herrera GMT, Pérez EE (2010). Alternativas para la producción de carne ovina en Nayarit, México. Revista Fuente. 2:12-16.

Martínez GS, Macías CH, Moreno FL, Zepeda GJ, Espinoza MM, Figueroa MR, Ruiz FM (2011). Análisis económico en la producción de ovinos en Nayarit, México Abanico Veterinario 1 (1)

New Zealand Sheepbreeders Association (2011). http://www.nzsheep.co.nz/. citado 04 04 12.

Nuncio OG, Nahed TJ, Díaz HB, Escobedo AF, Salvatierra YIB (2001). Caracterización de los sistemas de producción ovina en el estado de Tabasco. Agrociencia 35:469-477.

Realini CE, Duckett SK, Brito GW, Dalla Rizza M, De Mattos D (2004). Effect of pasture *vs.* concentrate feeding with or without antioxidants on carcass characteristics, fatty acid composition, and quality of Uruguayan beef. Meat Sci. 66:567-577.

Research into Lamb Meat Quality (2010). Alliance Group Limited http://www.alliance.co.nz/PDF/Lamb_Meat_Quality_booklet.pdf. citado 11/05/12.

Sánchez G (1999). Ciencia básica de la carne. 1ª edición. Editorial Guadalupe Ltda., Santa Fe de Bogotá.

Sañudo C (2008). Calidad de la canal y de la carne ovino y caprina y los gustos de los consumidores. Revista Brasileira de Zootecnia 37:143-160.

SAS (2004). User's guide: statistics 4th ed. Cary (NC): SAS Institute Inc.

Scerra M, Caparra P, Foti F, Galofaro V, Sinatra MC, Scerra V (2007). Influence of ewe feeding systems on fatty acid composition of suckling lambs. Meat Sci. 76:390-394.

Scheaffer R, Mendenhall W, Ott L (2007). Elementos de Muestreo. Edit., Thomson, España. P. 455.

Schreurs NM, Lane GA, Tavendale MH, Barry TN, McNabb WC (2008). Pastoral flavour in meat products from ruminants fed fresh forages and its amelioration by forage condensed tannins. Statgraphics, 1998. Anim. Feed Sci. Technol. 146:193-221.

Torrescano G, Sánchez A, Peñúñuri F, Velázquez J, Sierra T (2009). Características de la canal y calidad de la carne de ovinos pelibuey, engordados en Hermosillo, Sonora BIOtecnia. 11(1):41-50.

Whittington FM, Dunn R, Nute GR, Richardson RL, Wood JD (2006). Effect of pasture type on lamb product quality. 9th Annual Langford Food Industry Conference, 'New Developments in sheepmeat Quality 24-25th May, Bristol, UK. Proceedings of the British Society of Animal Science, pp. 27-31.

28

Comparative evaluation of milk yield and reproductive performances of dairy cows under smallholders' and large–scale management in Central Ethiopia

Nega Tolla

Adama Science and Technology University, School of Agriculture, P. O. Box 193, Asella, Ethiopia.

Comparative mik production and postpartum reproductive performances of Holstein Friesian cows under smallholder and large scale farmers' management was monitored in central Rift Valley of Oromia, Ethiopia. This study was conducted in three purposively selected districts of Arsi Negelle, Ziway, and Lume in Eastern Shoa Zone. Three large scale peri-urban farms having 170 to 195 and 21 small scale urban farms having 1 to 10 heads of dairy animals were identified during the initial exploratory survey. Based on the willingness of the farm owners, the presence of dairy cows of graded Holstein Friesian genotype, known parity and stage of pregnancy of the three large scale farms as a whole and 12 randomly selected small scale farms were considered for monitoring and data collection. A total of 59 animals from large scale (45 animals with average ± standard body weight 427±42 kg) and small scale (14 animals with average ± standard body weight 363±16 kg) were used for 28 weeks of data collection. Significantly (p<0.001) higher milk yield was recorded on large scale farms than on the small scale farms. The reproductive parameters measured were not statistically differed between farm scales. Although, the estimated amounts of crude protein and metabolizable energy consumed by animals were above requirements for the observed level of milk output, the productivity of animals in both farm scales were below their genetic potential particularly that of small scale farms were critically low even than other developing coutries with similar environment elsewhere. The quality of dietary nutrients in terms of the proportion of rumen degradable to undegradable protein sources, structural and non structural carbohydrate and sources of essential minerals needs further assessment for both farm scales.

Key words: Farm size, comparative, nutrients, intake, milk yield, reproduction.

INTRODUCTION

Inadequate and unbalanced nutrient supply is one of the major technical constraints of urban and peri-urban dairy production systems in Ethiopia (Abaye et al., 1991; Goshu and Mekonen, 1997). Dairy farms rely on varieties of feed materials. Feed resource markets provide primarily native grass hay, grain milling by-products and oil seed cakes to urban and peri-urban dairy producers. They also supply commercial mixed concentrates made up of mill by-products (Staal and Shapiro, 1996). Purchased crop residues are also important basal feed resources for small-scale farms in the secondary towns. Generally, the provision of feeds for dairy animals is based on availability than nutrient requirement for a particular productive state of animals.

Conserved native grass hay (mainly composed of *Digitaria decombens, Eragirostis pilosa, Trifolium repens* and *Trifolium prantense*), agro-industrial by-products and commercially formulated concentrate rations are the major feed resources used (Azage and Alemu, 1998) in the urban and peri-urban dairy production systems. However, there is no practice and skill of using nutritionally balanced concentrate diet in these production systems (Staal and Shapiro, 1996). In addition, there is no quality controlling system to regulate the nutrient compositions of commercially formulated concentrate in the way it can fulfill the nutrient requirements of dairy animals in different productive states. This can be one of the major factors to limit the expression of genetic potentials of exotic dairy cattle.

Generally, documented information on the nutrient composition of the available feed resources and its influences on the productive and reproductive potential of urban and peri-urban dairy farms are also limited and needs assessment.

MATERIALS AND METHODS

Study area

This study was conducted in the Central Rift Valley (CRV) of Oromia, at Arsi Negelle, Ziway, Wonji Kuriftu and Lume districts of East Shoa Zone. East Shoa Zone (Figure 1) is located between 38°00'E to 40°00'E longitude and 7°00'N to 9°00'N latitude. It is characterized with different altitude ranges of 1550 to 1900 meters above sea level and average minimum and maximum temperature of 20 and 27°C respectively. It has an erratic and unreliable rainfall, ranging from 500 to 900 mm per year.

Sampling and animal management

A rapid exploratory survey was undertaken to identify and locate the existing large and small scale dairy farms in the urban and peri-urban centers of the study area. Based on the willingness of the farm owners, the presence of dairy cows of graded Holstein-Friesian genotype, known parity and stage of pregnancy, 3 large peri-urban farms having 170 to 195 heads of dairy animals and 21 small scale urban farmers in the secondary towns having 1 to 10 heads of cattle were identified.

The farms were categorized based on the existing herd size as small scale (≤10 animals) and large scale (>10 animals). Accordingly, the three large scale farms and 12 randomly selected smallholder farmers were considered. A total of 59 animals from both large scale (45 animals with 426±85 kg average body weight) and small scale (14 animals with 363±78.5 kg average body weight) with the parity ranging from 1 to 6 and in the last trimester of pregnancy were used for data collection. Animals in different parities were classified as early parity (1 to 2 lactations) and advanced parity (3 to 6 lactations).

In the urban small scale production system the animals were entirely confined at home utilizing whatever space is available in the residential compounds. There were no sufficient exercising areas for animals. Animal houses and feeding facilities on small scale farms were poor relative to that of large scale farms. In the peri-urban large scale production sub-units animals were housed in sheds with well ventilated corrugated metal sheet roof and cemented floors. Their major basal feed sources was based on purchased or very few harvested native grass hay. Green feeds of

alfalfa and elephant grass were also used.

Data collection

Data collection was started from about one week postpartum. The utilization of available feed resources and daily milk yield of each selected farm was monitored and recorded every five days for 28 weeks. Daily milk yield of individual animal was monitored and recorded for both AM and PM using portable spring balance. The amount and type of feeds offered to individual animal was also weighed and recorded for each monitoring date. Both daily feed intake and milk yield for none collection days were estimated from average values of the preceding measurements.

Accordingly the refusal of any feed type offered was weighed and recorded. The amount of daily nutrient intake over a given period was estimated by multiplying the nutrient content of the feeds (per kg dry matter) by the daily dry matter intake of the respective feed. Dry feed samples from each farm were collected fortnightly and bulked. They were thoroughly mixed, sub sampled and delivered to International Livestock Research Institute (ILRI) laboratory for chemical analysis.

On the large scale farms, any signs of estrus manifestation was visually observed and recorded by barn attendants and the veterinarian daily in the morning and afternoon. On the small scale farms the enumerators assigned for data recording visually visited the farms and the farmers oriented to report any sign of estrus recorded accordingly. In both cases mating practice was by artificial insemination (AI). However, on the small scale farms several skipped mating were observed due to shortage of AI facilities and/or unavailability of AI technicians.

Chemical analysis

Feeds were analyzed for dry matter (DM), organic matter (OM) and crude protein (CP) using standard procedures of AOAC (1990). Neutral detergent fiber (NDF) was determined as described by Van Soest and Robertson (1985). The *in vitro* organic matter digestibility (IVOMD) was determined using the procedures described by Tilley and Terry (1963). Metabolizable energy (ME) content (MJ/kg DM) of feeds was estimated from *in vitro* organic matter digestibility (IVOMD (g/kg DM) × 0.016) as suggested by McDonald et al. (1988) and Barber et al. (1984). Metabolizable energy intake (MEI) was estimated by multiplying dry matter intake (DMI) of the feeds with the values of their respective energy concentration, that is, MEI (MJ/day) = DMI (kg/day) x ME (MJ/kg DM) according to Kearl (1982) and MAFF (1985). Calcium and sodium contents of the feeds were analyzed using atomic absorption spectrophotometers according to Perkins (1982), and phosphorus content was determined using auto analyzer of AOAC (1990). The daily crude protein (CP) and ME requirement for the animals was estimated based on actual average daily milk yield and fat content according to NRC (1989) recommendation.

Statistical analysis

Data on daily milk yield, milk compositions, postpartum reproductive efficiencies and nutrient intake were analyzed for farm scale, parity class and lactation period differences using the General Linear Model and multivariate analysis procedure of SPSS (1997). Mean differences between subjects under study were tested by pair-wise comparison and least significant difference (LSD) method. Since all the parameters measred were not significantly affected by parity it was excluded from the model.

Table 1. Means chemical compositions of available feed resources for Holstein Friesian cows in small scale farms of central Rift Valley, Oromia.

Feed types	DM	%DM								
		CP	ME (MJ/kg)	NDF	EE	Ash	IVOMD	Ca	P	Na
Grass hay	91	8.68	7.12	75.34	1.44	10.33	46.54	0.40	0.18	0.01
Wheat straw	92	2.56	5.66	81.58	0.93	8.21	36.33	0.09	0.05	0.01
Maize stover	91	5.63	8.74	81.24	0.69	8.21	55.30	0.35	0.10	0.02
Tef straw	92	3.75	6.04	83.49	1.13	6.64	38.63	0.17	0.08	0.01
Haricot bean straw	91	5.19	6.66	70.95	0.65	8.06	42.50	0.18	0.05	0.01
Maize forage	21	10.38	8.90	72.49	1.65	9.22	55.85	0.23	0.15	0.02
Grass forage	13	13.50	9.34	75.45	1.14	12.22	54.54	0.51	0.31	0.02
Nigerseed cake	91	29.75	10.31	36.43	6.22	10.91	58.07	0.69	0.99	0.00
Linseed cake	90	28.81	13.40	37.78	8.07	8.47	68.40	0.51	0.50	0.02
Cottonseed cake	90	22.31	12.06	48.73	6.63	5.56	47.50	0.19	0.74	0.01
Wheat bran	88	17.06	11.63	38.44	4.55	4.36	72.44	0.11	1.00	0.01
Atala[c]	13	21.40	11.00	57.22	-	4.02	69.00	0.61	0.59	0.00
Mixed concentrate	89	22.06	-	36.74	5.87	6.12	64.91	0.28	1.05	0.00

[c] Local brewers grain residue.

Table 2. Chemical compositions of available feed resources for Holstein Friesian cows in large-scale farms of central Rift Valley, Oromia.

Feed types	DM	%DM								
		CP	ME (MJ/kg)	NDF	EE	Ash	IVOMD	Ca	P	Na
Grass hay	92	4.75	5.91	76.90	1.19	8.50	38.94	0.30	0.15	0.05
Alfalfa forage	35	15.50	9.62	56.72	1.91	11.80	53.89	1.04	0.32	0.02
Elephant grass forage	15	13.10	9.02	68.00	1.94	17.06	50.33	0.21	0.28	0.01
Maize forage	35	6.25	9.90	62.32	1.32	9.22	62.24	0.23	0.33	0.00
Grass forage	13	9.13	6.70	75.85	1.37	9.32	44.96	0.43	0.29	0.01
Molasses intake	74	3.50	15.90	-	5.90	18.84	99.69	1.80	0.10	0.26
Mixed concentrate	89	20.16	-	36.62	4.66	7.91	66.56	0.36	1.03	0.42

RESULTS AND DISCUSSION

Chemical compositions of available feed resources

Chemical compositions of available feed resources on small and large-scale farms are presented in Tables 1 and 2 respectively. Crude protein (CP) content of hay was higher (8.7% DM) in small scale farms than in large scale farms, which may be due to proper harvesting time and preservation methods. In large scale farms, the hay purchased may be harvested after being over matured and exposed to sun for longer time as opposed to that in small scale farms.

In both small and large scale farms, Ca, P and Na contents of hay were deficient for dairy cattle requirement of 0.48 to 0.77, 0.25 to 0.48 and 0.06 to 0.25 respectively as suggested by McDowell (1997). Crop residues were used only in small scale farms (Table 1) and all types of crop residues were deficient in CP, Ca, P and Na

contents, and lower in IVOMD and higher in NDF contents. This was consistent with reports of Kabaija and Little (1989) in which levels of essential minerals in most commonly used fibrous feed resources were reported to be deficient to marginal.

Among the maize and native grass forages used in both farm scales, grass forage had sufficient Ca content in small scale farms (Table 1) while it was deficient in large sale farms (Table 2). Both forage types were deficient in P and Na contents in both farm scales. The variation in mineral compositions of grass forages may be due to variations in species compositions, soil types on which the forages were grown and stages of maturity at harvesting (Chesworth and Guerin, 1992).

Agro-industrial by-products and non-conventional feed resource of homemade liquor residue (atala) were used only in small scale farms (Table 1). Nougseed cake and linseed cake were sufficient in Ca and P content but deficient in Na contents and this agrees with Solomon

Table 3. Mean daily feeds and nutrient intake of Holstein Friesian cows under two different farmers' management scales in central Rift Valley.

Daily nutrient intake	Farm scales		SE	p
	Small	Large		
Number of animals	14	45		
Total DM intake (kg/day)	11.4	15.8	0.47	***
Roughage	4.9	6.3	0.33	**
Supplement	6.5	9.5	0.44	***
Total CP intake (g/day)	1704	2343	123..00	***
Roughage	259	477	45.85	**
Supplement	1445	1886	104.63	**
CP intake (%DMI)	14.5	15.4	0.59	NS
Total ME intake (MJ/day)	115	141	5.66	**
Roughage	34	41	2.26	*
Supplement	81	100	6.31	*
NDF intake (%DMI)	45.21	49.90	2.35	NS
EE intake (%DMI)	4.2	3.4	0.14	***
Roughage NDF (% total NDF)	71	53	2.26	***
Ca intake (%DMI)	0.40	0.43	0.02	NS
P intake (%DMI)	0.48	0.77	0.04	***
Na intake (%DMI)	0.03	0.26	0.001	**
CP requirement (g/day/head)	1238	1696	89.00	-
ME requirement (MJ/day/head)	103	133	3.78	-

***=$p<0.001$; **=$p<0.01$; *=$p<0.05$. NS, non significant; CP and ME Requirements were estimated based on the actual milk yield (kg/d) and body weights of animals (NRC, 1989).

(1992) who reported that locally produced oil seed cakes were deficient in Ca and Na contents, but sufficient in P, K and Mg contents. Cottonseed cake and wheat bran were deficient in Ca and Na contents but sufficient in P content. The mixed concentrate used was sufficient in CP, ME and P contents in both farm scales. But it was deficient in Ca and Na in small scale farms, while in large scale farms it was deficient in Ca but sufficient in Na content. The low Ca and Na content of concentrate diets in small scale farms was due to the fact that they used mainly the ingredients of wheat bran and nougseed cake or linseed cake which are both deficient in these mineral elements. However, large scale farms used reasonably sufficient common salt to supply sufficient Na but did not consider supplements for Ca sources such as limestone. Some of the small scale farmers used cereal crop residues after soaking them with home-made brewers' grain by-products. It was observed that soaking tef and wheat straws with brewer's grain by product substantially increased the CP, ME, Ca compositions and IVOMD of the feeds (Table 1).

Feeds and nutrient intake

Daily dry matter and nutrient intake for small and large scale dairy farms is presented in Table 3. Significant

differences were observed between the large and small scale farms in daily intake of DM, CP, ether extract (EE), P ($p<0.001$), ME, and Na ($p<0.01$). Animals on the large scale farms had higher intake of DM (39%), CP (38%), ME (23%), P (60%) and Na (33%) than those on small scale farms. Dietary calcium intake (%DMI) for animals in small scale farms was below the recommended range (0.43 to 0.77% DMI) of NRC (1989) while that of large scale farms was marginal. The intake of P was within the recommended marginal level of 0.33 to 0.48% of DMI for small scale farms while it was sufficiently higher for animals on the large scale farms.

The ratios of Ca: P was 0.8:1 and 0.6:1 in small scale and large scale farms respectively, while the recommended optimum level was 1: to 2:1 (McDowell et al., 1983). Dietary sodium intake was critically below the required level of 0.18% (NRC 1989) in small scale farms, but sufficiently high in large scale farms. Preserved crop residues of different types (maize stover, tef straw, wheat straw and haricot bean) were the major basal feed base for small scale farms in this study. These crop residues were low in digestible matter, nitrogen and true protein content which may limit the intake of DM and other nutrients. Chenost and Sansoucy (1991) and Chesworth and Guérin (1992) reported that voluntary feed intake of ruminants essentially depends on the rate of degradation of its digestible matter. About 43% of the total DM

Table 4. Mean daily milk yield, milk compositions and reproductive efficiency of Holstein Friesian cows under two farm scales.

Parameter	Farm scales		SE	p
	Small	Large		
Number of animals	14	45		
Milk yield (kg)	11.5	15.8	0.73	***
FC milk yield (kg)[a]	11.1	14.7	0.67	***
Calving to first sign of estrus (days)	96	115	14.60	NS
Days open	171	148	23.50	NS
Services/conception	1.6	2.1	0.24	NS

***=$p<0.001$; [a]FC = fat corrected.

consumed by animals on the small scale farms was roughage feed as compared to 38.6% on the large scale farms. Similarly about 71% of NDF intake of animals on the small scale farms was roughage feeds as compared to 53% on the large scale farms indicating that the smallhoder farmers mainly depend on poor quality roughages for their dairy animals.

Since the potential intake of forage is inversely related to its NDF content, the DM intake and consequently that of other nutrients were limited for animals in small scale farms. The utilization of agro-industrial by-products such as nougseed cake (Guizota abyssinica), linseed cake, cotton seed cake and wheat bran was minimal (7.76% DMI). A home-made brewer's grain by products (39% DMI) mixed with crop residues or alone was also used. Mixed concentrate in the daily dietary DM intake of animals in small scale farms was only 9.6%.

Milk yield and reproductive efficiency

The mean daily milk yield and postpartum reproductive efficiencies of Holstein Friesian cows on large and small scale farms are presented in Table 4. Difference in actual and fat corrected (FC) daily milk yield was highly significant ($p<0.001$) between the two farm scales. There were higher daily milk yield and FC milk yield (15.8 and 14.7 kg respectively) on the large scale farms than on small scale farms (11.5 and 11.1 kg respectively). The higher milk yield of animals on large scale farms may be attributed to higher intake of DM, CP, ME, Ca and P relative to the small scale farms.

It was also observed that smallholder farmers maily depend on poor quality roughages as a basal diets. With high technological inputs in terms of feeding management and health care milk production per cow is extremely very high than with low inputs (Leng, 1991). The utilization of conserved hay, green forages and mixed concentrate on large scale farms were about 26.6, 12 and 59% of daily DM intake respectively as compared to only 6.4, 5.4 and 9.6% respectively on small scale farms. This also reflects that the productivity of ruminants

is influenced primarily by quantity and quality of feed intake (Preston and Leng, 1987).

The quantity of CP and ME may not be the limiting factors for low performances of the animals under this study. Possible reasons for the low milk yield performances of animals in small scale farms may be due to poor nutritional qualities of crop residues used (about 36.4% of daily DM intake), the CP intake from home-made liquor residues which was about 39% of daily DM intake and provided about 76% of the total CP intake may be heat damaged during the long time boiling for alcohol distillation. Therefore, its CP may be unusable or poorly digested in the lower digestive tract as well as in the rumen. This may result in too low ammonia levels in the rumen which can not meet the requirement for efficient growth of rumen micro-organisms (Perston and Leng, 1987). Sources of supplemental CP are an important factor to influence the response of animals, due to their variation in type and levels of essential amino acid (EAA) contents (Christensen et al., 1993).

In addition, the minimal and inconsistent use of protein sources from mixed concentrate and agro-industrial by-products by small scale farmers may result in lower nutrient intake and consequently low productivity of the animals. On both small and large farm scales the ratio of Ca: P was observed to be very low (0.83:1 and 0.56:1 respectively). McDowell et al. (1983) reported that with dietary ratios below 1:1 and over 7:1 growth and feed efficiency decreased significantly. Simon (2005) in Tanzania reported that, there were high milk yield in large scale farms compared to smallholder farms. Differences in milk production between the two management systems were attributed mainly to the level of management. Like small scale farmers in other developing countries (Leng, 1991), smallholder farmers in this study could not be able to select quality basal diet, than using whatever was available at no or low cost.

The days from calving to first estrus (CFE), days open (DO) and the number of services per conception (S/C) was not significantly ($p>0.05$) different between the farm scales (Table 4). But the value of DO (171 days) on small scale farms was longer than on the large scale farms

(148 days). This is due to lack of regular AI services and several skipped services were observed. The days to first estrus (115 days) and SPC (2.1) were higher on the large scale farms than on the smallholder farms, (96 and 1.6 respectively), which may be due to poor heat detection practices on large scale farms. Since the number of animals on small scale farms was very few, farmers could be able to closely observe estrus manifestation of their animals than the large scale farms where there was relatively large number of animals and could not be easy to closely observe estrus manifestation of individual animal.

It was observed that commercial animal feed marking firms could not consider the inclusion of Ca sources which is demanding for the productive and reproductive performances of dairy cattle. In addition, the mixed commercial concentrate as well as the milling by-products of cereal brans were more of fiber and lack grains which can provide soluable carbohydrates. The protein source used was entirely negerseed cake (*G. abyssinica*) and there may be limited rumen by-pass protein sources.

Conclusion

Although, the estimated amounts of crude protein and metabolizable energy consumed by animals were above requirements for the observed level of milk output, the productivity of animals in both farm scales were below their genetic potential particularly that of small scale farms were critically lower than even those in other developing coutries with similar environment elsewhere. The quality of dietary nutrients in terms of the proportion of rumen degradable to undegradable protein sources, structural and non structural carbohydrate and sources of essential minerals needs further assessment for both farm scales.

ACKNOWLEDGEMENTS

Oromia Agricultural Research Institute (OARI) and Ethiopian Agricultural Research Organization (EARO) are highly acknowledged for funding this study.

REFERENCES

Abaye T, Tefera GM, Alemu GW, Bruk Y, Philip C (1991). Status of dairying in Ethiopia and strategies for future development, pp. 25–36. In Proceedings of 3rd National Livestock Improvement Conference, 24-26 May 1989. Institute of Agricultural Research (IAR), Addis Ababa.

AOAC (Association of Official Analytical Chemists) (1990) Official Methods of Analysis (15th ed)., Arlington.

Azage T, Alemu GW (1998). Prospects for peri-urban dairy development in Ethiopia, pp. 28-39. In Proceedings of 6th National Conference of Ethiopian Society of Animal Production (ESAP), 15–17 May 1997, Addis Ababa.

Barber WP, Adamson AH, Altma JFB (1984). New methods of forage evaluation, pp.161-176. In W. Hareisgn and D.J.B. Cole, (eds). Recent Advances in Animal Nutrition. Butterworth, London.

Chenost M, Sansoucy R (1991). Nutritional characteristics of tropical feed resources, pp. 66–81. In A. Speedy and R. Sansoucy (eds.). Feeding Dairy Cows in the Tropics. Proceedings of the FAO expert consultation held in Bangkok, Thailand, 7–11 July 1989. Food and Agricultural Organization of the United Nations, Rome.

Chesworth J, Guérin H (1992). Ruminant Nutrition. Macmillan Education Ltd. London. P. 170.

Christensen RA, Lynch GL, Clark JR (1993). Influence of amount and degradability of protein on production of milk and milk components by lactating Holstein cows. J. Dairy Sci. 76:3490-3496.

Goshu M, Mekonen HM (1997). Milk production of Fogera cattle and their crosses with Friesian at Gonder, northern Ethiopia. Ethiopian J. Agric. Sci. (EJAS) 16(2):61–74.

Kabaija E, Little DA (1989). Potential of agricultural by-products as source of mineral nutrients in ruminant diet, pp. 379-394. In A.N. Saic and A.M. Dzowela (eds), Overcoming Constraints to the Efficient Utilization of Agricultural By-products as Animal Feed. Proceedings of the Annual Workshop Held at Institute of Animal Research, Marken Station, Bamenenda, Cameron, 20-29 October 1987, ARNAB, ILCA Addis Ababa.

Kearl LC (1982) Nutrient Requirement of Ruminants in Developing Countries. International Feedstuff Institute, Utah Agricultural Experiment Station. Utah State University, Logan. P. 381.

Leng RA (1991). Feeding strategies for improving milk production of dairy animals managed by small farmers in the tropics, In A. Speedy and R. Sansoucy (eds.). Feeding Dairy Cows in the Tropics. Proceedings of the FAO expert consultation held in Bangkok, Thailand, 7–11 July 1989. Food and Agricultural Organization of the United Nations, Rome, pp. 82-104.

MAFF (1985). Energy Allowance and Feeding Systems for Ruminants. Ministry of Agriculture, Fisheries and Food, Reference Book No. 433, London. P. 79.

McDonald P, Edwards RA, Greenhalgh JFD (1988). Animal Nutrition. 4th edition, Longman Scientific and Technical, New York. P. 281.

McDowell LR (1997). Minerals for Grazing Ruminants in Tropical Regions. Third edition, University of Florida, Gainesville. P. 81.

McDowell LR, Conrad JH, Ellis GL, Loosli J.K (1983). Minerals for Grazing Ruminants in Tropical Regions. University of Florida, Gainesville. P. 86.

NRC (1989) Nutrient Requirement of Dairy Cattle. Sixth Revised Edition, National Research Council, National academy Press, Washington, D.C. 157 pages.

Perkins E (1982). AAS Manual Model 2380. Norwalk, Connecticut.

Preston TR, Leng RA (1987). Matching Ruminant Production Systems with Available Resources in the Tropics and Sub-Tropics. Penambul books Ltd. Armidale, New South Wales, Australia. 245 pages.

Solomon M (1992). The Effects of Method of Processing of Oil Seed Cakes in Ethiopia on Nutritive Value. PhD Thesis, University of Bonn, Germany.

SPSS (1997) Statistical Procedure for Social Science, version 10. SPSS Inc, Chicago.

Staal SJ, Shapiro BI (1996). The Economic Impact of Public Policy on Smallholder Peri-urban Dairy Producers in and around Addis Ababa. ESAP Publication No. 2. ESAP (Ethiopian Society of Animal Production), Addis Ababa. 57 pp.

Tilley JAM, Terry RA (1963). A two-stage technique for the in vitro digestion of forage crops. J. Brit. Grassland Soc. 18:104–111.

Van Soest PJ, Robertson JB (1985). Analyses of forage and fibrous foods. A Laboratory Manual for Anim. Sci. Cornell University, Ithaca, New York 613:98–110.

Permissions

List of Contributors

Pipat Arunvipas
Department of Large Animal and Wildlife Clinical Sciences, Faculty of Veterinary Medicine, Kasetsart University, Kamphaengsaen Campus, Thailand

Sathaporn Jittapalapong
Department of Parasitology, Faculty of Veterinary Medicine, Kasetsart University, Thailand

Tawin Inpankaew
Department of Parasitology, Faculty of Veterinary Medicine, Kasetsart University, Thailand

Nongnuch Pinyopanuwat
Department of Parasitology, Faculty of Veterinary Medicine, Kasetsart University, Thailand

Wissanuwat Chimnoi
Department of Parasitology, Faculty of Veterinary Medicine, Kasetsart University, Thailand

Soichi Maruyama
Laboratory of Veterinary Public Health, College of Bioresource Sciences, Nihon University, Japan

Bekele Dinsa
College of Agriculture and Veterinary Medicine, Jimma University, P. O. Box 307, Jimma, Ethiopia

Moti Yohannes
College of Agriculture and Veterinary Medicine, Jimma University, P. O. Box 307, Jimma, Ethiopia

Hailu Degefu
College of Agriculture and Veterinary Medicine, Jimma University, P. O. Box 307, Jimma, Ethiopia

Mezene Woyesa
School of Veterinary Medicine, Wollega University, Nekemte, P. O. Box 395, Ethiopia

Berhanu Kuma
EIAR, Holetta Agricultural Research Center, P. O. Box 2003, Addis Ababa, Ethiopia

Derek Baker
Research Economist, International Water Management Institute, P. O. Box 5689, Addis Ababa, Ethiopia

Kindie Getnet
Senior Agricultural Economist and Agricultural Marketing Program Leader, ILRI, Nairobi, Kenya

Belay Kassa
Haramaya University, P. O. Box 138, Dire Dawa, Ethiopia

V. M. Mmbengwa
North-West University (NWU), Potchefstroom Campus, Potchefstroom, Republic of South Africa

J. A. Groenewald
North-West University (NWU), Potchefstroom Campus, Potchefstroom, Republic of South Africa

H. D. van Schalkwyk
North-West University (NWU), Potchefstroom Campus, Potchefstroom, Republic of South Africa

P. J. C. Greyling
University of Free State (UFS), Bloemfontein Campus, Bloemfontein, Republic of South Africa

Saleem Ashraf
Institute of Agricultural Extension and rural Development, University of Agriculture, 38040- Faisalabad, Pakistan

Muhammad Iftikhar
Institute of Agricultural Extension and rural Development, University of Agriculture, 38040- Faisalabad, Pakistan

Ghazanfar Ali Khan
Institute of Agricultural Extension and rural Development, University of Agriculture, 38040- Faisalabad, Pakistan

Babar Shahbaz
Institute of Agricultural Extension and rural Development, University of Agriculture, 38040- Faisalabad, Pakistan

Ijaz Ashraf
Institute of Agricultural Extension and rural Development, University of Agriculture, 38040- Faisalabad, Pakistan

A. Reddy Varaprasad
Sri Venkateswara Veterinary University, Tirupati, India

T. Raghunandan
Livestock Research Institute, College of Veterinary Science, Rajendranagar, Hyderabad, Andhra Pradesh, India

M. Kishan Kumar
Department of Animal Genetics and Breeding, College of Veterinary Science, Rajendranagar, Hyderabad, Andhra Pradesh, India

M. Gnana Prakash
Department of Animal Genetics and Breeding, College of Veterinary Science, Rajendranagar, Hyderabad, Andhra Pradesh, India

Mehmet Arif Şahinli
Economics Department, Karamanoğlu Mehmetbey University, Karaman, Turkey

Ahmet Özçelik
Ankara Üniversitesi, Ziraat Fakültesi Tarım Ekonomisi Bölümü, Dıskapı Ankara, Turkey

W. A. Lamidi
Department of Agricultural Education, Osun State College of Education, Ila-Orangun, Osun State, Nigeria

J. A. Osunade
Department of Agricultural Engineering, Obafemi Awolowo University, Ile-Ife, Nigeria

L. A. O. Ogunjimi
Department of Agricultural Engineering, Obafemi Awolowo University, Ile-Ife, Nigeria

B. Okon
Department of Animal Science, University of Calabar, Calabar, Cross River State, Nigeria

L. A. Ibom
Department of Animal Science, University of Calabar, Calabar, Cross River State, Nigeria

O. J. Ifut
Department of Animal Science, University of Uyo, Uyo, Akwa Ibom State, Nigeria

A. E. Bassey
Department of Animal Science, University of Uyo, Uyo, Akwa Ibom State, Nigeria

Dawit Assefa
Adami Tullu Agricultural Research Center, P. O. Box 35, Ziway, Ethiopia

Ajebu Nurfeta
School of Animal and Range Sciences, Collage of Agriculture, Hawassa University, P. O. Box 5, Hawassa, Ethiopia

Sandip Banerjee
School of Animal and Range Sciences, Collage of Agriculture, Hawassa University, P. O. Box 5, Hawassa, Ethiopia

S. M. Hassan
Animal and Fish Production Department, King Faisal University, Al-Ahsa, Kingdom of Saudi Arabia

J. Garba
Department of Agricultural Education, Zamfara State College of Education, P. M. B. 1002 Maru, Zamfara State, Nigeria

A. Y. Yari
Department of Agricultural Education, Zamfara State College of Education, P. M. B. 1002 Maru, Zamfara State, Nigeria

M. Haruna
Department of Agricultural Education, Zamfara State College of Education, P. M. B. 1002 Maru, Zamfara State, Nigeria

S. Ibrahim
Department of Agricultural Education, Zamfara State College of Education, P. M. B. 1002 Maru, Zamfara State, Nigeria

Philippe Sessou
Laboratory of Research in Applied Biology, Polytechnic School of Abomey-Calavi, University of Abomey-Calavi, 01 P. O. Box 2009 Cotonou, Benin
Laboratory of Study and Research in Applied Chemistry, Polytechnic School of Abomey-Calavi, University of Abomey-Calavi, 01 P. O. Box 2009 Cotonou, Benin

Souaïbou Farougou
Laboratory of Research in Applied Biology, Polytechnic School of Abomey-Calavi, University of Abomey-Calavi, 01 P. O. Box 2009 Cotonou, Benin

Paulin Azokpota
Laboratory of Food Microbiology and Biotechnology, Faculty of Agronomic Sciences, University of Abomey-Calavi, 01 P. O. Box 526 Cotonou, Benin

Issaka Youssao
Laboratory of Research in Applied Biology, Polytechnic School of Abomey-Calavi, University of Abomey-Calavi, 01 P. O. Box 2009 Cotonou, Benin

Boniface Yèhouenou
Laboratory of Study and Research in Applied Chemistry, Polytechnic School of Abomey-Calavi, University of Abomey-Calavi, 01 P. O. Box 2009 Cotonou, Benin

Serge Ahounou
Laboratory of Research in Applied Biology, Polytechnic School of Abomey-Calavi, University of Abomey-Calavi, 01 P. O. Box 2009 Cotonou, Benin

Dominique Codjo Koko Sohounhloué
Laboratory of Study and Research in Applied Chemistry, Polytechnic School of Abomey-Calavi, University of Abomey-Calavi, 01 P. O. Box 2009 Cotonou, Benin

R. P. Naga
Department of Entomology, S. K. N. College of Agriculture (S. K. Rajasthan Agricultural University), Jobner (Rajasthan) - 303329, India

Ashok Sharma
Department of Entomology, S. K. N. College of Agriculture (S. K. Rajasthan Agricultural University), Jobner (Rajasthan) - 303329, India

K. C. Kumawat
Department of Entomology, S. K. N. College of Agriculture (S. K. Rajasthan Agricultural University), Jobner (Rajasthan) - 303329, India

B. L. Naga
Department of Entomology, S. K. N. College of Agriculture (S. K. Rajasthan Agricultural University), Jobner (Rajasthan)- 303329, India

Rene Maria Ignácio
Biochemical-Pharmaceutical Technology Department, Pharmaceutical Sciences Faculty, São Paulo University, Av. Prof. Lineu Prestes, 580/Cidade Universitária, São Paulo-SP, CEP 05508-900, Brazil

Suzana Caetano Da Silva Lannes
Biochemical-Pharmaceutical Technology Department, Pharmaceutical Sciences Faculty, São Paulo University, Av. Prof. Lineu Prestes, 580/Cidade Universitária, São Paulo-SP, CEP 05508-900, Brazil

Arse Gebeyehu
Adami Tulu Agricultural Research Center, P. O. Box 35, Batu, Ethiopia

Feyisa Hundessa
Adami Tulu Agricultural Research Center, P. O. Box 35, Batu, Ethiopia

Gurmessa Umeta
Adami Tulu Agricultural Research Center, P. O. Box 35, Batu, Ethiopia

Merga Muleta
Adami Tulu Agricultural Research Center, P. O. Box 35, Batu, Ethiopia

Girma Debele
Adami Tulu Agricultural Research Center, P. O. Box 35, Batu, Ethiopia

Manuel Murillo Ortiz
Faculty of Veterinary Medicine, Juárez University of the State of Durango, Castaña 106, Fraccionamiento: Nogales, CP: 34162, Durango, Dgo, México

Osvaldo Reyes Estrada
Faculty of Veterinary Medicine, Juárez University of the State of Durango, Castaña 106, Fraccionamiento: Nogales, CP: 34162, Durango, Dgo, México

Esperanza Herrera Torres
Faculty of Veterinary Medicine, Juárez University of the State of Durango, Castaña 106, Fraccionamiento: Nogales, CP: 34162, Durango, Dgo, México

José H. Martínez Guerrero
Faculty of Veterinary Medicine, Juárez University of the State of Durango, Castaña 106, Fraccionamiento: Nogales, CP: 34162, Durango, Dgo, México

Guadalupe M. Villareal Rodriguez
Faculty of Veterinary Medicine, Juárez University of the State of Durango, Castaña 106, Fraccionamiento: Nogales, CP: 34162, Durango, Dgo, México

A. E. Olatunji
Department of Animals Science, University of Abuja, Nigeria

R Ahmed
Niger State Agricultural Development Programme (ADP) Minna, Nigeria

A. A. Njidda
Department of Animal Science, Bayero University Kano, Nigeria

R. Trevor Wilson
Bartridge Partners, Bartridge House, Umberleigh EX37 9AS, UK

Katarzyna Ognik
Department of Biochemistry and Toxicology, Faculty of Biology and Animal Breeding, University of Life Sciences in Lublin, Poland

Anna Czech
Department of Biochemistry and Toxicology, Faculty of Biology and Animal Breeding, University of Life Sciences in Lublin, Poland

Bożena Nowakowicz-Dębek
Department of Animal Hygiene and Environment, Faculty of Biology and Animal Breeding, University of Life Sciences in Lublin, Poland

Łukasz Wlazło
Department of Animal Hygiene and Environment, Faculty of Biology and Animal Breeding, University of Life Sciences in Lublin, Poland

Phil Glatz
SARDI Pig and Poultry Production Institute, J. S. Davies Building, Roseworthy Campus, Roseworthy, South Australia 5371, Australia

Belinda Rodda
SARDI Pig and Poultry Production Institute, J. S. Davies Building, Roseworthy Campus, Roseworthy, South Australia 5371, Australia

R. Sendhil
Directorate of Wheat Research, Karnal (Haryana), India

D. Babu
National Academy of Agricultural Research Management, Hyderabad, India

Ranjit Kumar
National Academy of Agricultural Research Management, Hyderabad, India

K. Srinivas
National Academy of Agricultural Research Management, Hyderabad, India

F. M. El-Hag
Agricultural Research Corporation (ARC), Dryland Research Center (DLRC), Soba, Khartoum

M. M. M. Ahamed
Institute of Environmental Studies, University of Khartoum, Khartoum, Sudan

K. E. Hag Mahmoud
State Ministry of Agriculture, Animal Resources and Irrigation, Kordofan State, El-Obeid, Sudan

M. A. M. Khair
Agricultural Research Corporation (ARC), Wad Medani, Sudan

O. E. Elbushra
Agricultural Research Corporation (ARC), Dryland Research Center (DLRC), Soba, Khartoum

T. K. Ahamed
Food Research Centre, Shambat, Sudan

S. N. Goswami
National Bureau of Soil Survey and Land Use Planning, Amravati Road, Nagpur, Maharashtra 440 010, India

A. Chaturvedi
National Bureau of Soil Survey and Land Use Planning, Amravati Road, Nagpur, Maharashtra 440 010, India

S. Chatterji
National Bureau of Soil Survey and Land Use Planning, Amravati Road, Nagpur, Maharashtra 440 010, India

N. G. Patil
National Bureau of Soil Survey and Land Use Planning, Amravati Road, Nagpur, Maharashtra 440 010, India

T. K. Sen
National Bureau of Soil Survey and Land Use Planning, Amravati Road, Nagpur, Maharashtra 440 010, India

T. N. Hajare
National Bureau of Soil Survey and Land Use Planning, Amravati Road, Nagpur, Maharashtra 440 010, India

R. S. Gawande
National Bureau of Soil Survey and Land Use Planning, Amravati Road, Nagpur, Maharashtra 440 010, India

M. R. Hassan
Department of Animal Science, P. M. B 1044, Ahmadu Bello University; Zaria, Nigeria

F. A. Steenstra
Animal Production Systems Group, Wageningen University; P. O. Box 338, 6700 AH Wageningen, The Netherlands

H. M. J. Udo
Animal Production Systems Group, Wageningen University; P. O. Box 338, 6700 AH Wageningen, The Netherlands

Asifa Hameed
Cotton Research Station Multan, Pakistan

Muhammad Saleem
Entomological Research Institute Faisalabad, Pakistan

Haider Karar
Entomological Research Substation Multan, Pakistan

Saghir Ahmad
Cotton Research Station Multan, Pakistan

Mussarat Hussain
Cotton Research Station Multan, Pakistan

Wajid Nazir
Cotton Research Station Multan, Pakistan

Muhammad Akram
Cotton Research Station Multan, Pakistan

Hammad Hussain
Cotton Research Station Multan, Pakistan

Shuaib Freed
Bhauddin Zakariya University Multan, Pakistan

M. D. Mariezcurrena
Universidad Autónoma del Estado de México. Facultad de Ciencias Agrícolas. Código Postal 50200. Instituto Literario N0. 100 Col. Centro. Toluca, Estado de México

A. Z. M. Salem
Universidad Autónoma del Estado de México. Facultad de Medicina Veterinaria y Zootecnia, Código Postal 50200. Instituto Literario N0. 100 Col. Centro. Toluca, Estado de México

C. Tepichín
Universidad Autónoma del Estado de México. Facultad de Ciencias Agrícolas. Código Postal 50200. Instituto Literario N0. 100 Col. Centro. Toluca, Estado de México

M. S. Rubio
Laboratorio de Ciencia de la Carne, Secretaria de Producción Animal, Facultad de Medicina Veterinaria y Zootecnia. Universidad Nacional Autónoma de México. Código Postal 04510. Av. Universidad 3000, Del. Coyoacán, Ciudad Universitaria, Distrito Federal, México

M. A. Mariezcurrena
2Universidad Autónoma del Estado de México. Facultad de Medicina Veterinaria y Zootecnia, Código Postal 50200. Instituto Literario N0. 100 Col. Centro. Toluca, Estado de México

Nega Tolla
Adama Science and Technology University, School of Agriculture, P. O. Box 193, Asella, Ethiopia

www.ingramcontent.com/pod-product-compliance
Lightning Source LLC
Chambersburg PA
CBHW081710240326
41458CB00156B/4699